THE
PLANET SATURN

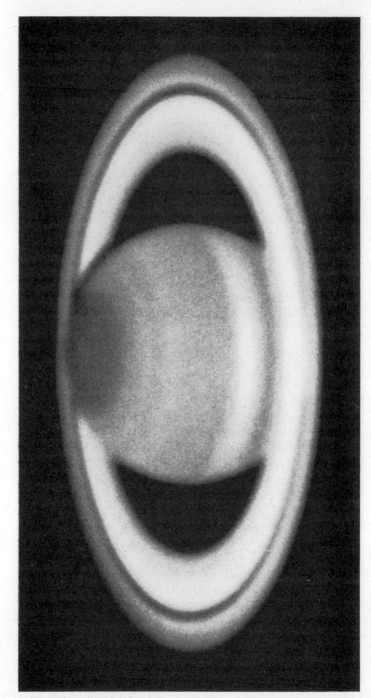

Saturn and ring system photographed (1943). 100-inch Hooker Telescope, Mount Wilson Observatory.

THE PLANET SATURN

A History of Observation, Theory and Discovery

A. F. O'D. ALEXANDER

Former Director 1946–51 of the Saturn Section and 1951–57 of the Jupiter Section of the British Astronomical Association

DOVER PUBLICATIONS, INC.
NEW YORK

Published in Canada by General Publishing Com-
pany, Ltd., 30 Lesmill Road, Don Mills, Toronto,
Ontario.
Published in the United Kingdom by Constable
and Company, Ltd., 10 Orange Street, London
WC2H 7EG.

This Dover edition, first published in 1980, is an
unabridged and corrected republication of the work
originally published in 1962 by Faber and Faber
Limited, London.
The Dover edition is published by special arrangement with Faber and Faber Limited, 3 Queen
Square, London WC1N 3AU England.

International Standard Book Number: 0-486-23927-6
Library of Congress Catalog Card Number: 79-055907

Manufactured in the United States of America
Dover Publications, Inc.
180 Varick Street
New York, N.Y. 10014

CONTENTS

7

CONTENTS

ILLUSTRATIONS

PLATES

DIAGRAMS

PREFACE

Efforts to observe and understand the planet Saturn can be traced back into the past for 2,600 years. Local and very sporadic in early times, such efforts have become in the last two centuries almost unceasing and world-wide. Many resounding successes have been gained, the whole story being an epic of human achievement worthy of the pen of a great writer. The author of this book, conscious of his lack of literary gifts, has tried to produce a clear and readable chronicle based on original sources. Assuming the role of editor he has often let the earlier observers convey the drama of their observations and discoveries by quoting their actual (or translated) words. He has tried to obtrude his personal opinions as little as possible, to avoid bias, and to present the evidence fairly for the reader's judgment. But since a true and complete chronicle must not omit the errors, illusions, false trails and controversies of the past, he has had to point out here and there what opinions seem in the light of later knowledge to be doubtful or mistaken, so that the reader may not be misled.

The attempt to select from and summarise the colossal mass of material (some of it available only in French, Latin or Spanish) has posed constantly the question of how much detail to include so that the book may be interesting, readable, comprehensive and useful as a work of reference. Brief, arid statements would usually satisfy none of these aims. What may seem too detailed to some readers may well be too scanty for those making a serious study of some aspect of the subject. The very frequent references to sources show where fuller information should be sought (the page number in a reference is usually to the page on which the paper or report begins). The aim is that the book should be one in or through which the general reader and the astronomer can find anything he wants to know about Saturn and its satellites.

Of the many instances that could be given of the importance and interest of details and of the value of using original sources, a few must suffice. Previous writers seem to have overlooked that Huygens himself stated that the discovery of Titan gave him the clue to solving the problem of Saturn's appendages; again, that Tuttle was the first to suggest an inner ring as the explanation of the appearance seen by the Bonds. Moreover, though any textbook will list the dates of discovery and names of the discoverers of the satellites, it is surely more satisfying to read (especially in their own words) how the observers first came to suspect, then to confirm, that certain star-like objects were really satellites of Saturn.

This book contains a very large amount of other information about the satellites, especially the difficult problems of their orbits, sizes and masses, based on original papers in the Monthly Notices and other journals and using also the recent excellent paper by Dr. J. G. Porter. Diagrams show the apparent paths of all the satellites, and drawings of Titan's markings made at the Pic-du-Midi Observatory are shown on Plate XVII, Fig. 1.

One of the author's aims was to try to present for the first time, and with some approach to continuity, the story of the naked-eye observations of Saturn during $2\frac{1}{4}$ millennia preceding the invention of the telescope. Glimpses of the historical background have been inserted to help to explain changes in outlook, and more especially shifts of the centre of observing activity, at certain epochs.

This book may help to rescue some past astronomers from being known merely as names of lunar craters. Something is said of the work of the great Syrian astronomer, Muhammad Ibn Jabir al-Battânî (Albategnius), and of the ingenious Moor, Azarquiel (Arzachel), contemporaries respectively of Alfred the Great and William the Conqueror. Drawings of Saturn by Gassendi, Fontana and Riccioli are reproduced. There is discussion of Saturn observations by Goldschmidt (Fabricius) and Sörensen (Longomontanus). It is explained why Boulliau (Bullialdus) was dubious about his friend Huygens's ring theory. Others, not even commemorated on the lunar map, are rescued from complete oblivion, e.g. Yĕhudá bar-Mošé and Rabiçag aben Cayut, the substance of whose account of how they set to work to produce the original Alphonsine Tables is given.

The author believes that readers will be interested in the actual or translated words of such great Saturn observers as Ptolemy, Tycho Brahe, Galileo, Huygens, Cassini, Herschel, Dawes, Lassell, Barnard, Antoniadi and others, both because of their sometimes vivid descriptions, and because the first-hand story of their observations is more realistic and illustrates their own remarkable personalities.

The adherence as far as possible to strict chronological order has the great advantage of showing approximately at any epoch the state which the knowledge of Saturn had then reached, but means inevitably that no part of the subject can be completely dealt with in one place. The section headings inside chapters and the index will help to compensate for this. Chapters 1 and 5 (latter parts), and 23 and

27 seemed necessary where they are placed to introduce, explain or summarise, and so stand somewhat apart from the strict sequence. The chronological arrangement has often resulted in some discovery of major importance being included in the same chapter with other (contemporary) observations that are merely useful and interesting; in such cases the author has endeavoured to draw special attention to the really important one.

It seems that the only previous books devoted exclusively to Saturn were Huygens's *Systema Saturnium* (1659) and R. A. Proctor's *Saturn and its System*, written nearly a century ago, and chapters 15 to 41 (nearly three-fifths) of the present work show the vast amount of progress since Proctor's time. But *The Planet Saturn* is no mere continuation of Proctor's classic, which was written from the geometrical and physical rather than the historical standpoint, and though very useful as a secondary source for some things (e.g. the orbit, the ring phases, the theories of Laplace and older ideas about the physical condition of Saturn) contained nothing of any consequence about observations before Galileo (since little was known about ancient observations in Proctor's time), and only one chapter on the observations from 1610 to 1863, and that one unfortunately not entirely free from error.

The Planet Saturn devotes several whole or part chapters to the occasional spot outbreaks such as occurred in 1876, the 1890s, 1903, 1932–33, 1947 and 1960, with two drawings, a diagram and a photograph to illustrate some of them. A full explanation and eight diagrams clarify the confusing but not difficult process of the periodical passages of the Earth through Saturn's ring-plane, when the ring may briefly disappear and Saturn be seen (in Herschel's words) 'stripped of its noble ornament, and dressed in the plain simplicity of Mars'.

Though written in England this book can hardly be considered insular, since it includes work accomplished in some thirty countries, embracing most of Europe and the Middle East, a huge contribution over the past 150 years from the United States, and in the present century notable observations made in the Southern Hemisphere: South Africa especially, and also Australia, New Zealand and Brazil. Important studies in the U.S.S.R. and in France, at the Pic-du-Midi Observatory, have marked the past two decades.

The vast progress of the past sixty years in photographic, spectroscopic, polarimetric and visual observation and theoretical studies, with the very recent suspicion of two forms of radio emission from Saturn, occupies more than one-third of the book. Physico-mathematical investigations, such as those of Sir Harold Jeffreys and Dr. W. H. Ramsey have revolutionised opinion and knowledge about the constitution of Saturn and its rings. Parts of the argument and the conclusions of mathematical papers have been stated, but it was felt unsuitable to include in this book the mathematical discussions: references show where the original papers can be consulted. A non-mathematical book obviously cannot do full justice to such work, or to the vital assistance rendered to observers by computers. The author has tried as far as possible to stress the importance of the predictions calculated by mathematicians such as Marth, Newcomb, H. Struve, Crommelin, Comrie, Levin and those responsible for the former *Nautical Almanac* and similar publications.

A brief reference here to some of the richest sources of original material seems desirable. For the early observations these include: A. Campbell Thompson's translations of the Ancient Babylonian reports; Halma's French translation of Ptolemy's *Syntaxis;* Millás's scholarly study of Azarquiel; and Dreyer's splendid complete edition in Latin of the observations of Tycho Brahe. For the seventeenth century the scholarly *Œuvres Complètes* provides not only complete French translations of the Huygens pamphlets but also a mine of information on contemporary astronomers. The *Philosophical Transactions* of the Royal Society are the outstanding source for the later seventeenth and eighteenth centuries, including the Saturn papers of Sir William Herschel, also collected and admirably edited by Dreyer. For the nineteenth and twentieth centuries the *Monthly Notices* and other publications of the Royal Astronomical Society are invaluable, and so from 1890 onwards are the Journal and other publications of the British Astronomical Association. Extremely useful sources for these last two centuries are also the *Harvard College Observatory Annals*, the bulletins of great Observatories such as Lick and Lowell, and such journals as the *Astronomical Journal*, the *Astrophysical Journal*, the *Publications of the Astronomical Society of the Pacific* and '*L'Astronomie*', together with the *Transactions of the International Astronomical Union*. Many other primary and secondary sources are referred to in the various chapters; of secondary sources one of the most generally useful has been Signor Abetti's *History of Astronomy*.

Being a detailed book *The Planet Saturn* is peculiarly prone to errors and misprints: the best the author can hope is that they may be few and minor.

ACKNOWLEDGMENTS

The author is deeply grateful to his friend, Mr. M. B. B. Heath, Director of the B.A.A. Saturn Section, for outstanding help and encouragement at every stage of the work; apart from the many useful data he has supplied to Chapters 1, 5, 27 and elsewhere (most of them specially worked out for the book), he has read the entire text and given much sound advice. Gratitude is gladly expressed to other friends for their expert aid: to Dr. W. H. Steavenson for reading the text and for many valuable suggestions and criticisms; to Dr. J. G. Porter for dealing similarly and equally thoroughly with chapters relating to orbits, masses, dimensions and other statistics, and for clearing up numerous queries; and to Mr. R. H. Mann, A.R.I.B.A., who has most generously performed the exacting task of redrawing for publication more than twenty of the diagrams (Figures 1, 2, 4–6, 8, 9, 11a, 12, 13, 15–18, 22).

The many useful notes and references supplied by Mr. Patrick Moore have been of great service, as have some important references found by Mr. C. A. Ronan. Mrs. J. E. Phillips and Mrs. B. Harrington (R.A.S. Librarians) and Mrs. M. L. Stanley (B.A.A. Assistant Librarian) have been most helpful in drawing up lists of references, arranging photography for illustrations, and forwarding many parcels of books. Certain rare books have been obtained by the Dorset County Library staff. Thanks have been expressed in the text to Dr. A. Dollfus and Professor R. R. de F. Mourão for sending prompt news of their recent observations.

The author wishes to express his great appreciation to the Councils of the Royal Astronomical Society and the British Astronomical Association for allowing the unrestricted use of material in their various publications, and to the Council of the Royal Society for a similar concession in regard to the older volumes of the *Philosophical Transactions* and the *Scientific Papers of Sir William Herschel*. Acknowledgment is also made to the Controller of Her Majesty's Stationery Office for many statistics quoted from the *Astronomical Ephemeris* (formerly *Nautical Almanac*); to Signor Abetti and Messrs. Sidgwick and Jackson for allowing translations of Galileo's writings to be quoted from the *History of Astronomy;* and to the Encyclopaedia Britannica for like permission for quoting translations of Copernicus's observations from *Great Books of the Western World,* vol. 16. Many similar acknowledgments should probably be made to other sources; the author hopes that his intention will be inferred from references made to them in the text and in the preface.

Acknowledgment for the use of illustrations is made as follows:

Frontispiece and Plates XII(1) and XIX to the Director of Mount Wilson and Palomar Observatories.

Plates I and II(1) to the Director of the British Museum for allowing them to be photographed from the copy of *Systema Saturnium* in the British Museum Library.

Plate II(2) to the Royal Society.

Plates III and IV(1) and Figures 9a, b to the Royal Society and Royal Astronomical Society.

Plates IV(2), V(2), V(3), VI to IX inclusive, and Figures 3, 13, 18a, b, and 19 to the Royal Astronomical Society.

Plate V(1) to the *Astronomical Journal*.

Plates X, XIII, XV, XVI(2) and Figures 11, 12, 20 and 21 to the British Astronomical Association.

Plate XI to the Director of Lowell Observatory.

Plates XII(2), XVI(2) and also VII and VIII to the Director of Lick Observatory.

Plate XIV to the late Sir Harold Spencer Jones, F.R.S., and the English Universities Press.

Plates XVI(1), XVII, XVIII, XX and XXII(1) to the Director of the Pic-du-Midi Observatory and Dr. H. Camichel and Dr. A. Dollfus. (Special thanks are due to Dr. Dollfus and Dr. Camichel for supplying prints of these items and of Figure 22).

Plate XXI(2) to the Director of Milano-Merate Observatory and Signor G. Ruggieri, and Plates XXI(1) and XXII(3) to Signor Ruggieri, who very kindly copied all three of his drawings specially for this book.

Plate XXII(2) to the Union Astronomer, Johannesburg, and Mr. J. H. Botham (who kindly sent the author a print of his photograph).

Also for Plates V(1), IX, X and XV to the Directors of the Observatories of Harvard, Juvisy, Yerkes and Meudon respectively, and for Plate XIII(1) to the Astronomer Royal.

Figures 4, 5 and 6 to the Dutch Society of Sciences.

24

Figure 7 to the Hutchinson Group.

Figure 10 to the late Sir Harold Spencer Jones and Messrs. Edward Arnold (Publishers) Ltd.

Figures 13 and 14 to the Astronomer Royal and the Controller of H.M. Stationery Office, respectively.

Figures 15, 16 and 17 to Mr. M. B. B. Heath.

Figure 22 to Dr. A. Dollfus.

Mr. J. R. Bazin has shown helpful interest in the book's progress: he made some suggestions for it, and so did Mr. A. P. Lenham.

Thanks are due to the Publishers, Messrs. Faber and Faber Ltd., and especially to Mr. M. Shaw for much helpful advice on the preparation of the book and particularly the illustrations.

The author is grateful to his wife not only for her encouragement, but also for practical help in the research and in revision of the text.

Dorchester, Dorset, England.
1961 November

ABBREVIATIONS

Astronomical Societies and Publications

Abetti	Giorgio Abetti: *The History of Astronomy*, 1954
A.J.	Astronomical Journal
A.L.P.O.	Association of Lunar and Planetary Observers
A.N.	Astronomische Nachrichten
Ann. d'Astroph.	Annales d'Astrophysique
Ap. J.	Astrophysical Journal
B.A.A.	British Astronomical Association
B.A.A.J., Hdbk.	B.A.A. Journal, B.A.A. Handbook
Harvard C.O.A.	Harvard College Observatory Annals
J.R.A.S. Canada	Journal of the Royal Astronomical Society of Canada
Lick O.B.	Lick Observatory Bulletin
Lowell O.B.	Lowell Observatory Bulletin
M.N.	Monthly Notices of the Royal Astronomical Society
McIntyre: *Translucency*	D. G. McIntyre: *The Translucency of Saturn's Rings*, 1935
Mém. de l'Acad. Roy. des Sciences	Mémoires de l'Académie Royale des Sciences
Millás	J. M. Millás Vallicrosa: *Estudios sobre Azarquiel*, Madrid 1943–50
N.A.	*Nautical Almanac* (from 1960 unified with American Ephemeris under title: *Astronomical Ephemeris*)

Nat.	*Nature*
Neugebauer	O. Neugebauer: *The Exact Sciences in Antiquity*, 1952
Occ. N.	Occasional Notes (Royal Astronomic Society)
Oeuvres Complètes	Œuvres Complètes de Christiaan Huygens, Société Hollandaise des Sciences, 1925 etc.
Pannekoek	A. Pannekoek: "The Origin of Astronomy" (*M.N.* vol. 111, p. 347)
P.A.S.P.	Publications of the Astronomical Society of the Pacific
Phil. Trans.	Philosophical Transactions of the Royal Society
Proc. Acad. Sci. U.S.S.R.	Proceedings of the Academy of Sciences of the Soviet Union
Procgs. Amer. Phil. Soc.	Proceedings of the American Philosophical Society
Proctor: *Saturn*	R. A. Proctor: *Saturn and its System*, 1882 edition
Publ. Obs. Nac.	Publicação Observatório Nacional, Rio de Janeiro, Brazil
R.A.S.	Royal Astronomical Society
Roy. Soc.	Royal Society
Sci. Papers	The Scientific Papers of Sir William Herschel, published by Royal Society and Royal Astronomical Society, 1912
Thes. Obs. T.B.	Thesaurus Observationum Tychonis Brahe Dani, edited by J. L. E. Dreyer, 1923–26
Trans. I.A.U.	Transactions of the International Astronomical Union
Webb	T. W. Webb: *Celestial Objects for Common Telescopes*, 1859 etc.

ABBREVIATIONS
General

Acad.	Academy, Académie	Mem.	Memoir(s)
Ann., A.	Annals, Annales	N.	Notes
App.	Appendix	O., Obs.	Observatory
A., Ast.,		Obs., Obsns.	Observations
Astron.	Astronomical	Proc.	Proceedings
Astroph.	Astrophysics,	p.	page; preceding
	Astrophysique	pub.	published
A.U.	Astronomical Units	Publ.	Publication(s)
B.	Bulletin	S., Soc.	Society
Circ.	Circular	Sci.	Science, Scientific
Contrib.	Contribution(s)	Trans.	Transactions
ed.	edited (by)	Univ.	University
f.	following	vol.	volume
J.	Journal		

Star names

The standard three-letter abbreviations for constellation names are used in star names, e.g. Cnc for Cancer, Sgr for Sagittarius.

CHAPTER 1

SATURN AS A WANDERING STAR: ITS ORBIT

Apart from the Moon there are only five planets which are easily visible to the unaided eye, namely (in order of distance from the Sun): Mercury, Venus, Mars, Jupiter and Saturn. Of these Saturn is much the farthest from the Earth also. Although they do not shine by their own light but merely by reflecting that of the Sun, the phases of Mercury and Venus and the slighter phase of Mars cannot be detected without optical aid. Except, therefore, that their light is somewhat steadier than that of the stars, these five planets might seem to give no indication in their mere appearance which would lead a casual naked-eye observer to distinguish them from stars. How was it, then, that the five planets were recognised as being different from the other stars as much as three thousand years or more before the invention of the telescope? What (if anything) can be found out about them by observing them merely with the unaided eye? The short answer to both questions is: their brightness and their peculiar movements in the sky. But so much of interest and importance is covered, particularly by the second half of this answer, that a detailed treatment of it is clearly necessary.

Stellar magnitude

Venus is always far more brilliant than any star; Jupiter is comparable with the brightest star (Sirius) but at opposition is always appreciably brighter than Sirius. Mars varies much in apparent brightness owing to the considerable variation in its distance from the Earth; at a close opposition it can outshine Jupiter and is then all the more conspicuous owing to its fiery red colour, whereas the dull cream or pale yellow of Saturn and the whiteness of the other three planets are tints much more commonly found among the brighter stars. Mercury (when seen with the naked eye) and Saturn are never as brilliant as Sirius, but at maximum light can be brighter than any other star in the northern sky or any but Canopus in the southern.*

* This high rating of Mercury's apparent brightness may surprise British naked-eye observers, accustomed to see that planet always low on the eastern or western horizon when visible at all, dimmed by the dust and haze of the horizon atmosphere and by the bright background of a dawn or sunset sky. But in the Mediterranean region Mercury looks much brighter and more conspicuous, and so was noticed and recognised as a planet in remote antiquity.

M. B. B. Heath, Director of the Saturn Section, has written papers (*B.A.A.J.*, vol. 66, p. 166, 1956 April) giving the dates of oppositions of Mars, Jupiter, Saturn and Uranus between the years 1956 and 2000. His paper on the Oppositions of Saturn is reproduced as Appendix I of this book. From the table in that paper it will be seen that the stellar magnitude of Saturn at opposition can be as low as +0·8, which is somewhat fainter than Vega, Capella, Arcturus, Rigel and Procyon, or as high as −0·3, brighter than any star except Sirius and Canopus. There are therefore several bright stars with whose apparent brightness that of Saturn could be compared with the unaided eye provided the planet happens to be sufficiently near them in the sky for a satisfactory comparison to be made.

Apparent motion

It must, however, have been their movements in the sky rather than their brilliant light which really led to the discovery thousands of years ago that there was something quite different from all other stars about these five planets, i.e. 'wandering stars'. Though the stars have their nightly apparent movement across the sky from east to west (now known to be a mere effect of the Earth's rotation on its axis from west to east) and the planets seem to share in this movement, an attentive watcher will find that though the stars appear to maintain their exact positions relative to each other from year to year, and a succession of watchers will see that they do so from century to century, in a matter of weeks or even days the planets will be found to change their positions in relation to the so-called 'fixed stars'. The main interest of naked-eye planetary observation lies in trying to follow their complicated movements.

Mercury and Venus revolve round the Sun in elliptical orbits inside that of the Earth and are therefore known as 'inferior planets'. They are always to be found in the same part of the sky as the Sun: they appear to recede from the Sun, then approach it, disappear more or less behind it, recede again on the other side, again approach and pass more or less in front of it, the whole performance appearing to an observer on the Earth to take place for Mercury in about 116 days, and for Venus in about 584 days. During part of their receding and approaching movement they are seen as evening stars after sunset in the west, and during the other part as morning stars in the east before sunrise. This has been mentioned to make it clear that the apparent movement is entirely different in the cases of Saturn, Jupiter and Mars, which are 'superior' or 'exterior' planets, revolving round the Sun in elliptical orbits outside that of the Earth. Hence to an observer in the northern hemisphere they appear, when visible in the night sky, to move like most of the stars across all or some part of the southern sky each night from east to west. Periodically the Earth, moving faster in a smaller orbit, crosses almost in line between one or other of them and the Sun, and then there is an opposition. This happens in the case of Saturn at an average interval of 378 days and this is called the synodic period; from the tables of dates of opposition in Appendix I and in this chapter it will be seen that the length of the interval varies a little, from about 377 to 380 days. When Saturn is in opposition it is considerably nearer the Earth than at other times, and is diametrically opposite to the Sun in the sky, rising at sunset in the eastern sky, crossing the south meridian at midnight, remaining above the horizon all night, and setting towards the west at sunrise. It is therefore most favourably placed for observation round about the dates of oppositions, but although at the Equator all oppositions are almost equally favourable to the observer, in higher latitudes some oppositions are much more favourable than others owing, as will be explained further on in this chapter, to differences between them in the planet's altitude above the horizon.

Though they appear to move nightly with the stars from east to west, in actual fact the superior planets, like the Moon, are moving slowly from west to east in relation to the background stars. As Saturn takes nearly 29½ years to revolve once round the Sun, its slow west to east movement round the sky takes that length of time, and the change of position is so gradual that it is only when Saturn happens to be close to the direction of a bright star such as Regulus or Spica that the planet's motion can be readily perceived in the course of a few days. As the inclination of the plane of Saturn's orbit to that of the Earth is only about 2½°, Saturn is always near the ecliptic, the great circle round the sky which represents the Earth's orbit plane and along which the Sun appears to move round the sky from west to east in the course of a year. Owing to the small inclinations of their orbits to the ecliptic plane, all the major planets (except sometimes Pluto whose inclination is greater) are always to be found somewhere within a belt of sky extending for 9° north and 9° south of the ecliptic and known as the zodiac.

Regression and stations (Heath)

The fact that the Earth is moving round the Sun faster than the exterior planets in a smaller orbit causes a peculiar complication in the apparent motion of Saturn and the others with reference to the background stars. As the Earth is overtaking and passing Saturn (or any other superior planet) around the time of opposition, the latter appears to reverse its motion for a time and travel *westwards* among the stars. This phenomenon has been well described by E. W. Maunder (*Astronomy without a Telescope*, 1902, p. 148) as follows: 'For the greater part of the year Saturn is moving eastwards among the stars at an average rate of a degree in about eight days . . . but gradually as the time of opposition draws on, the speed . . . diminishes, until Saturn comes to a stop about 70 days before opposition. . . . Then for 143 days Saturn moves westward . . . , becoming stationary again at the end of the period, and then resuming once more (its) eastward march. This period of westerly movement or retrogression marks the time when the planet is nearest the Earth and therefore brightest, and it is at the middle of this period that the planet is in opposition, that is to say, on the meridian at midnight.' In quoting the paragraph references to Jupiter have been omitted.

By a substitution of appropriate periods the description could apply to any of the superior planets.

It may be as well to point out that the periods given by Maunder for Saturn's retrogression are approximate only and require some adjustments according to the planet's position in its orbit. Thus, when, as in 1944, Saturn was in perihelion the first station occurred 66 days 21½ hours before the opposition and the second station 66 days 19½ hours after it; but in 1959, when Saturn was in aphelion, the first stationary point was reached 70 days 12 hours before opposition and the second station 70 days 22 hours after it. Thus in 1944 the planet's regression lasted only 133 days 17 hours, but in 1959 it continued during 141 days 10 hours.

Year	Conjn.	Staty.	Oppn.	Staty.	Period of Regression
1960		Apr. 27	Jul. 7	Sep. 15	141 days 7 hours
1961	Jan. 11	May 9	Jul. 19	Sep. 27	140 days 22 hours
1962	Jan. 22	May 22	Jul. 31	Oct. 9	140 days 10 hours
1963	Feb. 3	Jun. 4	Aug. 13	Oct. 21	139 days 19 hours
1964	Feb. 15	Jun. 15	Aug. 24	Nov. 2	139 days 6 hours
1965	Feb. 26	Jun. 29	Sep. 6	Nov. 14	138 days 15 hours
1966	Mar. 10	Jul. 12	Sep. 19	Nov. 27	137 days 22 hours
1967	Mar. 23	Jul. 26	Oct. 2	Dec. 10	137 days 4 hours
1968	Apr. 4	Aug. 7	Oct. 15	Dec. 22	136 days 11 hours
1969–70	Apr. 18	Aug. 21	Oct. 28	Jan. 4	135 days 20 hours
1970–71	May 2	Sep. 4	Nov. 11	Jan. 18	135 days 6 hours
1971–72	May 17	Sep. 19	Nov. 25	Jan. 31	134 days 17 hours
1972–73	May 30	Oct. 2	Dec. 8	Feb. 13	134 days 4 hours
1973–74	Jun. 14	Oct. 16	Dec. 22	Feb. 27	133 days 17 hours

Mr. M. B. B. Heath has kindly calculated and supplied to the author particulars of Saturnian stations, periods of regression, and dates of oppositions and conjunctions with the Sun for the years from 1960 to 1973–74. As these particulars may well be both interesting and useful to readers they are given in the preceding table.

R. A. Proctor has pointed out how very slow is Saturn's retrograde motion: only three degrees in the period of over two months from opposition to the second stationary point, on reaching which the planet resumes its eastward progress. In the course of the synodic period of about 378 days Saturn advances during nearly eight months and retrogrades slowly for about four and a half months, so that on balance the planet does manage to progress eastward along the zodiac by some twelve degrees per annum.

All this is illustrated by Fig. 1 which shows Saturn's path through parts of the constellations Cancer and Leo during the synodic period from the conjunction of 1947 August 5 to that of 1948 August 19, plotted from the daily positions given in the *N.A.* for those years. The direct apparent motion for some time after (and before) conjunctions is comparatively rapid, the apparently retrograde motion through opposition between stations slower; when approaching a stationary point the planet seems to slow to a halt and at the station make a sharp turn. In this example a long narrow loop in the track (on the side away from the ecliptic) was traced by Saturn having appeared at the beginning of 1948 July to cross its former path near where it had been at the end of 1948 January. There are no bright stars in this small area of sky; the places at epoch 1950·0 of a few of the nearby fifth and sixth magnitude stars of Cancer and Leo are shown. (Proctor: *Saturn*, plate 3, embraces on a smaller scale the planet's apparent path for seven synodic periods.)

Inferences from naked-eye observation (Proctor)

Proctor (*Saturn*, chap. 1) thoroughly investigated, with much geometrical detail, the inferences that might be drawn by an observer without optical aid, from Saturn's slow retrograde motion around opposition and long period of retrogression, and the small arc passed over by the planet in that period. For the full details Proctor's chapter should be read; here only a summary of the main points will be given. The parallax effect of Saturn's apparent backward movement as the faster-moving Earth is overtaking and passing it can easily be understood by anyone who, from a car overtaking and passing a bicycle moving in the same direction, has noticed the

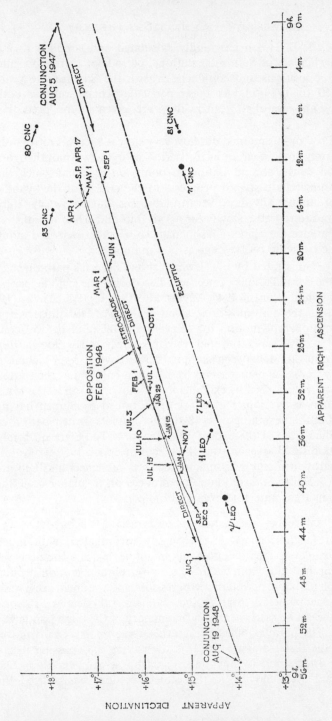

FIGURE 1. Apparent path of Saturn 1947–48. Based on daily apparent places of Saturn given in *Nautical Almanac*. (Arrows parallel to Saturn's path show direction of motion; others merely indicate position).

apparently backward movement of the bicycle with respect to the distant scenery. The stations are accounted for by the fact that the Earth, in a small part of its elliptical orbit some time before opposition, will be moving almost directly towards Saturn, and similarly some time after opposition almost directly away from the planet; in each case the Earth's direction of motion will exactly annul its superior velocity and Saturn will appear to be stationary. After the second station has been passed, the Earth moves on towards conjunction (when Saturn will be placed on the other side of the Sun). During this period the Earth's motion, with reference to the line of sight to Saturn at any instant, will be in a contrary direction to the planet's motion, which it therefore seems to augment on the celestial sphere. Proctor then pointed out, with a geometrical proof, that if Saturn were visible when in conjunction its motion would appear faster then than at any other time, but of course no planet is visible to the unaided eye for some time before and after as well as at conjunction because it is lost in the light of the Sun. The effect of this swifter motion can, however, be detected, because when Saturn is next seen it will be found to have moved farther along the zodiac than the observer might have expected from its customary slow progress. The progressive motion then gradually diminishes until the next stationary point is reached, and the whole cycle of movement begins again.

Proctor further suggested that the observer could infer from the slowness of Saturn's retrogression when in opposition, that this must arise either from Saturn's velocity being very nearly equal to that of the Earth, or else from the planet's orbit being very large compared with that of the Earth. He went so far as to suggest that an observer (if aware that both planets revolve round the Sun) could have deduced these alternative possibilities within a week of Saturn's discovery! But from the observation that Saturn retrogrades for nearly two months and a half after opposition, the observer might have been led to the correct alternative: during this time the Earth would have gone over nearly a quarter of its orbit, and hence the Earth's direction of motion would have become inclined at a very small angle to the line of sight to the planet, and only then does Saturn appear stationery. It follows from this that the Earth's motion is much faster than Saturn's; hence the slowness of the planet's retrograde movement must be due to its being at a vast distance from the Earth.

Having reached this conclusion that Saturn's orbit is much larger than the Earth's, but without knowing the size of that orbit, the observer could (in Proctor's opinion) deduce, either by geometry or a trigonometrical calculation based on naked-eye observations extending over a few months, that Saturn's velocity is in the proportion of approximately 11 : 34 compared with that of the Earth. Proctor then gave two alternative methods for determining from this that Saturn's distance from the Sun is about $9\frac{1}{2}$ times the Earth's distance from the Sun. From these results it follows at once that the ratio of Saturn's period of revolution to that of the Earth (one year) is as $\frac{19}{2} \times \frac{34}{11} : 1$, or as $29\frac{1}{2} : 1$ very nearly. Hence Saturn's year is equal to about $29\frac{1}{2}$ terrestrial years.

Nodes of the orbit

A further complication of Saturn's motion through the zodiac arises from the fact that the planet's orbit plane does not coincide with the ecliptic plane but is inclined to it at a small angle, about $2\frac{1}{2}°$. As two planes intersect in a straight line, it follows that there will be two points common to the ecliptic and Saturn's orbit, or, in other words, two points on opposite sides of its orbit at each of which Saturn crosses the ecliptic; these are called the nodes. About one-half of the planet's orbit lies north of the ecliptic and the other half to the south of it. The ascending node, the point where Saturn crosses the ecliptic from south to north, is at the present time in ecliptic longitude about 113°, in the constellation Gemini, fairly near the star 79 Geminorum. The descending node, where Saturn crosses the ecliptic to go south of it, is therefore at about longitude 293°, in Sagittarius fairly near 56 Sagittarii. Although, as Proctor said (*Saturn*, p. 26), the nodes regress right round the ecliptic very slowly in a period of about 66,000 years, this would only hold good if they were referred to a fixed equinox and ecliptic. In a footnote he pointed out that precession must be added, since everything is referred to a moving equinox; this completely swamps the regression, and causes the nodes to advance by half a minute of arc a year. The inclination of Saturn's orbit to the ecliptic at the present time decreases very slightly but steadily year by year, but may at some time in the distant future reach a minimum and begin to increase again.

Loops in the track

The most striking effect of the inclination of Saturn's orbit plane to the ecliptic is that when the planet regresses and then again moves directly, the two movements do not take place to and fro on the same line but result in the formation of a long narrow loop in that part of the planet's path, so that Saturn's track through the zodiac consists of a series of lines and loops alternating (see Fig. 1). When the planet is near a node the path is twisted but without a loop; elsewhere in the orbit there are definite loops in the track between each pair of stationary points. The loops or twists are always turned away from the ecliptic, lying to the north of the planet's main line of advance when it is north of the ecliptic, and on the south side of the part of Saturn's track that lies south of the ecliptic. Proctor in his first chapter gave a geometrical explanation of the causes of these phenomena. The looping of the track in connection with retrograde motion applies to all the superior planets, as their orbit planes are all inclined to the plane of the ecliptic.

Opposition positions (Heath)

For a few months in each year round about conjunction, Saturn is accompanying the Sun in the daylight sky and is not available for observation. The apparitions (periods before, during and after opposition) are not all equally favourable to the observer in latitudes such as that of the British Isles. Winter oppositions, when Saturn is high in the night sky and observable most of the long night, are the best: the planet is then to be seen in the direction of the stars of such constellations as Taurus, Gemini, Cancer or Leo. At summer oppositions when Saturn is in Libra, Scorpio, Ophiuchus, Sagittarius or Capricornus, the planet is much lower in the sky, and is above the horizon at night for a more limited period, the interval between sunset and sunrise being much shorter, so the observing conditions are less favourable. Another of Mr. Heath's valuable contributions to this book is the following table of the approximate place of Saturn at every opposition for the 31 years 1943–73 covering slightly more than a Saturnian year. The table shows that Saturn in the course of $29\frac{1}{2}$ years travels right round the zodiac, passing through not only the twelve well-known zodiacal constellations but also a thirteenth (Ophiuchus). The opposition positions of Saturn given in the table are sufficiently accurate to enable a naked-eye observer with a star atlas and some familiarity with the stars to find the bright planet quite easily round about each opposition for the next dozen years, and also to get a rough idea from his star maps of the course of Saturn among the stars throughout a Saturnian year (1943–72). The ring aspect of each year noted in the last column can only be seen with a telescope, and will be discussed in Chapter 5.

The opposition of 1973 should be a particularly favourable one in Great Britain, given clear skies, as it occurs at mid-winter with Saturn crossing the meridian some 60° above the horizon. Moreover, as the planet will be near η Geminorum it will also be close to its perihelion, and so at minimum distance from Sun and Earth, and as at the same time the rings will be widely open, the stellar magnitude of Saturn will have the maximum value of $-0\cdot3$, as stated in Appendix I.

Date of Opposition	Position of Saturn in sky	Ring aspect
1943 Dec. 16	Taurus, near β	Rings (South face) widest open during year
1944 Dec. 29	Gemini, near ϵ	
1945 (No opposition in year)		
1946 Jan. 12	Gemini, near Pollux	South face of rings on view
1947 Jan. 26	Cancer, near η	
1948 Feb. 9	Between Cancer and Leo	
1949 Feb. 21	Leo, near Regulus	
1950 Mar. 7	Leo, near χ	Rings edgewise to Earth September 14
1951 Mar. 20	Virgo, near β	
1952 Apr. 1	Virgo, near γ	
1953 Apr. 14	Virgo, near Spica	
1954 Apr. 26	Between Virgo and Libra	North face of rings on view
1955 May 9	Libra, near α	
1956 May 20	Scorpio, near β	
1957 Jun. 1	Ophiuchus, near ω; not far from Antares	
1958 Jun. 13	Ophiuchus, near θ	Rings (North face) widest open during year
1959 Jun. 26	Sagittarius, near λ	
1960 July 7	Sagittarius, near π	
1961 July 19	Sagittarius, near ω	
1962 July 31	Capricornus, near ρ	North face of rings on view
1963 Aug. 13	Capricornus, near ι	
1964 Aug. 24	Aquarius, near θ	
1965 Sept. 6	Aquarius, near λ	
1966 Sept. 19	Pisces, near λ	Rings edgewise to Earth thrice during 1965–66
1967 Oct. 2	Pisces, near δ	
1968 Oct. 15	Pisces, near ζ	
1969 Oct. 28	Between Pisces and Aries	South face of rings on view
1970 Nov. 11	Aries, near δ	
1971 Nov. 25	Taurus, near Pleiades	
1972 Dec. 8	Taurus, near ι	
1973 Dec. 22	Gemini, near η	Rings (South face) widest open during year

Distance and orbital speed

It may surprise the reader to find in the next chapter evidence that oppositions, retrograde motion, stations and the synodic period were familiar to naked-eye observers more than two thousand years ago, and that much of the explanation given in this chapter is really necessary to the understanding of the observations of the ancient Babylonians and Ptolemy. It was, however, impossible for the ancient and mediaeval astronomers, working on the assumption that the Sun and planets revolved round the Earth, to get any idea of the true distances of the planets or of their orbital speeds, though they deduced correctly from Saturn's slow apparent motion with respect to the stars that it was the most distant of the planets known to them.

It is now known that Saturn's orbit is much less elliptical than those of Pluto and Mercury, and appreciably less so than that of Mars, but has a greater eccentricity than those of the other major planets. Hence its distance from the Sun has a range of almost 100 million miles, from about $837\frac{3}{4}$ million at perihelion to about $936\frac{1}{2}$ million miles at aphelion (for 1960, assuming the astronomical unit to be 93 million miles, but subject to large variations at different dates). Saturn's mean distance from the Sun is therefore more than 1·8 times that of Jupiter and more than 9·5 times that of the Earth. Saturn's mean distance even at opposition is so great that its light, travelling at over 11 million miles a minute, takes about one hour and twelve minutes to reach the Earth.* A simple calculation will give the further illustration that if a space probe reaches the Moon in $1\frac{1}{2}$ days, one starting at a similar speed would take over 13 years to reach the orbit of Saturn, but perhaps only $7\frac{1}{2}$ years if a space vehicle can reach Venus in three months.

Owing to its remoteness from the Sun, the planet's linear velocity in its orbit is comparatively low: the range is from about $5\frac{2}{3}$ to $6\frac{1}{3}$ miles per second, the mean being 6 miles per second. This is less than three-quarters of the mean linear velocity of Jupiter, less than one-third that of the Earth, and not much more than one-fifth that of the innermost major planet, the fast-moving Mercury.

Orbit particulars

It seems appropriate to end this chapter by giving a list of statistics relating to Saturn's orbit, reserving for later chapters (especially

* Saturn's least possible distance from the Earth, at a perihelion opposition (towards the end of December), is about 744 million miles; its greatest distance, at an aphelion conjunction, is about 1,028 million miles (Spencer Jones: *General Astronomy*, 1951, p. 249).

Chapter 23) others of the planet (mass, diameter, volume, density, etc.), the ring system and the satellites.

These statistics are all derived from the *N.A.* Most of them appear in the *B.A.A. Hdbk.* 1960, but those of linear orbital velocity and daily motion at perihelion and aphelion have been calculated by M. B. B. Heath. The assumed value of the astronomical unit is 92,900,000 miles.

Saturn's mean distance from Sun :

Millions of Kilometres	1,425·7
Millions of Miles	886·1
Astronomical Units (Earth = 1)	9·538843
Mean Sidereal Period of Revolution in Tropical Years	29·45772
Sidereal Mean Daily Motion	0°033460
Daily Motion at Perihelion	0°0374649
Daily Motion at Aphelion	0°0299762
Mean Synodic Period (Opposition to Opposition) in Days	378·09
Mean Linear Orbital Velocity (miles per second)	6·00
At Perihelion	6·340
At Aphelion	5·671

The following quantities vary a little from year to year, the values stated below being approximations for the years round 1960:

Mean Longitude of the Perihelion (direction at present near η Gem.)	92°
Mean Longitude of the Ascending Node (direction at present near 79 Gem.)	113°
Inclination of Orbit to Ecliptic	2° 29′
Eccentricity of Orbit	0·0557

CHAPTER 2

ANCIENT UNAIDED EYE
OBSERVATIONS OF SATURN

(A) ANCIENT BABYLONIAN DATA

The earliest known records of observations of Saturn were made in Mesopotamia and can be reliably dated to the mid-seventh century B.C., and the way in which this astronomical activity arose and developed can only be explained by fairly frequent reference to the historical background. Very much earlier, long before the rise of the first civilisations some 5,000 years ago, primitive man must have been forced to use the Sun and stars for direction-finding on journeys, and when he began to engage in agriculture, to watch the variation in the Sun's declination and look for the heliacal rising of stars just before sunrise so as to ascertain the seasons, and to use the Moon's synodic period of $29\frac{1}{2}$ days as the basis of his calendar.

The late Professor A. Pannekoek (*M.N.*, vol. 111, p. 347) said that the transition from barbarism to civilisation was marked by the invention of script and the rise of a ruling class; about 5,000 years ago the first civilisations arose in various parts of the world on fertile plains watered by great rivers: agriculture was organised and city states were established. In Mesopotamia these developments seem to have coincided with the settlement, in the southern part of that country, of the Sumerians about 3000 B.C. The regulation of the calendar became one of the chief functions of the Sumerian priesthood; the priests looked from their *ziggurat* (terraced pyramid) temples for the first appearance of the thin crescent Moon so as to be able to announce the new month, and they had to decide when, every two or three years, to add a thirteenth month to make their calendar conform with the seasons for the purposes of agriculture and religious festivals.

Omens from observation of planets

When the bright planets first came to be distinguished from the stars, and when they began to be observed, are unknown. The earliest evidence of planetary observation is the description of the phenomena of Venus on the *Nindar-anna* tablet, believed to date from about 1600 B.C. By that time the Sumerian race had become merged in the

Semite population, and the independent, often warring, city states had come under the rule of the Babylonian kingdom. This in turn became, in the centuries after 1000 B.C., a frequently rebellious province of the despotic military empire of Assyria. It was during the turbulent centuries of Assyrian domination over all the neighbouring states that the services of the astrologer-priests came into great demand, and this in turn caused more attention to be paid not only to the Moon but also to the planets. As Pannekoek pointed out, their irregular motions among the stars and the almost endless variety of their phenomena were thought to provide the astrologers with signs from the gods which they could use to foretell the future. It was in this way, and for this purpose, that, for the first time in history, a large number of data came to be accumulated about the planets, and knowledge of facts about their motion increased.

After the collapse of the first Assyrian empire and the end of the dynasty, the second Assyrian empire was founded in 745 B.C. by a military adventurer who assumed the title of Tiglath-Pileser III. He was an able organiser, who, according to Dr. Sayce (*Assyria*, 1926, p. 44), tried to consolidate his conquests by carrying out large scale centralisation for the first time in that region, and replacing the old feudal nobility of Assyria by a bureaucracy. The astrologer-priests became an influential branch of the civil service, making astronomical observations for the purposes of calendar adjustment, weather forecasting and above all the production of omens, and sending frequent reports to the king. Such reports seem to have been especially numerous, or at least more carefully preserved, during the reigns of Esar-haddon (681–669) and his son Asshur-bani-pal (669–626 B.C.). The latter, unlike his predecessors, was unwarlike and chiefly interested in collecting ancient Sumerian and Babylonian literature for his palace library at Nineveh, presumably relying on his astrologers to warn him of imminent dangers, and on his generals to overcome them.

The source for the reports of the astrologer-priests is A. Campbell Thompson's *The Reports of the Magicians and Astrologers of Nineveh and Babylon in the British Museum*, 1900, which contains transcriptions, transliterations and in many cases translations of nearly 300 reports found in the ruins of the royal library at Nineveh. A great many of these concern the Moon and the other bright planets, but of Saturn there are not more than a dozen.

The omens were not all of war and political affairs; many concerned daily life, trade, agriculture and weather. This is illustrated by the following rather startling statements about Saturn:

'When a planet changes colour opposite the Moon and enters the Moon, lions will die and the traffic of the land will be hindered (or) cattle will be slain. Saturn has entered the Moon. . . . '(No. 175)

'When a halo surrounds the Moon and a planet stands within it, robbers will rage. Saturn stood within the halo of the Moon. . . . When Mars reaches the path of the Sun, there will be a famine of cattle. . . . Mars reached Saturn.' (No. 103)

Sayce stated that the system of augury was based on the false assumptions: (a) that an event was caused by a phenomenon in the sky that had immediately preceded it, and (b) that therefore if the phenomenon recurred the event would again follow. Hence the astrologers seem to have carefully recorded any terrestrial event that had happened shortly after each observation, and thus had compiled collections of precedents. Once an observation was made the production of an omen was pure routine and childishly simple: find a similar observation in the records and copy out the supposed result! The futility of this method is underlined by the fact that often recurring phenomena would have several different omens in the collection. They sometimes tried to cover this by giving two alternatives, as in No. 175 above. They may have taken more trouble over the astronomical observations than over the forecasts: one report says the observer had himself roused seven times during the night to try to observe a lunar eclipse.

Saturn about 650 B.C.

The reports of the astrologer-priests must all be earlier than 612 B.C., when the Assyrian empire was finally overthrown and the palace and library at Nineveh were burned: the huge collection of literary, grammatical and other works, being inscribed in cuneiform characters on kiln-baked bricks, on the whole escaped serious damage in the fire. References to events appear to date Nos. 22 and 264 to 660 and 655 B.C. respectively, and there is a purely astronomical report about the observation of an expected lunar eclipse being prevented by clouds which exceptionally contains an exact date corresponding in modern reckoning to 649 B.C. Hence, it seems fair to presume that most of these reports belong to the mid-seventh century B.C., and as each begins with a reference to an earlier observation of the same phenomenon, that similar observations were being made and recorded in the early part of that century, if not before.

The purely astronomical parts of the few Saturn reports are as follows:

'Saturn stood within the Moon's halo.' (Nos. 90, 98, 103, 180)

'Last night a halo surrounded the Moon; Saturn stood within it near the Moon.' (No. 144)

'Last night Saturn drew near to the Moon. Saturn is the star of the Sun.' (No. 176)

'When a planet changes colour opposite the Moon and enters the Moon. . . . Saturn has entered the Moon.' (No. 175)

'Saturn has not once approached Venus.' (No. 67)

'Mars left an interval of four (degrees ?) away from Saturn; it did not approach.' (No. 88)

'When Mars reaches the path of the Sun. . . . Mars reached Saturn.' (No. 103)

'Saturn in front of Regulus.' (No. 180)

'Saturn has appeared in Leo.' (No. 216)

These primitive observations are, as might be expected, very inexact: Saturn being inside the *tarbatsu* (smaller halo) of the Moon, while sufficient for the omen, means nothing astronomically, as a radius of 22° from the Moon covers a large area of sky. 'Saturn is the star of the Sun' is explanatory; a legendary association between these two bodies was an excuse for making reports even more ambiguous by using the word for Sun when Saturn was meant. A further ambiguity was caused by using *lubad* ('planet') sometimes for Saturn, sometimes for Mercury. According to R. Brown (*Researches into the Origin of the Primitive Constellations* vol. 1, 1899, p. 346), *lubad* meant 'old sheep', and the name mostly used for Saturn in these reports, *lubadsagush*, meant 'the eldest old sheep', presumably because it wandered very slowly among the stars. Perhaps the most interesting of the reports is No. 175, recording an occultation of Saturn by the Moon and referring to the dimming of the planet by contrast when close to the Moon's bright limb.

It is interesting that Regulus ('prince') was called *sharru* ('king') by the Assyrians. The reports show that they had names for at least all the zodiacal constellations, for the Pleiades, and for several bright stars, such as Spica and Aldebaran, while there were six alternative names for Mars and at least five for Jupiter, a different one being used for the latter when rising, another when two hours after rising, and a third when on the meridian. They also had names for the phases of the Moon in five-day periods, and called its four quarters after the four main divisions of the empire, so as to describe the incidence of partial eclipses; the poetic term *atal mâtâti* (darkness of the countries) therefore signified a total eclipse of the Moon.

State of astronomy 650 B.C.

Some of the reports on other planets are far more revealing than those on Saturn as to the state of astronomical knowledge at that period. Nos. 205 and 208 refer to Venus disappearing at sunrise and at sunset respectively; Nos. 221 and 226 state respectively that Mercury was visible at sunrise, at sunset. Hence they knew that Venus and Mercury were sometimes morning, sometimes evening stars. Then consider the following:

'. . . Mars has turned and proceeded into Scorpio.' (No. 70)

'. . . Mars stands and waits in the sting of Scorpio. . . .' (No. 236)

'. . . Jupiter has stood for a month over its reckoned time. . .' (No. 183)

These not only indicate knowledge of stations and regression, but the last one strongly suggests computations being made about these phenomena. No. 272 gives a description of an apparent regression by Jupiter whose path seemed to form a loop near Regulus. No. 112 contains the following definition of 'planet': 'The planets are those whose stars pass on their own road over themselves'. This, while far from clear, suggests a reference to alternate direct and retrograde motion. No. 170 contains the intriguing sentence '. . . The ?? stopped last night. . . .'; the translator suggested the word 'clock' for the unknown sign, and this seems not unlikely, because, as will be seen presently, the priests at this period were able to measure time both day and night in 2-hour units.

Pannekoek considered that there is no reason to think that these ancient observers compared observations to find regularities; there is a much greater likelihood 'that the regularities in the planetary phenomena gradually forced themselves upon the attention of the observers, arousing expectations, which developed into (astronomical) prediction.' It is quite evident that by this period they had found a way of predicting lunar eclipses, e.g.: 'The night of the 13th, the night of the 14th a watch was kept but no eclipse occurred', 'when the Moon appears on the 30th of Sebat (January–February), a total eclipse will happen', 'on the 14th day a watch of the Moon we kept—the Moon was eclipsed.'

The last of these quotations is from one of a few astronomical reports (without omens) of the period, reproduced in L. W. King's *First Steps in Assyrian* (1898). Sayce (p. 137) quoted a report of an unsuccessful attempt to observe a solar eclipse at new Moon. Pannekoek has pointed out that the prediction of a lunar eclipse is very simple: 5 or 6 follow each other regularly at intervals of six months, and then a new series starts 11 or 17 months later. Hence if an eclipse of the Moon has been observed the chances are 5 to 1 in favour of another six months later, though about half of them occur in the daytime when the Moon is below the horizon.

Sayce has pointed out that the gnomon or dial (referred to in 2 Kings xx. 11) set up at Jerusalem by Ahaz, who was reigning about 735 B.C., was probably obtained from Assyria, and it seems almost certain that by about the seventh century B.C. the astrologer-priests in the Assyrian empire were using not only the gnomon but also the clepsydra (water clock) to measure time. A determination of the date of the vernal equinox (King: *First Steps in Assyrian*, p. 282) states: 'On the 6th day of Nisan (March-April), the day and the night were equal. Of six *kasbu* the day, of six *kasbu* the night. . .' The *kasbu* was equivalent to two hours. In later times when more exact measurements of the positions of stars and planets were made, distances were sometimes expressed in degrees, sometimes by the weight of water that had passed through the clepsydra between the culmination of one star and that of another, which is equivalent to the time interval (G. Contenau: *Everyday Life in Babylon and Assyria*, 1954, p. 304).

In the mid-seventh century B.C. purely astronomical reports without omens were fairly numerous, most of them relating to the appearance of the Moon or eclipses (see King pp. 282–85, Sayce pp. 136–38, and especially Leroy Waterman: *Royal Correspondence of the Assyrian Empire*, pub. 1930–36), so the difficult question arises whether a scientific interest in the phenomena was already arising. Pannekoek thought not: that the sole interest in the observations lay in their use for omens or weather forecasts or calendar adjustment. On the other hand, A. T. Olmstead ("Babylonian Astronomy," *American Journal of Semitic Languages and Literatures*, vol. 55, p. 113) traces the evolution of a more scientific attitude in the eclipse reports and the correspondence about them between king and priests in the reigns of Esar-haddon and Asshur-bani-pal. The most striking example is a peremptory letter from the former king asking for a definite reply whether an eclipse of the Sun will take place or not. The observer has to confess that he is not sure: '. . . The sign is not clear, I am cast down, I do not understand it. . . .' But there is no clear indication of the motives of either kings or observers, though the accumulation of data inevitably led to more exact astronomical knowledge.

Transition to exact science: 550–300 B.C.

In 612 B.C. the unstable Assyrian empire finally collapsed and for the next seventy years the kingdom of Babylon again became the leading power. The regular recording of eclipses is said to have begun there 135 years before, and Olmstead has given particulars of two tablets, supposed to date from 577 and 568 B.C., the first giving the approximate synodic periods of the planets, and the second very precise distances for them from certain bright stars. The exactness of the information in these tablets raises serious doubts whether they really belong to this period and seems to imply a very much later date, especially as the general opinion among authorities seems to be that no spectacular advance towards mathematical astronomy occurred until the late sixth century, after the Persian conquest.

The absorption of Babylonia and Assyria from 539 B.C. into the far greater Persian empire, the end of local wars, and the consequent decline in the demand for omens, was, in Pannekoek's view, the main cause which brought about the study of astronomy for its own sake, and the beginning of the production of tables consisting mostly of figures and resembling the modern ephemeris. Olmstead has given an example of one of the very earliest of such tables, reliably dated to the seventh year of Cambyses (522 B.C.). He also states that an early third-century B.C. textbook on the construction of lunar and eclipse tables attributes the system used to an eminent Babylonian astronomer and mathematician called Nabu-rimanni who flourished about 490 B.C. This agrees well with the opinion of O. Neugebauer (*The Exact Sciences in Antiquity*, 1952), who considers that the development into systematic mathematical theory probably took place rather rapidly about or shortly after 500 B.C. Until about 480 B.C. there was no regularity in the adjusting of the lunar calendar; a century later a systematic adjustment seems to have been used. He mentions a text of 419 B.C. which for the first time showed a division of the zodiac into twelve parts each 30° in length. Positions could be accurately determined by degrees and zodiacal signs, although it seems that actual positions in the sky continued generally to be expressed with reference to bright stars. It is very unfortunate that so few texts are available for this crucial transition period, for by 300 B.C. at latest, mathematical astronomy in Babylonia was, in Neugebauer's opinion, fully developed.

Saturn tablets in late Babylonian times

The Persian empire was overthrown by Alexander the Great in 331 B.C., and for a century and a half from 312 B.C. Babylonia was under the rule of the Greek Seleucid dynasty, and then for the last century and a half of the pre-Christian era it formed part of the neo-Persian Parthian kingdom. It was during those three centuries that Babylonian knowledge of planetary phenomena reached the stage of pure arithmetical theory.

Ptolemy has referred in his *Syntaxis* (book XI, chap. vii) to a Babylonian observation of Saturn made in the evening of a date corresponding to 228 B.C., probably January 9* when Saturn was described as being two 'fingers' below the fixed star of the southern shoulder of Virgo. This star was presumably η Virginis*, and the width of two fingers held at arm's length would cover on the sky about two degrees.

Neugebauer has pointed out that the backbone of Babylonian mathematical astronomy was the equating of x intervals of one sort with y intervals of another sort, e.g. 2 complete journeys of Saturn round the zodiac take approximately 59 years. A very good example of the application of this principle is found in a tablet cited by Pannekoek (from F. X. Kugler, S.J.: *Sternkunde und Sterndienst in Babel*, 1907, vol. I, pp. 59, 85), entitled: 'The first day, the phenomena, the motions and the eclipses for the year 140' (of the Seleucid era, presumably corresponding to 173–72 B.C.). It gives the phenomena of Saturn of the Seleucid year 81, since they would be repeated at approximately the same places in the year 140; those for Venus of the year 132, since 13 revolutions of Venus are nearly equal to 8 terrestrial years, and so after every 8 years the elongations and conjunctions of Venus are repeated at about (2 or 3 days earlier than) the same dates. The tablet gives similar information for the other three planets.

According to Neugebauer the tablets of this period fall into two classes: 'procedure texts' which give the rules for the computation of 'ephemerides', which give the lunar and planetary positions at regular intervals for a particular year or series of years. Available tablets from the Seleucid period are not numerous: out of a total of about 100 for the planets the great majority relate to Jupiter and comparatively few to Saturn. Neugebauer suspects that there may be thousands in museums still awaiting transcription and publication.

Pannekoek said that 'a still higher and more perfect' kind of theoretical knowledge was achieved by Babylonian astronomers during the Parthian period of the last 150 years before the Christian era. Badly damaged fragments of tablets of that time dug up from the ruins of city temples contain nothing but rows of numbers in columns with names of months and zodiacal signs. Father Kugler, an able astronomer as well as an Assyriologist, discovered that each

* Hind (*Solar System*, p. 117) has 'γ Vir.' and 'March 1.'

column gives the successive values of the longitude and date of one of the special phenomena, e.g. opposition position, heliacal rising, heliacal setting. There was no description but just columns of very exact numbers in sexagesimal fractions, showing periodical increases and decreases of the successive intervals. It is thought probable that these computations may have been derived from comparatively few original observations, and that as the Babylonians had now reduced all the planetary phenomena to pure arithmetical theory which could be calculated back into the past or forward into the future, it is likely that they felt that reference to actual observations was no longer necessary.

It is worth noting that modern astronomy owes to the Babylonians among other things the division of the zodiac into 360 degrees, and the sexagesimal system which still survives in the reckoning of time and angles, e.g. 9^h 25^m 32^s is really the sexagesimal number

$$9; \ 25, \ 32 \quad \text{or} \quad 9 + \frac{25}{60} + \frac{32}{60^2} \quad \text{thinly disguised.}$$

Tribute has been paid by Neugebauer to the high level of mathematical theory of lunar and planetary motion attained by the Babylonians in this latest period: the last known of their tablets relates to A.D. 75. At the same time he suggests that the value of their observations has been overrated. They were most interested in observing phenomena close to the horizon, such as the appearance and disappearance of planets and the determination of the date of opposition by observing their rising at sunset or setting at sunrise. Ptolemy, complaining of the lack of reliable planetary observations by them, remarked that the inaccuracy was due to such phenomena and those of stationary points being of their nature difficult to observe. Moreover, Pannekoek's view is that the final stage of Babylonian astronomy, admirable as it is, amounts to 'merely formal theory, entirely devoid of any physical interpretation. . . . Its mathematics is only arithmetic, with no geometry.' It would seem that the Babylonian astronomers, unlike the Greeks, had no conception in their astronomy of three-dimensional space; to them the planets were merely luminous objects moving (as Pannekoek put it) 'with unravelled regularity along the firmament'. This shows how wide of the mark were Proctor's suppositions in his Note A on Chaldean Astronomy; he wrote at a time when practically no first-hand information was available, he imagined Babylonian priests of two millennia ago as having a nineteenth century outlook and perhaps anticipating the work of Copernicus, and he thought the Persian conquest brought a decline, whereas it probably freed astronomy at least for a time from the shackles of astrology, and certainly was coeval with the advance in mathematical theory.

(B) ANCIENT GREEK THEORIES AND INVESTIGATIONS

Egyptian calendar

Unlike those of Babylonia, the vast written records of Ancient Egypt seem to contain hardly anything about astronomy and nothing at all about planetary observations. Saturn is said to have been known to the Egyptians as 'Horus the Bull', Mars as 'the red Horus', and the planets in general as 'the stars who never rest', which, though appropriate in the wide sense, does not suggest careful observation of stationary points. In Neugebauer's opinion the best Egyptian contributions to astronomy were the division of the day into 24 hours, and the adoption of a solar year of 365 days, divided conveniently into 12 months of 30 days each, plus 5 extra days. In the last period of Egypt as an independent kingdom the correction of one additional day every fourth year was introduced. This simple and excellent arrangement was, however, the eventual outcome of extraordinary complications, and many changes in the date of new year, as is shown by J. W. S. Sewell (chap. 1 of *The Legacy of Egypt*, ed. S. R. K. Glanville, 1942). Sewell (p. 10) mentioned as one new year day August 29 (Julian) of the Alexandrine calendar of $365\frac{1}{4}$ days. It has been assumed in this book for the purpose of interpreting the dates of Ptolemy's observations that this was the calendar he used, his '1 Thoth' corresponding to 29 August. This also agrees with the Ancient Egyptian calendar given by E. A. Wallis Budge (*First Steps in Egyptian*, 1923, p. 36) according to which the month Thoth began, not at the beginning of the inundation of the Nile, but at its height.

'Homocentric spheres' (Eudoxus)

The only people of the ancient world who visualised the planetary system in depth and tried to think out geometrical models to explain its working, were the Greek philosophers. But during the sixth and fifth centuries B.C., when the Greek city states reached the greatest heights of political and artistic achievement, their astronomy was still in its infancy. The Ionian and Pythagorean Schools of philosophers which flourished during those centuries showed insatiable curiosity about the Earth and the heavenly bodies, and were prolific in ideas, mainly of a fantastic kind; they relied on theory and pure reason rather than observation, and, in fact, Plato, who was living towards the end of that period, though credited with having recorded

Saturn's yellow colour, showed a strong preference for reasoning and is said to have discouraged observation and experiment.

Plato had supposed that every body in the sky was carried around by a rotating sphere, and one of his pupils, Eudoxus of Cnidus who lived in the earlier part of the fourth century B.C., developed this idea into a mathematical basis for the planetary motions. Eudoxus also studied under an Egyptian priest from Heliopolis, from whom he is thought to have learned something of the planetary observations of the Egyptians. Apparent irregularities of motion could be explained (according to Eudoxus) by supposing the poles of the sphere carrying a planet to be carried by a slightly larger concentric sphere revolving about different poles, that in turn being carried by a third sphere and the third by a fourth. The motions of the third and fourth spheres combined caused a planet to describe a sort of elongated figure-of-eight on the surface of the second sphere, and the combined motions of the four spheres seemed to give a satisfactory explanation of the motions of the planet, including retrogression and stationary points. Thus the theory of Eudoxus required three spheres each for the Sun and Moon, four for each of the planets, and one for the fixed stars, making 27 in all. (For a somewhat fuller explanation see G. Abetti: *The History of Astronomy*, 1954, p. 29, and M. Davidson: *The Stars and the Mind*, 1947, p. 44). Callippus added a few more spheres, and Aristotle, to counteract the disturbing effect of the spheres on one another, increased the grand total to 55, making the scheme so hopelessly complicated that when, half a century or so later, it was found not to fit the facts, it had to be abandoned. It was at about this time, the mid-fourth century B.C., that Heraclides Ponticus suggested that the Earth rotates on its axis and that Venus and Mercury revolve round the Sun, though he still supposed the latter to revolve round the Earth.

Aristotle, who rightly concluded from the phases of the Moon and the shape of the Earth's shadow on the partially eclipsed Moon that the Earth and Moon are spherical, laid down the principles that the circle is the perfect figure, and that the motions of the heavenly bodies are uniform circular motions. These principles, which for many centuries were regarded as fundamental, caused serious difficulties to the exponents of the epicycle theory, and had a very adverse effect on astronomical progress, even hampering Copernicus and, for a time, Kepler.

Alexandrian School and epicycle theory

By the end of the fourth century B.C. the Greek city states had lost their independence, having come under Macedonian, later to be replaced by Roman, domination. The conquests of Alexander the Great, however, brought about the establishment of a famous library and observatory at Alexandria in Egypt, henceforth the centre for the most eminent Greek philosophers, astronomers and mathematicians. Another result of those conquests was that the Greeks for the first time gained acquaintance with the statistical results of Babylonian astronomy. Though it may be rather over-enthusiastic to refer, as Pannekoek did, to the 'observing practice' and the 'abundance of observational knowledge' of the later Babylonians, nevertheless the fact remains that from the very first the Greeks of the school of Alexandria were observers, especially of stars, much less so of planets. Aristyllus and Timocharis about 300 B.C. were observing and cataloguing the positions of stars; Aristarchus of Samos, who attempted to measure the sizes and distances of the Sun and Moon, even asserted that the Sun was the centre of the system, and the Earth moved round it, a suggestion which, however, found hardly any support from others; somewhat later Eratosthenes from observation attempted to measure the size of the Earth and obtained quite a good result for the obliquity of the ecliptic.

It is not certainly known who first put forward the epicycle theory: the suggestion of Heraclides that Mercury and Venus revolved round the Sun, while the Sun revolved round the Earth, contained the germ of the idea, while the Babylonian planetary statistics and the observations of the Alexandrian Greeks themselves must have forced them to seek a replacement for the revolving spheres of Eudoxus. The problem facing the inventors of the epicycle theory was to devise a system of uniform circular motions which would represent the variable velocity, the stations, the direct and retrograde movements, and the motion in latitude observed in planets such as Saturn. When the new theory is first met with in its discussion by Apollonius of Perga, in the later third century B.C., he was already introducing the device of viewing the uniform circular motion of each planet from a point slightly eccentric to the centre of the orbit. According to Neugebauer, Apollonius proved that this kind of eccentric movement is equivalent to an epicyclic motion, in which the planet P moves uniformly round a small circle (epicycle), whose centre X moves uniformly round a large circle (eccentric) with the observer at its centre, the radius XP of the epicycle being equal to the eccentricity. The motions of P and X are so chosen that at the near side of the epicycle the planet appears to have slow retrograde motion (the difference between the motions of P and X); at the far side swifter direct motion (the sum of the motions of P and X);

52

and at the points on the epicycle where the planet is directly approaching or receding in the line of sight, it will appear stationary in the sky. Movement in latitude can be provided for by allowing the plane of the epicycle a slight tilt.

Abetti states that Apollonius explained how to deduce the ratio of the radii of the epicycle and eccentric, and that while the path along the epicycle corresponds to the synodic period, the path along the eccentric circle corresponds to the sidereal period of the planet. On this basis Ptolemy succeeded in working out the theory of the planets, thus bringing to completion the astronomy of the Alexandrian School.

An explanation of the epicycle theory such as the one just given is necessary for the understanding of Ptolemy's observations of Saturn. Assuming uniform motions in circular orbits, the theory could be deduced from a heliocentric one by deducting from all movements the motion of the Earth, and it gives a correct description of the angular motions of the planets as they appear from the Earth. It had the great advantage of being extremely adaptable, and, as Neugebauer has pointed out, when the theory had been brought to perfection by Ptolemy, 'only greatly refined observations could eventually disclose' its defects. One of its chief drawbacks was that it could afford no help in determining quantitatively the relative distances of the planets. Of course Galileo was easily able to prove with the telescope that the heliocentric theory of Copernicus is the correct one, but it is not difficult to understand why the Ptolemaic theory should have held the field for many centuries of the pre-telescope era.

The epicycle theory was adopted by Hipparchus, the greatest of all the ancient astronomers, who lived in the second century B.C. Among his many achievements were tables of the Sun's positions, the determination of the Moon's parallax and the tilt and eccentricity of its orbit, the invention of the planisphere and the classification of the stars in apparent magnitudes, the cataloguing of the positions of 1,080 stars, and the discovery of the precession of the equinoxes. He made this last discovery by comparing his stellar observations with those of Aristyllus and Timocharis made a century and a half earlier. Most of his work, including his star catalogue, was preserved through being recorded by his great admirer, Claudius Ptolemaeus (Ptolemy), the last great astronomer of the Alexandrian School, who lived nearly three centuries later.

Ptolemy's book, *Mathematike Syntaxis*, had an earlier Greek title meaning 'great composition' from which is derived the Arabic name

Almagest ('the greatest') by which it is popularly known. It embraced all the astronomy of the ancient Greeks, and in numerous ways showed the influence of the later Babylonian astronomy: in the planetary tables with a Babylonian date as their starting point, in the use of Babylonian observations, of arithmetical as well as geometrical methods, and of sexagesimal fractions. Mr. M. B. B. Heath having drawn attention to the fact that the *Syntaxis* recorded some observations of Saturn made during the first half of the second century A.D., the remainder of this chapter will be devoted exclusively to the information given by Ptolemy about Saturn, in book IX and onwards of vol. II of the *Syntaxis* (Greek text with parallel French translation by Abbé Halma, Paris, 1816).

Saturn observations and data from Ptolemy's *Syntaxis*

Ptolemy wrote (*Syntaxis*, book IX, chap. 1) that the planets are nearer to the Earth than the stars, and that Saturn is the farthest planet from the Earth and moves on the largest sphere around the centre of the zodiac, but that there was no means of proving the true position of the planets, because none of them had a sensible parallax which would be the sole means of determining their distances. He placed Mercury and Venus in orbits inside the Sun's and the other three further out, as he recognised a difference between the movements of the two former and those of the others. He went on to state (chap. 2) that he intended to prove for the planets, as he had done for the Sun and Moon, that the apparent irregularities in their movements could all be explained on the basis of uniform circular motions. He mentioned, as one of the hindrances in the task, the lack of really accurate observations in the past, but he recognised the difficulties of former observers, e.g. in the observation of stationary points, remarking: 'the change of position is insensible for several days before and after the station'.

He said that the memoirs of Hipparchus showed that he had not begun to work out a theory for the planets, partly through lack of reliable observations and partly because Hipparchus had realised the complications more clearly than had previous mathematicians, but he had laid down the right lines for the solution. Ptolemy proposed to use a new method which he felt might make the task easier without vitiating the results, namely, to assume that the planets were moving in circles in the plane of the zodiac.

Ptolemy (chap. 3) gave some definitions and planetary statistics, based on the observations of Hipparchus and corrected by his own researches. He defined 'movement in longitude' as the movement

54

of the centre of the epicycle on the eccentric circle, and 'anomaly' as the movement of the planet in the epicycle.

He stated that he found that 57 anomalies of Saturn make 59 solar years and nearly ¾ of a day. (To take a modern example for comparison, there will be 57 synodic periods of Saturn from the opposition of 1940 November 3 to that calculated to take place on 1999 November 6 —see Appendix 1—a period of 59 years, 3 days. Hence the disagreement with Ptolemy's period amounts to 2¼ days in 59 years, about 0·001 per cent).

Ptolemy gave the mean anomaly (i.e. the angular movement in the epicycle) for Saturn in an Egyptian year of 365 days as

$$347° \ 32^{\text{I}} \ 0^{\text{II}}, \ 48^{\text{III}}, \ 50^{\text{IV}}, \ 38^{\text{V}}, \ 20^{\text{VI}}.$$

This rather fearsome figure in sexagesimal fractions appears to be equivalent to a value taken to seven places of decimals of a second. (A simple calculation based on the modern mean synodic period of Saturn of about 378 days, given at the end of Chapter 1 of this book, would make the anomaly, if epicycles were now used, about 347° 37′, a difference from Ptolemy's value of about 0·024 per cent).

Ptolemy gives for the mean annual movement in longitude of Saturn another figure in sexagesimals of a second which, to the nearest second, is 12° 13^{I} 24^{II}. (This corresponds to the sidereal mean annual motion of Saturn, and the modern value, based on the figure for mean daily motion given at the end of Chapter 1, would be 12° 12′ 46″44, the disagreement with Ptolemy being about 0·08 per cent.)

These examples will show how accurately the Ptolemaic system represented the angular motions of Saturn; similar statistics were given by him for all the other planets then known. He then gave very detailed tables for the movements in longitude and anomaly for all five planets, choosing as his epoch or starting point the planetary positions (doubtless calculated back) at noon on the first day of *Thoth* in the first year of Nabonassar, i.e. Nabû-nâzir, who, contemporary with Tiglath-Pileser III of Assyria, was a king of Babylon who reigned from 747 B.C., the year from which the Babylonians are said to have kept a complete record of eclipse observations. Ptolemy's starting date was therefore probably 29 August 747 B.C. According to Ptolemy the direction of *Kronos* (Saturn)* at noon on that date was Capricornus 26° 43′, the direction of the apogee of the orbit was Scorpio 14° 10′ and the Sun was at 0° 45′ in Pisces. He appears to have measured longitudes from the preceding point of each zodiacal constellation, each constellation being 30° in length.

* *Saturnus* was the name of the Roman god identified with the Greek *Kronos*.

After general discussions and chapters dealing with the other planets Ptolemy wrote (book XI, chap. 5): 'We have also taken (for Saturn) as for the other planets, in order to determine its apogee and eccentricity, three *akronykt* (opposition) positions of that planet diametrically opposite to the mean place of the Sun.

Opposition of 2 May 128 A.D.* 'We observed the first by means of the astrolabe (at 6 hours) in the evening of the 7th of the Egyptian month *Pachon* of the 11th year of Hadrian, (Saturn being) in 1° 13' of Libra;'

Opposition of 12 July 134 A.D.* 'and the second the 18th of the Egyptian month *Epiphi* of the 17th year of Hadrian. We calculated from our observations at that opposition that the exact time of opposition was at 4 hours after midday and the place in 9° 40' of Sagittarius.'

Opposition of 17 August 137 A.D.* 'Finally we observed the third on the 24th of the Egyptian month *Mesori* in the 20th (? 21st) year of Hadrian, and we found in the same way the true time of opposition was noon on the 24th, and the place in 14° 14' of Capricornus.'

[*Notes:* (*a*) They had to find the date on which Saturn came nearest to acronical rising, i.e. at the instant of sunset, and calculate from that the true time of opposition.

(*b*) 137 is evidently the year of the third opposition observed, from the time intervals stated below, but as the Roman Emperor Hadrian is considered to have succeeded to the Empire on 11 August 117, the date of the observation would come very early in his 21st regnal year.]

* *Great Books of the Western World*, vol. 16 (edited R. M. Hutchins, Encyclopaedia Britannica, Inc.) contains complete English translations of *Revolution of the Heavenly Spheres* by Copernicus, and of Ptolemy's *Syntaxis*.

Copernicus (book V, chap. 5) interprets the dates of Ptolemy's opposition observations as follows: A.D. 127 March 26; 133 June 3; 136 July 8.

These dates appear to disagree by one year with the regnal years of the emperors, and would make '1 Thoth' fall respectively on 23, 21 and 20 July.

Ptolemy continued: 'Of these two intervals, that from the first opposition to the second comprises 6 Egyptian years, 70 days and 22 hours, and for the apparent motion of the planet 68° 27'. The interval from the second to the third includes 3 Egyptian years, 35 days and 20 hours, and 34° 34' of motion. But the mean movement in longitude during the first interval, taken in round numbers, is found to be 75° 43', and during the second 37° 52'.'

[*Note.* 1° 13' Lib. to 9° 40' Sgr. = 30° − 1° 13' + 30° + 9° 40' = 68° 27'.

As the result of a very long mathematical discussion of the data he appears to have found that at the third opposition Saturn was, according to its mean motion, at 19° 30' of Capricornus, and 56° 30' in longitude from the apogee of the eccentric and 174° 44' by anomaly from the apogee of the epicycle. The apogee of the eccentric was therefore at Scorpio 23° and the perigee at Taurus 23°.

[*Note.* 23° Sco. to 19° 30′ Cap. = 30° − 23° + 30° + 19° 30′ = 56° 30′.]

Observation of 31 *January* 140 A.D. (book XI, chap. 6): 'To find the size of the epicycle we have taken the observation which we made in the second year of Antoninus, four hours before midnight 6th–7th of the Egyptian month *Mechir*, the astrolabe then showing the end of Aries on the meridian, and the mean Sun being at the same moment in Sagittarius 28° 41′. Saturn compared with the bright star in Hyades (Aldebaran), appeared to be at 9° 4′ of Aquarius and it was half a degree behind the centre of the Moon, for it was at that distance from the northern horn of the crescent. But at this moment the Moon, according to its mean motion, was at 8° 55′ of Aquarius and at 174° 15′ of anomaly from the apogee of the epicycle. Consequently its true place must have been at 9° 40′ of Aquarius, and it appeared at Alexandria to be at 8° 34′. Thus Saturn, left about half a degree behind the centre of the Moon, must have been at 9° 4′ of Aquarius, and it was at 76° 4′ from the apogee of the eccentric (Scorpio 23°), provided that the latter had not advanced appreciably in so short a time (as less than three years). Since the time interval from the third opposition to this observation is 2 Egyptian years, 167 days and 8 hours, and since Saturn advances approximately 30° 3′ in longitude during that length of time, and in anomaly 134° 24′, if we add these quantities to the places given by the third opposition, we shall have for the time of observation in question (4th observation): 86° 33′ in longitude from the apogee of the eccentric, and 309° 8′ of anomaly from the apogee of the epicycle.'

After a mathematical discussion, Ptolemy drew the inferences that at the beginning of the reign of the Emperor Antoninus the apogee of Saturn was at 23° of Scorpio, and that the ratio of the radius of the epicycle to that of the eccentric was as 6½ : 60, which gave him the value of the eccentricity; these were the two results he had set himself to find from the observations.

[The place of the apogee of the epicycle might be regarded as corresponding, in modern heliocentric theory, to the ecliptic longitude of Saturn at conjunction in the year in question; that of the apogee of the eccentric as corresponding to the ecliptic longitude of the aphelion of Saturn's orbit, which changes very slowly by precession, and, according to Ptolemy's calculation, had only altered by 3° 40′ in 366 years—in modern reckoning the change would be about 5° 7′.

According to Newcomb, annual precession slightly exceeds 50″; Ptolemy evidently supposed it to be 36″, a very erroneous determination.]

Ptolemy (in book XI, chap. 7) introduced a further discussion about Saturn with the words: 'For the correction which we have still to give to the periodic movements we have chosen one of the most precise of the ancient observations.' In this investigation he needed two observations separated by a long interval of time; therefore he took as the first a Babylonian observation (already mentioned) which was made in the evening of the 14th of the Egyptian month *Tubi* in the 519th year of (the era of) Nabonassar, which would probably correspond to 9 January 228 B.C., when Saturn was seen to be due south of a star presumed to be η Virginis. He proceeded:

'. . . the mean Sun was in Pisces 6° 10'. But the fixed star of the southern shoulder of Virgo was, at our observation at 13° 10' of Virgo and at the observation in question, as in 366 intervening years the fixed stars have advanced nearly 3° 40', it must have been at 9° 30' of Virgo, and so was Saturn . . . Moreover, having shown that in our time the apogee of Saturn was in 23° of Scorpio, it must have been at 19° 20' of Scorpio at the time of that old observation. Hence one concludes that at the time (of the old observation) the apparent planet was distant from the apogee by 290° 10' on the zodiac, and that the mean Sun was at a distance of 106° 50' (from the apogee) . . .'

[*Note.* From 19° 20' Sco. to 9° 30' Vir. = 30° − 19° 20' + (9 × 30°) + 9° 30' = 290° 10'.]

After a mathematical discussion he continued:

'. . . at the time of the observation in question Saturn according to its mean motion was 283° 33' from apogee, that is at 2° 53' of Virgo. But as the mean movement of the Sun is supposed to be 106° 50', if we add to it 360°, and from the sum, 466° 50', subtract 283° 33' of longitude, we have for this same time 183° 17' of anomaly from the apogee of the epicycle.

'Since it is proved that at that observation (9 January 228 B.C.) Saturn was 183° 17' away from the apogee of the epicycle, and that in the third opposition (17 August 137 A.D.) it was at 174° 44', it is evident that in the length of time elapsed between those two observations, which comprises 364 Egyptian years and 219¾ days, Saturn has traversed, in addition to entire circumferences, 351° 27' of anomaly, which agrees very nearly with the surplus that our tables of mean motion give.'

The brief extracts from the *Syntaxis* which have been given in translation will leave no doubt of Ptolemy's ability in arithmetic, and a study of his mathematical discussions would afford an even more impressive demonstration of his mastery of geometry and trigonometry. Moreover, what has been related by no means covers the aspects of Saturn he discussed, since further chapters of his book covered true places, anomalies, longitudes, retrogression, stations, latitudes, and appearances and disappearances for all the five known planets, with tables for some of these matters. For example, his table of latitude for Saturn gave its limiting values as 3° 2' north and 2° 59' south of the central plane of the zodiac, showing that he had arrived at a very fair idea of the tilt of the planet's orbit.

He dealt with the retrogression of Saturn (in book XII, chap. 2), making use of the ratio he had found between the radii of the epicycle and eccentric, $6\frac{1}{2} : 60$, and working out by trigonometry the periods of retrogression as follows: at apogee $140\frac{2}{3}$ days, at an intermediate position on the orbit 138 days, and at perigee 136 days, values in quite good agreement with those of the present time given in Chapter 1 of this book.

It may seem scarcely credible, but it is a fact, that all the discoveries that have been recounted in the present chapter were based on naked-eye observations with instruments that would now be considered very primitive. The progress of mathematical astronomy in the later centuries of antiquity, culminating in Ptolemy's remarkable book, is even more astonishing and is underlined by Neugebauer's remark: 'The ancient astronomers rightly had greater confidence in the accuracy of their mathematical theory than in their instruments.' The proportions of the historical time-scale too are rather striking: the few observations of Saturn mentioned in this chapter spanned eight centuries, and another fourteen and a half centuries separated Ptolemy (the last ancient astronomer) from Galileo (the first modern one), and yet a mere three and a half centuries have gone by from Galileo's first use of the telescope to observe Saturn until the present day.

CHAPTER 3

MEDIAEVAL AND SIXTEENTH CENTURY UNAIDED EYE OBSERVATIONS

(A) ISLAMIC AND SPANISH TABLES

Ammonius; the 'Sind Hind' and the Islamic revival of astronomy

Ptolemy lived in the Age of the Antonines, later regarded as a golden era of peace and prosperity for the Roman Empire, but his work marked the virtual end of ancient astronomy. The last director of the declining School of Alexandria, Ammonius Hermiae (who lived in the fifth century A.D.) is said, however, to have produced an ephemeris of planetary positions, now lost. Millás (p. 236) considers that the lost ephemeris was probably that known to the Islamic astronomers of the middle ages as the Almanac of 'Humeniz' or 'Aumatius' which formed the basis of Azarquiel's Almanac. Also, about the year 400 a lawyer at Carthage called Martianus Capella wrote a book on human knowledge which renewed the suggestion that Mercury and Venus revolve round the Sun, and was quoted a millennium later by Copernicus (*R.A.S. Occ. N.*, vol. 2, pp. 17, 31). But it would be vain to look for anything new or original in astronomy in the West for nine centuries after Ptolemy, and the revival occurred in the Middle East, under the successors of the Prophet ruling at Baghdad.

The revival seems to have originated in the compilation in India, early in the fifth century A.D., of the fundamental work called *Surya-Siddhanta*, derived from ancient Indian astronomical doctrines and Greek science (Millás, p. 23). In A.D. 772, during the reign of Caliph Al Mansur, a patron of learning, a copy of this Indian treatise was brought to his Court, translated into Arabic, and under the name 'Sind Hind' not only became the standard textbook for half a century but influenced Islamic writers long afterwards. This first step was followed in the late eighth and early ninth centuries under such enlightened caliphs as Harun al Rasid and Al Mamun by the translation of Greek works including Ptolemy's *Syntaxis*, thenceforth known under the more familiar name *Almagest*, the building of observatories at Damascus and Baghdad, and the making of astronomical observations.

The new observations soon showed that the positions of Saturn and the other heavenly bodies did not agree with those calculated from Ptolemy's tables, a fact no doubt puzzling to the astronomers of Islam but in the light of modern knowledge not at all surprising. Of the factors that in the course of centuries were bound to produce discordances, by far the most important was the Ptolemaic attempt to represent by a complicated system of circular motions round a nearly central Earth the planetary orbits which are in reality elliptical with the Sun at a focus. Other subsidiary causes were: Ptolemy's underestimate of precession, the inexactitude of naked-eye observations, slight inaccuracies in the instruments used, the effect of perturbations (e.g. of Jupiter and Saturn on each other's motion) which were completely unsuspected until the time of Laplace, and errors in the copying of astronomical tables.

To bridge the gap in the study of Saturn between the observations of Ptolemy and Copernicus account should be taken of the successive mediaeval astronomical tables and almanacs, and of the astronomers who, in order to compile them, had to make fresh observations of the planets.

Al-Khwârizmî; Al-Battânî

F. J. Carmody (*Arabic Astronomical and Astrological Sciences*, 1956) says that 'the fountainhead of Arabic astronomical tables and rules for their use' were those of Al-Khwârizmî (fl. 820), a Persian who used the 'Sind Hind' as one of the bases of his tables and is also credited with having conferred on all later astronomers and mathematicians the inestimable boon of introducing (from India) the so-called arabic numerals, including a sign for zero.

One of the most celebrated of all the Islamic astronomers was Al-Battânî (Albategnius), son of a skilful constructor of astronomical instruments and born at Battan in western Mesopotamia. He made many observations about A.D. 877–918, mostly at Rakkah on the Euphrates, some at Antioch.

His correct name seems to have been Muhammad Ibn Jabir al-Battânî, and the period of his life from before 858 to 929 A.D. His fine book on astronomy was translated and edited under the title *Opus Astronomicum* by C. A. Nallino (Milan – Reale Osservatorio di Brera, No. 40, 1899–1907, British Museum 8752. h. 1).

Abetti (pp. 45–47) gives an account of the book, which includes in its later chapters particulars of the positions of the five major planets on the ecliptic at the various epochs. In the preface Al-Battânî stated that he had written the book to correct differences regarding the celestial motions found in the works of other authors, some of whom had been wrong in the fundamentals. To do so he had used the methods and followed the precepts of Ptolemy, whose work he evidently much admired. He called upon others to observe and investigate after him, as the subject was too great for any one man to cover it completely. He said he had sought to explain difficult principles and had corrected the movements of the celestial bodies on the ecliptic from observations, calculations of eclipses and other operations.

Volume 2 of the *Opus Astronomicum* consists of a vast number of tables of the Sun, Moon and planets, including conjunctions, oppositions, eclipses, stations, latitudes, etc., dates being given in Arab, Roman and Coptic years and months, with tables for the conversion from one era to another. His table of mean motion of Saturn in longitude in Roman years shows a variation from 12° 13′ to 12° 16′ per annum, and a period of approximately 29½ years for a complete circuit of the zodiac. His table of latitudes of Saturn indicates maximum latitudes north and south of the ecliptic of 3° 2′ and 3° 5′, respectively. A calculation from his table of the anomalies of Saturn's stations, for all directions of Saturn from the Earth at 6° intervals, appears to give the following results for the periods of Saturn's regression: at apogee 141 days, at an intermediate position 138 days, and at perigee 135½ days. These agree quite closely with Ptolemy's figures (see Chapter 2), and reasonably well with modern values (see Chapter 1).

Al-Battânî also gave a very long list of star positions, determined very accurately the obliquity of the ecliptic, and the positions of the equinoxes, introduced sines into trigonometry, and used new formulae for the solution of spherical triangles.

In the late tenth century Ibn Junis, observing at Cairo, compiled the Hakimite tables of the planets.

Azarquiel's *Toletan Tables* and Almanac

Though Albategnius was not the last of the writers on astronomy under the caliphate of Baghdad, both learning and political power declined there after his time, culminating in the fall of the city to the Seljuk Turks in 1055. The hub of science had moved west, to Cairo and much further: to the caliphate of Cordova in Spain, and especially to the academies in that city and Toledo. There, though the Moorish power was already declining, flourished in the eleventh century the brilliant Moor astronomer known as Ibn-al-Zarqellu ('son of Zarqellu') or more simply, Azarquiel, less correct forms of the name being Zarkali and Arzachel. J. M. Millás Vallicrosa (*Estudios sobre Azarquiel*, Madrid 1943–50, chap. 1) has culled not a few biographical details from contemporary writings and Azarquiel's own works. From these sources it seems that he was a native of Toledo and originally a skilled engraver and metal worker, without a scientific education but with great natural gifts. The excellent astronomical instruments* he made brought him to the notice of the learned men. With their encouragement he easily mastered the works of the ancient authors, becoming eventually the most eminent astronomer, mathematician and inventor of apparatus of his time, and in the opinion of one writer, the greatest since the coming of the Moors to Spain. He himself wrote that he had been dissatisfied with the disagreement between former observations of the stars and of the Sun and between the Persian and Indian theories of the Sun's mean position, so with a group of learned and reliable men he had devoted himself to observing the Sun, Moon and stars as often as possible for 25 years. About 1061 he was determining precession, and in 1081 the longitude of Regulus and the apogees of the planets, correcting the positions determined by Albategnius 202 years earlier. Then he had to retire to Cordova because of the outbreak of strife which culminated in 1085 in the conquest of Toledo by Alfonso VI of Castile. At Cordova he made his last observations in 1087–88 and died on 15 October 1100.

The original Arabic version of his famous *Toletan Tables* is lost, existing MSS. deriving from the Latin text of the celebrated twelfth century translator, Gerard de Cremona. Other authentic works by him include: the Almanac; a treatise on stellar motion (in Hebrew translation only); two on instruments; a mainly astrological one on the planets; and a lost work on solar motion. Many other writings have been attributed to him, some doubtfully, others falsely. After detailed examination and comparison of canons, Millás (p. 71)

* A finely engraved astrolabe, said to have been made and used by Azarquiel, is still preserved in the Science Museum at Barcelona. Also, according to Millás (pp. 6–9), Azarquiel installed two large tanks, in a house on the bank of the Tagus near the Gate of the Tanners on the outskirts of Toledo, in such a way that they were completely filled with water, and emptied, by hidden conduits according to the waxing and waning of the Moon. People who had seen them said that they filled to the extent of 1/14 in each 24 hours from the first appearance of the new Moon, were completely full by the fifteenth night, and then the water fell at the same rate so that they were completely empty by the 29th day; also that if anyone put in or drew off any water, they automatically adjusted to the correct level. In the mid-twelfth century, King Alfonso VIII of Castile, wanting to find out how they worked, authorised a Jewish astronomer, who thought he could improve on the invention, to dismantle one of these remarkable water-clocks; he did so but was unable to restore it!

concludes that the most influential sources of the Toletan Tables were the Tables of Al-Khwârizmî and the 'Sind Hind', with Al-Battânî's book a growing influence on Azarquiel's works, the later ones of which show a tendency to break away from the less desirable teaching of the ninth-century Baghdad astronomer Tabit ibn Qurra (Thebit) and his incorrect theory of a variation in precession known as 'trepidation'. The widespread influence of the Toletan Tables and the other works of Azarquiel on later mediaeval European astronomy has been shown by Millás in three long chapters. His chapter 4 reproduces the Arabic text of the canons of the Almanac with Spanish translation, and the whole of the Almanac's tables.

The Almanac's solar, lunar and planetary tables all start with the month of September because Azarquiel was adapting the ephemeris ascribed to Ammonius, and the Alexandrian School probably started the year on 1 Thoth (29 August). Hence Thoth roughly corresponds to September, and in fact Azarquiel, in his planetary tables, showed the Egyptian months starting with Thoth alongside the Roman ones. The Saturn tables in his Almanac give the constellation and longitude in degrees of the planet's place for days 1, 11, 21 of each month for 59 years, after which the places would approximately repeat on corresponding dates. The Egyptian months of 30 days each with 5 extra days at the end of the year lend themselves admirably to an almanac giving positions at intervals of 10 days; Azarquiel explained (Millás p. 120) that to adapt this to the Roman months he had to allow for the extra day in 31-day months and the 2 days short in February.

The epoch he chose was the year 1400 of the era of Alexander, by which he meant the Seleucid era used by the later Babylonians; it commenced on 1 October 312 B.C.* His year 1400 Alexander therefore corresponds to 1088–89 A.D., and as he said that this was the sixth year of his Saturn table, the first year would be 1083–84. Hence the first five or six years of the table fell within his observing period, the places of Saturn in the remaining 53 years being predicted ones. Since the table shows that Saturn had autumn oppositions in Pisces, Aries and Taurus in those first six years, it seems reasonable to suppose that Azarquiel actually observed the planet around at least some of those oppositions to ensure that his table would be based at the start on actual current observations. It happens that 17

---

* The date when Seleucus, son of one of Alexander the Great's generals, recovered Babylon and established a hereditary monarchy in Mesopotamia after a decade of struggle among the generals over the partitioning of the empire following the death of Alexander the Great.

Saturnian years approximate to 500 terrestrial ones, and that Tycho Brahe carefully determined Saturn's opposition places for the years 1583 etc., exactly 500 years later than the first six of Azarquiel's table. The approximate dates and places of the oppositions of Saturn for the years 1083–88 can be deduced from Azarquiel's solar and Saturn tables, and tested by comparing them with the corresponding ones of Tycho's time, as follows:

Approx. Oppositions (Azarquiel)			Equivalent Ecliptic Longitude (1950)		Oppositions (Tycho Brahe)		
1083 Sept.	6	Pisces 12°	353½°	355°	1583 Sept.	1	
1084	19	25°	6½°	8°	1584	13	
1085 Oct.	4	Aries 9°	20½°	21°	1585	26	
1086	17	22°	33½°	34°	1586 Oct.	10	
1087 Nov.	1	Taurus 7°	48½°	48°	1587	24	
1088	14	21°	62½°	62°	1588 Nov.	7	

(Allowance has been made in both cases for precession to epoch 1950; further details of the Tycho oppositions are given later in this Chapter.)

In the rules preceding his Almanac corrections for precession were given for the planets by Azarquiel, and also directions for determining latitude from another table and for deciding whether it was north or south. He pointed out that Saturn's mean position in direct motion is approximately that of conjunction with the Sun, and that its mean position in retrograde motion is approximately that of opposition (Millás, p. 137).

In his treatise on the planets Azarquiel showed how to deduce the anomaly for Saturn (and the other planets) and also how to find the anomalies of the stationary points. He determined the position of Saturn's apogee in 1081–82 as: 8 signs, 7° 24′, i.e. longitude Sagittarius 7°4 (Millás, pp. 470–478). Ptolemy had found Saturn's apogee about 9½ centuries earlier to be in Scorpio 23° which, allowing for precession, would be equivalent to about Sgr 5°4 at the time of Azarquiel's observation.

Nassir Eddin; Ulugh Begh

Before dealing with the further development of astronomy in Spain and its spread through western Europe there are two further meritorious though short-lived revivals in the Middle East which deserve mention. When the Mongol Hulagu Khan, a grandson of Genghis Khan, captured Baghdad (1258), he established a fine observatory at Meraga in Persia, said to have been equipped with a mural quadrant of 12 feet radius and other instruments which were the best to be constructed prior to those of Tycho Brahe three centuries later. Under the superintendence of Nassir Eddin the Meraga astronomers produced the *Ilkhanic Tables* containing data for the calculation of planetary motions as well as a star catalogue, they made a very accurate determination of the precession of the equinoxes, and translated Ptolemy's and other Greek scientific works. Unfortunately the School of Meraga barely survived Nassir Eddin's death (1273).

Tamerlane seems to have resembled the earlier Genghis Khan not only in being a conqueror and scourge of humanity but also in having a grandson devoted to astronomy, Ulugh Begh, who personally supervised the work of the Tartar astronomers at Samarkand Observatory in the early fifteenth century. They compiled planetary tables and a star catalogue, derived from Ptolemy's, giving positions in minutes of arc as well as degrees. Ulugh Begh's assassination (1449) seems to have ended mediaeval Tartar astronomy.

Abetti (p. 48) and Davidson* have pointed out that though Islamic and Tartar astronomers brought little in the way of new theories or discoveries to astronomy, they were very skilful observers and computers, accumulated observations, greatly improved mathematics and helped to preserve the writings of the ancient Greeks. Many star names and astronomical terms still in use derive from Islamic writings, which filtering through at points of contact such as Sicily and Spain, were translated into Latin and helped to found the astronomy of western Europe.

Alfonso X's *Libros del Saber*: Aben-Ragel's views about Saturn

An important step was taken by Alfonso X *el Sabio* (the 'wise' or 'learned') when, soon after succeeding to the throne of Castile in 1252, he set many scholars to work on translating Arabic writings into Spanish, collecting information and making observations so as to produce a sort of encyclopaedia of astronomical (and astrological) knowledge called the *Libros del Saber de Astronomía* (edited by Manuel Rico y Sinobas, Madrid, 5 vols., 1866 etc.). The Castilians had conquered from the Moors in the late eleventh century central Spain, including Toledo, but had then been severely checked by the Berbers, called in by the Moors. Then early in the thirteenth century Alfonso's predecessor had added by conquest much of southern Spain, including Cordova and Seville. As all these cities were centres of learning of the Moors and Jews, it is not surprising to find that

* *Stars and the Mind*, p. 78.

Alfonso employed on his project Moors, Jews and Spaniards as well as learned men from Italy and other countries, but the *Tables* were the work of two Jews.

The *Libros del Saber* were written in old Spanish and contained not only the Alphonsine Tables and canons for their use but also treatises on stars and constellations, some of them translated from Arabic, and others explaining with diagrams how to make sundials, water-clocks and other instruments for measuring time and observing. Alfonso's personal contribution seems to have consisted of prefaces and also precepts, philosophical reflections and nature notes included in the descriptions of the constellations. Information on the planets, apart from the Tables, was very meagre, but the editor quoted (vol. 5: Saturn on p. 256) writings on them by Aben-Ragel, a thirteenth-century Moor astrologer of Toledo whose works (esteemed by Alfonso X) had been translated into Spanish by a Jew, Ihuda el Cohen. It is interesting to know what was thought about Saturn in that period but Aben-Ragel, like many another mediaeval Islamic scholar, dwelt mainly on the astrological influence of the planets, and the small modicum of astronomy is of very uneven quality.

Aben-Ragel compared Saturn to an old, weary recluse, large and slow-moving in the depths of space. 'Large' would be a fair deduction from the brightness in spite of great distance. He expressed the opinion that Saturn 'has neither light nor splendour in itself but receives it from the other planets and most from the Sun, which makes it hot and bright.' This would have been rather a good guess if he had omitted the word 'hot' and had not made the wrong suggestion that Saturn receives light to any but a negligible extent from bodies other than the Sun.

He also made the following surprising statement: 'Saturn does not submit to any of the planets nor is guided by any, rather by the Sun, because it burns by him, and retrogrades and is made eastern and western by him, and all this in approaching him and elongating itself from him.' Some may be tempted (as was apparently the editor of the *Libros del Saber*) to read into this dictum of Aben-Ragel a heliocentric idea anticipating that of Copernicus by 2½ centuries. But Aben-Ragel made no suggestion whatever of solar control over the motion of Mars or Jupiter. Therefore the idea prompting his remark about Saturn remains obscure, and may stem from nothing more original than the ancient Babylonian notion of Saturn's being 'the star of the Sun'.

It was probably known to Aben-Ragel from observations or from earlier writings that Mars appears brightest (when at what is now known as an opposition) in late summer, July–September; Jupiter in autumn, September–November; Saturn in winter, November–January (see *B.A.A.J.*, vol. 66, p. 166). Hence he was roughly correct in giving Aquarius–Aries as the part of the sky where Mars attains greatest brightness; and Aries–Gemini for Jupiter; and also in stating that they show their minimum brightness in the constellations in the respective opposite parts of the zodiac. Applying the same principle to Saturn, Aben-Ragel fell into the dreadful error of supposing Saturn to appear brightest (at oppositions) in Gemini–Cancer–Leo and faintest in Sagittarius–Capricornus–Aquarius. In actual fact, the stellar magnitude at oppositions in Taurus, Gemini and Cancer is around maximum, but the *minimum* occurs at oppositions of Saturn in Leo and Virgo and in Pisces, because of the closing of the ring (see list of opposition places in Chapter 1, and list of stellar magnitudes in Chapter 5 and Appendix I). Aben-Ragel's error might have been avoided by careful naked-eye observations, but only if they had been carried out over a long period of years.

Alphonsine Tables (Yĕhudá bar Mošé and Rabiçag Aben Cayut); Regiomontanus; Santritter

Though Alfonso X showed little political wisdom and most of his ambitious schemes ended in failure, his undoubted love of learning and his astronomical project brought him lasting fame. His Tables, copied, recopied, and later printed, continued to be a *vade mecum* of astronomers for three centuries, and for most of that time it is doubtful whether any astronomical tables would have found a market without the hallmark of the name 'Alphonsine'.

Although many of the later versions were in Latin it seems that the original tables with 54 canons for their use were drawn up in old Spanish in 1272 by two astronomers: 'Yhuda fi. de Mose fi. de Mosca' and Rabiçag Aben Cayut. Millás (p. 452) has shown the probability that Ihuda, the translator of Aben-Ragel, and Yhuda, the part author of the Tables, were really the same man, evidently a Jew, whose full name may have been Yĕhudá bar Mošé ben Mosca ha-Kohén, and that he was a medical man, physician to the Infante Don Alfonso before the latter came to the throne as Alfonso X. Dreyer appears to have identified Rabiçag with Isaac ibn Sid, a Jewish astronomer of Toledo who was reputed to have observed eclipses 1263–66 (*M.N.*, vol. 80, p. 245).

In their preface to the canons Yhuda and Rabiçag pointed out that astronomy needs constant corrections that could only be made by a succession of men over a long period, since some of the movements in the heavens are so slow that they take thousands of years to complete a circumference. After giving the date (in the era of Caesar) of their revision, they stated that King Alfonso had ordered the work and had caused instruments to be made for it as prescribed by Ptolemy in the *Almagest*, and that about two centuries had elapsed since the last corrections had been made by Azarquiel, observing at Toledo, where they also carried out their observations. They said that they had observed the Sun's motion for a whole year, had corrected conjunctions, eclipses and other things considered doubtful, had constructed tables and drawn up explanatory directions. Their canons show great analogy in both arrangement and content with those of Azarquiel, but in some respects are more elaborate; for example, they went into considerable detail as to the reconciliation of year and month dates as between the various eras used then and in earlier works (those of Alexander, Caesar, Mahomet, the Persian era, etc.). They made it clear that they were adopting a new one—the Alfonsine era—the starting point being noon on the day preceding the first day of January of the first year of Alfonso's reign (i.e. A.D. 1252).

In view of the comprehensiveness and precision of the canons of Yhuda and Rabiçag it is very disappointing to find that the tables following them in Rico y Sinobas's vol. 4 are very incomplete, in no way coincide with the Alfonsine canons he had unearthed, and have numerous copyist's errors—there are 18 obvious ones in the Saturn tables alone. This discrepancy has been explained by Millás (pp. 407–8) and Dreyer (*M.N.* vol. 80, p. 243). Rico y Sinobas claimed that he had found an authentic early text of the Alphonsine Tables but what he published was really a Portuguese almanac. Dreyer found complete agreement between certain tables (mostly of the fourteenth century), based on the Alphonsine, and the Alphonsine canons.* It would appear that the genuine Alphonsine Tables, which deserved their great reputation, were based on those of Azarquiel,

* In his interesting paper *On the Original Form of the Alphonsine Tables* Dreyer, using various MSS. mostly preserved in the Bodleian Library, Oxford, reconstructed and reproduced parts of the Alphonsine Tables in almost their original form. His version gave, e.g. for Saturn, the longitude (in zodiacal 30° signs, degrees, minutes and seconds) for the end of every 20-year period from 1320 to 1600 A.D., with tables of the adjustments to be made for single years, months, days and hours, and for precession. He showed that most continental MSS., and the early printed versions, of the Tables differed from the original form in several ways, especially by using a peculiar sexagesimal system of time reckoning which greatly abbreviated the tables but involved more calculation for their users.

as their authors indicated, with the positions of the heavenly bodies amended in the light of the thirteenth-century observations.

In view of the inevitable failure of all tables based on the Ptolemaic system to accord exactly with later observations, it is not surprising to find in the mid-fifteenth century complaints being made of the 'falsity of the planetary motions' as given in the then current versions. These criticisms came from George Purbach, professor of astronomy at Vienna, and his pupil, Johann Müller of Königsberg (hence known as 'Regiomontanus'), who was a renaissance man, keen to get back to the original Greek of Ptolemy, and disliking the Islamic tendency to mingle astrology with astronomy. He published in 1471 the first astronomical ephemerides* to aid navigators, preferring the Greek term to the Arabic word 'almanac'. Purbach and Müller made a revision of the Alphonsine Tables; Müller established an observatory at Nuremberg where printing was just beginning, and was the first astronomer to take advantage of the new invention.

Part of his complaint about the Alphonsine Tables was directed against the various non-Alphonsine items, some of them of an astrological character, which had in course of time become attached like barnacles, and were believed to be part of the original. In spite of the criticism such items were included by Santritter (see *Libros del Saber*, p. 79 of vol. 5 where many MSS. and some early printed editions of the Tables were discussed). John Lucilius Santritter published one of the earliest printed versions of the Tables at Venice in 1492 (British Museum No. 1A 23354) and prefaced his edition by 33 explanatory canons, some by himself, some by others: he called his book 'Tables of the celestial motions of Alfonso of blessed memory, most illustrious King of the Romans† and Castille.' Careful successive revisions and corrections by various scholars enabled the Alphonsine Tables still to be of service a century later in the time of Tycho Brahe.

(B) RECORDED SATURN OBSERVATIONS IN SIXTEENTH CENTURY

Copernicus

The sixteenth century marked the transition from mediaeval to modern astronomy by two great advances: the promulgation of a heliocentric system by Copernicus, and the systematic recording of

---

* A vital aid to the ocean voyages of Columbus and others.

† Referring to Alfonso's unsuccessful claim to election (on a minority vote) as Holy Roman Emperor.

accurate observations by Tycho Brahe. This revolution was completed early in the following century by the discovery of Kepler's laws and the invention of the telescope.

Niklas Koppernigk (Copernicus), Polish and German by descent, was an outstanding renaissance scholar and thinker, who spent ten years at Italian universities, studying mathematics, astronomy and Greek and Latin literature, and acquiring degrees in canon law and medicine. By occupation administrator of a bishopric, his absorbing interest was in mathematical astronomy and especially cosmology. His life and work have been fully described in other books (e.g. Abetti's *History of Astronomy*, and A. Armitage's *World of Copernicus*); here it is proposed only to deal with how he came to write *De Revolutionibus Orbium Coelestium*, why he delayed publishing it, certain points and limitations in his theory, and his observations of Saturn.

He stated in the preface to his book (*Occ. N. R.A.S.*, vol. 2, no. 10, pp. 3–6) that he had found that the mathematicians were so unsure of the movements of Sun and Moon that they could not explain the constant length of the seasonal year, and that they did not use principles consistently in determining the motions of these and of the other five planets. He had been unable to find agreement on any one certain theory of the mechanism of the universe, and had therefore looked in the ancient writings for some alternative, finding that Philolaus had suggested that the Earth revolves, and Heraclides that it rotates on an axis. He had therefore assumed motions of the Earth and had found by long and frequent observations that he could on this basis explain the general system of the universe. He gave 'fear of ridicule' as his reason for reluctance to publish his theory, but the postponement of publication until the end of his life may well have resulted in part from qualms as to the reaction of the Church to such a revolutionary hypothesis.

In the first book of his treatise he asserted, without real proof, that the universe is spherical, gave some proofs that the Earth is spherical, and then hampered his theory by adhering to the Aristotelian principle that the motions of the heavenly bodies are uniform and circular, or composed of circular motions. As the result of this he had to retain epicycles, and to use 34 circles in all, including three for the Earth and five for Saturn, though it was an improvement on the Ptolemaic system which at that time postulated 79. He showed that the Earth is a planet, rotating, and revolving round the Sun: it was more reasonable to assume the Earth to be rotating in 24 hours than that the vast universe should do so. He said that any apparent motion of the Sun could be better explained by motion of the Earth. He pointed out that the outer planets are always nearer the Earth at their evening rising, i.e. at opposition, and more distant at the time of their evening setting, when the Sun is between them and the Earth. He showed that since the outer planets could even appear to move in a direction opposite to the Sun, they must have orbits larger than that of the Earth, and for the first time in history he was able to calculate their relative distances approximately. From observations of their oppositions he was able to obtain fairly accurate values of their synodic and sidereal periods. He stated that his system explained why progression and regression appear greater for Jupiter than Saturn, but less than for Mars, and why such oscillation appears more frequently in Saturn than in the other outer planets.

Copernicus went to very great trouble to try to reconcile all previous observations, many of which were inaccurate, with his system and his own observations, which were made, as Dr. Armitage says, with rather crude instruments and in which Copernicus apparently did not expect to attain an accuracy to within less than ten minutes of arc, although it is possible to measure an angle in the sky without a telescope to within one or two minutes of arc. In his book V, chapter 5 he gave particulars of Ptolemy's observations of Saturn and then stated (chap. 6) that the computation of Saturn's motion handed down by Ptolemy showed considerable discrepancy in his time, and as it was difficult to see where the error lay, he had been forced to make new observations. He then gave particulars, just as Ptolemy had done, of four Saturn observations, three of them at oppositions. His treatise, in fact, as to subject matter and order of treatment followed very closely the lines of Ptolemy's *Syntaxis*; he used in it some 27 of his own observations altogether, though records of others are shown in his papers.

Saturn observations. Translations (from *Great Books of the Western World*, vol. 16) are given below of his four observations of Saturn (from his book V, chaps. 9 and 6) in order of date. It is evident, especially from his statement on the 1527 opposition, that he measured ecliptic longitudes, not from the first point of Aries (then as now in Pisces) but from β and γ, the two bright preceding stars of Aries, and the equivalent positions on the 1950 star map have been given on this basis:

1. *Observation of 1514 April 26*: 'A.D. 1514 on 6 day before kalends of May, 5 hours after the preceding midnight, Saturn was seen to be in a straight line with the stars in the forehead of Scorpio, namely with the second and third stars, which have the same longi-

tude and are at 209° of the sphere of fixed stars.' (1950 equivalent: 241°, near β and δ Sco.).

2. *Opposition of* 1514 *May* 5: 'A.D. 1514 on 3rd day before nones of May, $1\frac{1}{5}$ hours before midnight; Saturn at 205° 24'.' (1950 equivalent: $237\frac{1}{2}$°, Libra near κ).

3. *Opposition of* 1520 *July* 13: 'A.D. 1520 on 3rd day before ides of July at midday; Saturn at 273° 25'.' (1950 equivalent: $305\frac{1}{2}$°, near o Cap.).

4. *Opposition of* 1527 *October* 10: 'A.D. 1527 on 6th day before ides of October, 6 2/5 hours after midnight; Saturn at 7' from the horn of Aries, 0° 7'.' (1950 equivalent: 32°, in a barren region between γ Ari. and ξ' Cet.; this agrees well with the opposition place given by Tycho for October 1586). (See p. 79).

Copernicus found the mean motion of Saturn between 1514 and 1520 oppositions to be 75° 39' whereas the difference of place from his observations was 68° 1'; between the oppositions of 1520 and 1527 the mean motion was 88° 29' while the observed difference of place was 86° 42'. To try to reconcile these differences between theory and observation, and his own observations with those of Ptolemy nearly 1,400 years earlier, he went into long mathematical discussions in the Ptolemaic manner.

His deduction regarding the distance of Saturn was stated as follows: 'the altitude of the apogee of Saturn is 9^p 42', where the radius of the orbital circle of the Earth equals 1^p, and the altitude of the perigee is 8^p 39'.' This would seem to mean in modern parlance that he found the greatest distance of Saturn from the Sun to be 9·7 A.U. (i.e. astronomical units) and the least distance 8·65 A.U., giving a mean distance of about 9 A.U. (modern value 9·539 A.U.).

He deduced (book VI, chap. 3) that the inclination of Saturn's orbit to the ecliptic had a maximum value of 2° 44' and a minimum of 2° 16'. (The mean of these is very close to the value for 1960 which is 2° 29'.)

The elaboration of a complete mathematical system for the universe on the basis of a central Sun with the Earth as one of the planets was a work of outstanding genius: *De Revolutionibus* of Copernicus ranks with Ptolemy's *Syntaxis* and Newton's *Principia*, as Abetti has said, as representing three basic stages in the development of astronomy. More than a century was to pass before Copernicus's work (published 1543) gained general acceptance: it was too difficult and too far removed from common experience for most people to understand, and for a time it was also regarded as contrary to religious doctrine.

Copernican Tables (Reinhold, Stade, Maestlin)

One of the very few to accept the Copernican system quickly was Erasmus Reinhold, professor of mathematics and astronomy at Wittenberg University, who worked out tables in 1551 based on the work of Copernicus from which the positions of the heavenly bodies could be determined at any date. As the expense of their publication was met by Duke Albert of Prussia they were known as the Prutenic (Prussian) Tables; they were the first non-Alphonsine tables to appear and though they showed a certain improvement over the Alphonsine (based on the Ptolemaic system), the difference was not very great. Another set of Copernican tables for the period 1554 to 1606 was worked out by Jan Stade (Stadius), a Belgian astronomer and mathematician, and called *Tabulae Bergenses* in honour of Robert de Berg, Prince-Bishop of Liège. Michael Maestlin, professor of mathematics at Tübingen, who taught Kepler the Copernican system, also produced tables or ephemerides on the Copernican basis. The Prutenic Tables and those of Stadius and Maestlin were all used and referred to by Tycho Brahe, who also made use of then current versions of the Alphonsine Tables (of Cyprianus 'Ludovicus'* and of Carellus).

Kepler reported to Tycho (*Thes. Obs. T.B.* vol. 4, p. 247) a conjunction of Saturn with Mars on 17 March 1576 as having been observed by Maestlin, who gave the positions for 10 h as: Saturn at 0° 36′ Cap. and Lat. + 1° 40′; Mars at 0° 34′ Cap. and Lat. + 0° 24′. The equivalent place of Saturn on the 1950 star map, allowing for precession, would be ecliptic longitude about 276°, in Sagittarius between 21 Sgr and λ Sgr; latitude was measured north or south of the ecliptic.†

Tycho Brahe

Tycho Brahe came of an aristocratic Danish family settled in southern Sweden, then under Danish rule, and was born in 1546,

* Cyprian Leowitz (or Livowski), a well-known German or Czech astronomer-astrologer who predicted the end of the world for 1584, was visited about 1569 by Tycho who asked him whether he checked his ephemerides by observations. Leowitz said his only instruments were clocks which he occasionally used for observing eclipses, and while admitting the superiority of Copernican tables for solar eclipses and the outer planets, claimed that the Alphonsine were the more accurate for inner planets and lunar eclipses. The Ephemerides of Carellus were published at Venice, 1557. (Dreyer: *Tycho Brahe*, pp. 29, 18).

† Mr. Heath has verified by calculation that there certainly was a conjunction of the two planets near λ Sgr. at that date and time. Maestlin could not have made the observation at 10 a.m. because of daylight, but he may have seen the two planets rather near each other at (say) 4 a.m. and computed the time of closest approach; he seems to have made the latitude of both 0°·6 too far north.

three years after the death of Copernicus. At the age of 14 he was so greatly impressed by the accuracy of the forecast of a solar eclipse that he developed a passion for astronomy and mathematics, was eventually allowed to study science instead of law, and from the age of 17 was making and recording astronomical observations, comparing them with calculations made from Alphonsine and Copernican tables, and finding the tables to be inaccurate. His careful observations of position and brightness of the supernova in Cassiopeia in 1572 and his skill in designing and constructing observing instruments showed his great promise as an astronomer, and in order to keep him in Denmark King Frederick II in 1576 granted him for life the island of Hveen (in the Sound between Denmark and Sweden) and one per cent of the royal revenues, so that he might build and maintain an observatory there. By the beginning of 1578 observations were being made by Tycho and his assistants at Uraniborg, which he equipped with the most accurate instruments of wood and iron made up to that time and organised as the first modern observatory; a second set of buildings and equipment was later added especially for the observation of stars.

M. André Danjon, in a most interesting address on Tycho's life and work (*L'Astronomie*, July 1947), described fully with illustrations the huge and splendid equipment, and instanced the adjustments and corrections Tycho worked out to give the utmost precision to these naked eye observations. Danjon stated that the places of nine fundamental stars were ascertained with an error less than 25″ of arc in their co-ordinates, and that errors rarely reached 2′ in the whole catalogue of about a thousand stars compiled at Hveen. He summarised the scientific programme of Tycho's observatory as: (i) getting accurate positions of a thousand stars in various parts of the sky; (ii) relating to them the places of Sun, Moon, planets and comets, so as to study their motion; (iii) determining the elements of the members of the solar system with much greater precision so that really accurate tables might be drawn up. Tycho realised the first two of these aims, but only commenced work on the tables towards the end of his life, so their completion was mainly due to his assistant, Kepler. Unlike Kepler, Tycho did not subscribe to the Copernican theory but in 1577 proposed a compromise of his own which he held to be more in accord with common sense, observation and the teaching of the Bible. The Tychonic system kept the Earth immobile at the centre with Moon and Sun revolving round it, but conceded that the other planets and the comets revolve round the Sun.

The massive output of observations and calculations, all recorded in Latin, by Tycho and his staff, was published 1923–26 in four large volumes with introduction and notes (also in Latin) by another distinguished Danish astronomer, Dr. J. L. E. Dreyer, well-known for his New General Catalogue of stars. Dreyer's *Thesaurus Observationum Tychonis Brahe Dani* covers in vol. 1 observations 1570–85; vol 2, 1586–89; vol. 3, 1590–95; vol. 4, 1596–1601 and observations of comets. For Saturn alone these records contain many hundreds of observations made on more than 400 dates. As a rule at least three readings of each position were taken using different instruments; distances of the planet from several bright stars, sometimes as many as six, were given in degrees, minutes and seconds of arc. Elaborate calculations and corrections of Saturn's place were shown, some in Tycho's own hand. In addition to ecliptic longitude and ɪatitude he introduced determinations of right ascension and declination. Comparisons with computations from tables were usually shown. Time was accurately recorded, various types of clocks being used; in connection with one Saturn observation it was noted that the observatory clock was found to be one minute slow and had been adjusted. This profusion of data on Saturn's place can perhaps best be represented by giving a list of determinations of opposition places and stations, which were calculated with special care, partly by Tycho himself, from the nearest observations to the date and time of opposition or station, and by giving some particulars of a few other Saturn observations of various kinds.

Conjunction of Saturn with Jupiter. The following seems to be Tycho's earliest recorded observation of Saturn made when a student at Leipzig aged 17: 1563 *August* 18, 13^h 34^m. 'The distance between Jupiter and Saturn is a little greater than between the kids (η and ζ Aur.) and less than between the two in the front right foot of the Great Bear (ι and κ UMa) but nearer to the interval of those in the Bear . . . The place of Saturn is south of the straight line from Jupiter to Venus, Saturn being further south than Jupiter. The distance between the kids is 30', the two in the Great Bear 50'; whence Saturn and Jupiter are 40' or 45' distant from each other.'

This is typical of a number of his observations of planetary conjunctions made in the early years by sighting along the legs of an instrument like a large pair of compasses. He recorded observations of Saturn in Leo on 17 dates in 1563 and 1564, and noted that it was near its first station on 20 November 1564.

Conjunctions with the Moon. (1) 1587 *January* 6. 'The Moon came very near Saturn this evening. At 10^h 4^m when the eye of the Bull (Aldebaran) was distant from the meridian towards the west 27° 50', then the two horns of the Moon being in quadrature were in exact alignment with Saturn as well as could be seen. For Saturn was distant from the lower limb of the Moon and southern horn towards the south 12′, for it was seen to be more than a third and less than half the diameter of the Moon from the lower horn.' (Figure 2a is a careful copy of the rough sketch that accompanied the record).

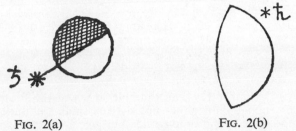

FIG. 2(a)   FIG. 2(b)

FIGURE 2. Tycho Brahe: sketches of conjunctions of Saturn with the Moon: (a) 6 January 1587; (b) 12 February 1592. ♄ is the symbol of Saturn, which is shown as a star. (Redrawn after Uraniborg sketches reproduced in Dreyer's *Thesaurus Observationum*, vol. 2, p. 142 and vol. 3, p. 191).

He then stated that two of his assistants, Peter Jacobi and Gellio Sasceride,* using better instruments, had estimated Saturn's minimum distance from the limb at 13′, but they all agreed a distance slightly less than one semi-diameter. He decided that the time of conjunction with the Moon was 10ʰ 10ᵐ, and that Saturn's place was then Long. 26° 1′ Ari., Lat. 2° 27¾′ S. (The equivalent ecliptic longitude for epoch 1950 would be 31°.)

(2) 1592 *February* 12. Saturn's nearest distance from the limb of the gibbous Moon was estimated at 9′ or 10′ at most (see Figure 2(b)).

Saturn's diameter. On 19, 20 and 23 January 1591 Tycho and his assistants attempted the impossible by making naked eye estimates with various instruments of Saturn's diameter. He took as the mean value 2′ 15″, which is more than three times too great, since it was then some seven weeks after opposition, and at mean opposition distance Saturn's apparent diameter is only 44″.

Alignment with stars. On 29 November 1594, 2½ hours after midnight, 'Saturn was seen to be stationary and beginning to regress

* Sasceride, born at Copenhagen 1562, was a pupil and assistant to Tycho 1582–88, and then went abroad to study medicine, later becoming professor of medicine at Copenhagen.

Peter Jakobsen of Flemlöse, also a Dane who studied and later practised medicine, was the first to join Tycho as assistant and worked for him for ten years (Dreyer: *Tycho Brahe*, p. 117).

on the straight line leading from Cor Leonis (Regulus) to that which is *informis et lucidior infra caudam Vrsae majoris*. Saturn was distant from Regulus not much more than a degree towards the north-east.' The 'unformed and brighter object below the Great Bear's tail' is evidently Cor Caroli (α CVn), which in Ptolemy's and also in Tycho's time was one of a number of bright stars that had not been formed into constellations.

Comparisons with tables. Here are three examples out of very many such comparisons: (1) 23 September 1586:

	By observation	Calc. Alph.	Calc. Coper.
Long.	0° 38′ 10″ Tau.	3° 3′ Tau.	0° 42′ Tau.
Lat.	2 47 20 S.	3 0 S.	2 58 S.

(2) 10 December 1594, an example in which the Prutenic tables are good enough in longitude but give too small a latitude, while the Alphonsine are 2° out in longitude but give a better approximation to the latitude:

	Longitude		Latitude
Observation	24° 30′ 20″ Leo.		1° 18′ 30″ N.
Prutenic, Calc.	24 45	,,	0 29 N.
Alphonsin, Calc.	26 46	,,	1 15 N.

(3) Comparison given for ten dates in April 1600, those nearest to opposition being:

	Observation	Alphon.	Coper.
Longitude on			
April 14	25° 12′ 50″ Lib.	25° 50′ Lib.	25° 52′ Lib.
17	24 58 31 ,,	25 35 ,,	25 38 ,,
Latitude on			
April 14	2 49 15 S.	3 2 S.	2 56 S.
17	2 49 14 S.	3 2 S.	2 57 S.

In this case the Alphonsine tables came a little nearer to observation in longitude, but the Copernican were closer in latitude.

Opposition places 1570 *to* 1601. After calculating Saturn's true opposition with the mean Sun to have occurred at 4½ʰ on 25 March 1599 in longitude 12° 54′ Lib., latitude 2° 49′ N., Tycho added the note: '29 years before, viz. in 1570, I observed Saturn in opposition with the Sun almost in this same place.' His record of the 1570 opposition gave the date as March 31 and the longitude 18° 30½′ Lib., so that Saturn was within 6° of its place at the opposition of 29 years earlier and would reach it in another half year.

The gap 1571–81 in the list on p. 79 is due to Saturn's being too far south for effective observation from northern Europe; no observations of that planet were recorded by Tycho in the years 1571–77 inclusive. From about 1586 the data of distances from bright stars become more plentiful and fix Saturn's position very well on the star map. Particularly good sets of these data are those for dates near the oppositions of 1594, 1596 and 1597:

1594 Jan. 14 Pollux 21°−; Procyon 24°+; Regulus 16½°
1596 Feb. 13 Denebola 15°+; Regulus 10½°+; Arcturus 50½°
1597 Feb. 28 Regulus 23½°; Denebola 10°; γ Virginis 17°; Arcturus 41°−
(Tycho of course gave these distances in minutes and often seconds as well as degrees.)

Saturn's opposition places, 1570, 1582–1601 (Tycho)

Date and Longitude from Observations	Lat. or Dec.	Approx. Ecliptic Longitude and Place epoch 1950
1570 Mar. 31 Lib 18° 30½' (Spica <5°)	—	204° Virgo near Spica
1582 Aug. 21 Psc 7° 34'	—	343° Aquarius near λ
1583 Sep. 1 Psc 20° 1½'	D 6¼° S.	355° Aquarius near 20 Psc
1584 Sep. 13 Ari 2° 46'	—	8° Pisces-Cetus near 14 Cet
1585 Sep. 26 Ari 15° 50'	—	21° Pisces near μ
1586 Oct. 10 Ari 29° 12'	L 2¾° S.	34° Aries-Cetus near ξ' Cet
1587 Oct. 24 Tau 12° 54½'	D 13° N.	48° Aries between δ Ari and λ Cet
1588 Nov. 7 Tau 26° 49' (Aldebaran < 8°)	D 17° N.	62° Taurus between Pleiades and Hyades
1589 Nov. 21 Gem 10° 56' (Aldebaran < 8°)	L 1° 47' S. D 20° N.	76° Taurus near ι and 106
1590 Dec. 6 Gem 25° 10'	—	90° Between 1 Gem and 139 Tau
1591 Dec. 21 Cnc 9° 24' (1592 No opposition)	—	104½° Gemini near ω
1593 Jan. 3 Cnc 23° 32½'	—	118½° Cancer near μ
1594 Jan. 18 Leo 7° 27'	L 0° 49½' N.	132½° Cancer between o and ξ
1595 Feb. 1 Leo 21° 11½'	—	146° Leo near ν
1596 Feb. 15 Vir 4° 34' (Regulus 10½°)	—	159½° Leo between ρ and χ near 53
1597 Feb. 27 Vir 17° 39' (Denebola 10°	L 2⅓° N. γ Vir 17°)	172½° Between ν Vir and τ Leo
F 1598 Mar. 13 Lib 0° 27½' T 14 0° 19'	L 2° 41' N. D 2° 19' N.	}185° Virgo near η
1599 Mar. 25 Lib 12° 54' (γ Vir <8½°)	L 2° 49' N.	198° Virgo near θ
F 1600 Apr. 6 Lib 25° 4' T 16 25° 5½' (N.S.) (Spica 8⅓°)	} L 2° 49' N.	210° Virgo near κ
1601 Apr. 17 Sco 6° 58'	—	222° Libra near α

F = data supplied to Tycho by Fabricius; T = Tycho's own data.
Tycho used new style dating for 1600 opposition.
Distances of rather near bright stars given by Tycho are shown.

In April 1597 Tycho Brahe with his family, assistants and most of his observing instruments left Denmark for good, observing for a year or so at Wandesburg, near Hamburg, and afterwards near Prague. Partly through envy of certain courtiers at his privileges and partly through complaints of his tenants against his oppressive treatment of them, Tycho's grants were withdrawn by the new King of Denmark, but in Austria he was given a castle and some financial aid by Emperor Rudolph II, a patron of the arts. The additional data sent him by Fabricius were no doubt very welcome to Tycho in these last years.

Stations of Saturn determined by Tycho

Data, not always complete, were given by Tycho for a few stations.

Date, Longitude and other data from observations	Approx. place of station for epoch 1950
1586 Dec. 26 Ari 25° 43'; Dec. 7½° N., Lat. 2½° S. Aldebaran 38°, Markab 43°	31° Pisces near ξ' Cet
1587 Aug. 18 Tau 16° 19' Aldebaran 18°, Hamal 19°	51¼° Aries between δ Ari and 5 Tau
1591 Oct. 11 Cnc 12° 49½'; Lat. ¼° S. Pollux 8½°, Aldebaran 39°	108° Gemini near δ
1594 Nov. 27 Leo 24° 50'; Dec. 14½° N. Regulus 1°, Nov. 29: Pollux 37½°	150° Leo near Regulus
1595 Apr. 12/13 −; Dec. 17° N. Regulus 6½°; Pollux 30½°	— Leo near ψ
1595 Dec. 9 −; — Regulus 14°−; Denebola 13°; Procyon 50½°	— Leo near χ
1596 Apr. 25 Vir 1° 13' Pollux 44°; ε Vir 35½°	156° Leo near ρ
1599 Jan. 15 −; — Arcturus 27°+; β Librae 26½°	— Virgo near 66 (?)
1601 Jun. 29 Sco 3° 45'; Lat. 2½° N. β Lib 11½°	219° Between μ Lib and λ Vir

Fabricius and Longomontanus

David Goldschmidt, who translated his surname into 'Fabricius,' was a Frieslander, pastor of a village in N. Holland, and an amateur astronomer. He had visited Uraniborg, studied Tycho's methods, and tried with slender resources to construct similar instruments. He was a very good observer and is credited with the discovery of the variable brightness of Mira Ceti. His calculation of the opposition place of Saturn in 1598 was based on star distances taken on March 9 and 12, and the meridian altitude of Saturn on March 13, when he noted: ' I was not able to take distances (from stars) because of intervening clouds.' His data are shown (*Thes. Obs. T.B.*, vol. 4, pp. 151–2) with a marginal note by Tycho pointing out a difference between the observations of Fabricius and his own of $2\frac{1}{2}'$ in longitude and $\frac{1}{2}'$ in latitude. Fabricius sent with his results the following note: 'If therefore your excellency has observed or has had observed the same things in Wandesburg or Denmark, I should like to be advised as soon as convenient as to what discrepancies there are in the observations. I hardly expect my opposition observations to differ from your most exact and infallible ones by more than one minute, if only you have had observations made. When I construct some new instruments in the form of your excellency's instruments, applying to them also that subtle graduation by transverse points, I hope the accordance will be absolutely complete. This I aspire to do later. . . .'

Fabricius was referring to Tycho's use of 'transversals': he divided into ten equal parts the hypotenuse of each little triangle whose base represented on his instruments an angular distance of ten minutes of arc. Hence by very careful sighting it was possible to read off angular distances to the nearest half minute of arc or sometimes even more precisely.

Fabricius sent Tycho observations of 13 dates in 1600, basing his determination of the opposition of April 6^{d} 5^{h} 43^{m} p.m. on the following star distances of April 8: Spica $8\frac{1}{4}°$, Denebola $39\frac{1}{2}°$, the third of the wing of Virgo (γ Vir) $20\frac{1}{4}°$, the N. lance of Libra (β Lib) $19\frac{1}{2}°$, the S. lance (α Lib) $15°-$.

Christian Sörensen, who called himself 'Longomontanus' because he came from Longberg in Denmark earned money to pay for his own grammar school education, was engaged as an assistant by Tycho in 1589, worked at Hveen until Tycho left Denmark, studied for three years at German universities, rejoined Tycho at Prague in 1600, and after Tycho's death returned to Denmark, where for 40 years he was professor of mathematics at Copenhagen. Among many treatises on astronomy, mathematics, etc., Sörensen published in 1622 *Astronomica Danica*, a large book with mathematical discussions of the planetary motions based on the Uraniborg observations. On p. 380 of that book he gave the following sample observation of Saturn:

'1591 17 March 7.30 p.m. at Uraniborg Saturn was observed at a distance from Aldebaran $19°$ $11\frac{1}{2}'$. But at the same time from Pollux $25°$ $54'$ with declination $22°$ $23'$ N. Hence it is found that the longitude of Saturn is $22°$ $43'$ Gem with latitude $0°$ $56\frac{1}{2}'$ S.' Then as the result of a long calculation he decided that Saturn's true longitude was Gem $22°$ $45'$ $35''$, and he gave the following

81

comparison with tables:

	Alphonsine	*Copernican*
Longitude	Gem 25° 7′	Gem 22° 23′
Latitude	0° 56′ S.	1° 35′ S.

Sörensen's sample observation would correspond to an ecliptic longitude (1950) of 88°, a position in Orion near χ' Orionis.

Kepler

It would seem that the friendship and co-operation between the brilliant young German astronomer, Johann Kepler, and the famous Danish exile, Tycho Brahe, belonged only to the last year or two of the latter's life. In August 1600 Kepler sent from Styria a report (*Thes. Obs. T.B.*, vol. 4, pp. 246–52) about certain records of former observations of Saturn and other planets and giving his own observations of 1594–99. These included a few of Saturn, not very precise, as presumably Kepler's equipment was rather limited:

1597 November 5, 6^h. He saw Saturn $1\frac{1}{2}$ diameters of the Moon from 6 Vir; the triangle between Denebola, Saturn and 6 Vir had a slightly obtuse angle at Saturn.

1598 February 8. Saturn in line with 6 and 7 Vir.

 March 4. The planet equidistant from 6 and 7 and about one-third of a degree N. of that line.

 April 6. Saturn with 6 and 7 Vir formed a right angle at itself.

 November 25, $6\frac{1}{2}^h$. He saw Saturn, Spica and the centre of the Moon in one and the same vertical.

Kepler's observations of the spring of 1598 suggest a position for Saturn of ecliptic longitude (1950) of 180° or rather less, whereas Tycho and Fabricius gave Saturn a position at mid-March of that year corresponding to about (1950) 185°.

Tycho, however, was evidently impressed with Kepler's mathematical skill and his inquiring theorising mind: they formed an ideal combination, Tycho being the practical man with a great store of very precise and accurate observations, Kepler having the imaginative insight to make the best use of that first-rate material. On Tycho's death in 1601 Kepler inherited the store of observation records and undertook the task of drawing up new planetary tables, but did so on the basis of the Copernican, not (as Tycho wished) the Tychonic, system. He struggled long and hard, trying many different expedients in the effort to solve the problem of the planetary motions. Mars gave him the most trouble (because of course of its very elliptical orbit); in the attempt to find a combination of circular motions for Mars, he even considered the assumption that Tycho's observations were 8 minutes of arc in error, but he rejected it and decided instead to try some orbit other than a circle. By this bold stroke he lifted planetary astronomy out of the rut in which it had been confined for two thousand years.

Kepler thus discovered his three Laws of Planetary Motion, stated by Davidson (*Stars and the Mind*, p. 101) as follows:

1. A planet moves in an orbit which is an ellipse, the Sun being at one focus.
2. The area swept out by the line joining the Sun and a planet is proportional to the time.
3. The squares of the periods of revolution are proportional to the cubes of the mean distances from the Sun.

Though the first two laws were made originally for Mars he quickly recognised that they must apply to all the planets. He had discovered the third law by 1619. By thus perfecting the Copernican theory, as well as by his friendships with Tycho, David Fabricius and Galileo, Kepler linked the old naked eye observing with the new era of the telescope.

Being court mathematician to the Emperor Rudolph, Kepler called his new tables, based on the Uraniborg observations, the *Rudolphine Tables*. Published in 1627 they remained the standard guide to planetary motions for a century. After that the observation of planetary positions and the further refinement of the theory of their movements became one of the continuing tasks of the great observatories. The truly remarkable thing is that it was the work of naked eye observers after two thousand years of struggle which, through the skill of Tycho Brahe and the genius of Copernicus and Kepler, solved in the main essentials the motions of the planets.

CHAPTER 4

THE FIRST TELESCOPES, THE STRANGE APPENDAGES AND TITAN (1610–1660)

The telescope has long been such a familiar tool of the astronomer that the revolutionary change wrought by its invention is hard to realise and apt to be forgotten. The observations referred to in Chapters 2 and 3, made at intervals during the long period from 650 B.C. to A.D. 1601, all with the unaided eye, could do no more than solve some of the problems of Saturn's motion and enable its future positions and movements to be predicted with a fair degree of success. The shape and appearance of the planet were as impossible to discover for Tycho Brahe as for the ancient Babylonians. The year 1610 can well be said to mark the end of mediaeval and the beginning of modern astronomy.

Even those familiar with Saturn from drawings and photographs can scarcely refrain from a gasp of astonishment at the first sight through a telescope of the exquisite planet with its encircling rings, unique in the solar system, and bearing no true resemblance to any other observable object in the whole sky. Until the advent of the telescope that wonder was hidden from the world. Its spectacular charm and its many enigmas have so fascinated astronomers that the history of its observation has, except for a break in the mid-eighteenth century, been practically continuous for the past 350 years.

First view with telescope (Galileo)

That history began on the night in July 1610 when Galileo Galilei at Padua, using his largest telescope, saw for the first time the planet's disk. But the aperture of his telescope was very small and the magnification it gave was only 32 diameters, and more important still, the lenses were very imperfect and afforded only a distorted view. Furthermore, he was unlucky in happening to make his first observation in a year when the ring was only narrowly open, and though his telescope could show its broad extremities the instrument was not powerful enough to reveal the narrower parts of the ring nearer the globe. Therefore although Saturn was approaching opposition and was favourably situated for observation, the result of his inspection was by no means satisfactory. The impression he obtained was that there was something quite unexpected and peculiar in the aspect of this planet.

Galileo's letter of 30 July 1610 announcing his discovery to Belisario Vinta, counsellor and secretary of state to the Grand Duke of Tuscany, is a landmark in the history of Saturn observation, since it describes the first view ever obtained of Saturn's disk; it also explains why he was making the announcement and why he wished it to be treated as confidential. The following extract from this letter is as quoted in the translation given by G. Abetti (*The History of Astronomy*, pub. by Sidgwick and Jackson, 1954, p. 104):

'. . . I have discovered a most extraordinary marvel, which I want to make known to Their Highnesses and to Your Lordship, but I want it kept secret until it is published in the work which I am going to have printed. But I wanted to announce it to Their Most Serene Highnesses so that, if someone else should discover it, they would know that no one observed it before I did. Yet I believe that no one will see it before I inform him. The fact is that the planet Saturn is not one alone, but is composed of three, which almost touch one another and never move nor change with respect to one another. They are arranged in a line parallel to the zodiac, and the middle one is about three times the size of the lateral ones. . . .'

To his friends in Germany and Italy he disclosed the news in the form of an anagram, the solution of which read: '*Altissimum planetam tergeminum observavi*' and which may be translated: 'I have observed the most distant planet to be a triple one.'

As only a few months previously he had discovered four satellites revolving round Jupiter, he naturally thought that the two lesser globes were very close satellites which revolved around Saturn very rapidly. In fact he wrote (as quoted by Abetti) to Giuliano de'Medici: 'So! we have found the court of Jupiter, and two servants for this old man, who help him to walk and never leave his side.' But continuing his observations for several months, he was much surprised and probably disappointed to find that the two attendant disks remained in the same position and appeared not to alter in brightness. The fixity of their position was inexplicable; it showed that they were different from Jupiter's moons. Galileo was confronted with a phenomenon unlike any hitherto revealed by his telescopes.

The missing globes

A greater surprise and setback was to come. Galileo had stated that Saturn's two attendants never left his side, but two years later, when he observed the planet again, he saw a single golden-hued disk more like that of Jupiter, and no trace whatever of the two lesser globes. According to R. A. Proctor, the ring became edgewise to the Sun on 28 December 1612; hence for several weeks before and after that date it would have presented much too thin a face to be visible in Galileo's telescope. One can imagine that he must have tried again and again to detect the lesser disks, but without avail. His perplexity, and his justifiable apprehension that his opponents and critics would not be slow to seize this chance of making capital out of his discomfiture, are strikingly expressed in a letter he wrote on 4 December 1612 (quoted in *Opere di Galileo*, vol. ii, p. 152, Padua edition, 1744):

'What is to be said concerning so strange a metamorphosis? Are the two lesser stars consumed after the manner of the solar spots? Have they vanished or suddenly fled? Has Saturn, perhaps, devoured his own children? Or were the appearances indeed illusion or fraud, with which the glasses have so long deceived me, as well as many others to whom I have shown them? Now, perhaps, is the time come to revive the wellnigh withered hopes of those who, guided by more profound contemplations, have discovered the fallacy of the new observations, and demonstrated the utter impossibility of their existence. I do not know what to say in a case so surprising, so unlooked for, and so novel. The shortness of the time, the unexpected nature of the event, the weakness of my understanding, and the fear of being mistaken, have greatly confounded me.'

The greatness of Galileo as an astronomer is shown by the fact that, in spite of this blow and the confusion he felt, he stuck to his belief in the reality of the 'two lesser Saturnian stars', and even predicted that they would reappear for a short time in the summer of 1613 and again in the winter of 1614, the phenomenon being repeated several times until they became more distinct and larger and brighter, and that then they would remain continually visible for many years. Though incorrect in detail, this prediction is a truly amazing forecast when it is remembered that Galileo was attempting to sketch the future course of a hitherto unknown phenomenon, the cause of which he himself was unable to understand. As he continued his observations he found that the supposed satellites began to resemble less two minor globes than two great arms or handles stretched towards the planet. In fact his drawing made in 1616 was, in the opinion of Signor G. Abetti, so accurate that had he then seen Saturn for the first time he might well have guessed the true explanation.

Before the observations of Galileo, the planets Saturn, Jupiter and Venus had merely appeared as they do today to the unaided eye as bright stars slowly changing their positions in relation to the other stars. Galileo's revelation, by means of the telescope, of Saturn's variable shape was an advance hardly less important than his discoveries of the brightest moons of Jupiter and the phases of Venus, though those discoveries had more immediate importance in relation to the Copernican conception of the solar system and universe. In bringing to light the strange behaviour of Saturn he presented astronomers with a problem which took them almost half a century to solve.

Other pioneer observers (Fontana, Gassendi, Hevel, Riccioli)

Among the earliest to follow up Galileo's observations of Saturn may be mentioned a German Jesuit, Christopher Scheiner, whose main contribution to astronomy was, however, a prolonged and systematic study of sunspots; Giuseppe Biancani, S.J., professor of mathematics at Parma; and Francesco Fontana, a Neapolitan lawyer and amateur astronomer, who was one of the pioneers in the making of telescopes and their use for planetary observations. Fontana was making drawings of Saturn during the period 1630–50 and so was Pierre Gassendi, a Provençal and, like Galileo and Scheiner, a professor of mathematics. There were other famous pioneer lunar observers besides Gassendi who were also active in the observation of Saturn about the middle of the seventeenth century: John Hevel (or Hevelius), a city councillor of Danzig who published the first complete work on the Moon with maps and drawings, and the Jesuit professor Joannes Riccioli of Bologna, whose map of the Moon laid the foundations of lunar nomenclature and contained many of the names still in use for lunar formations. Also, according to Riccioli, his friend Francesco Grimaldi, another Jesuit professor and lunar observer at Bologna, was the first to remark, in 1650, on the flattening of Saturn at the poles. Eustachio Divini, an Italian physician and expert telescope-maker, also observed and made drawings of Saturn.

Examples of the impressions of most of these observers are shown in Plate I, a set collected, studied and published by the famous Dutch astronomer, Christiaan Huygens in his epoch-making treatise *Systema Saturnium*. Figure 3 shows a set of tracings made by C. L. Prince (*M.N.*, vol. 36, pp. 108–9) from some of the engravings, probably the first of Saturn ever published, in a copy of Gassendi's works printed in 1658.

These early drawings, referred to derisively by the Rev. T. W. Webb as 'queer-shaped misrepresentations', seem much more meri-

A
June 19
1633

B
April 13
1634

C
Nov. 20
1636

D
Jan. 11
1645

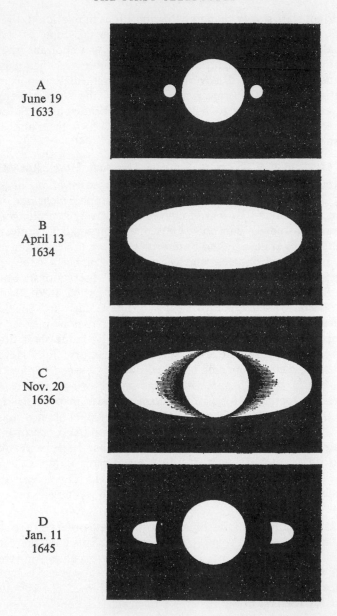

FIGURE 3. Gassendi: eight drawings (with telescope)

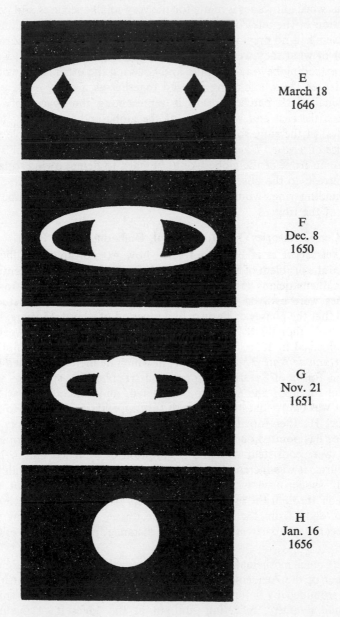

E
March 18
1646

F
Dec. 8
1650

G
Nov. 21
1651

H
Jan. 16
1656

of Saturn 1633–56 (from *M.N.*, vol. 36, pp. 108–9).

torious when allowance is made for the very small apertures and poor definition of the telescopes used and for the fact that these pioneer observers had no previous knowledge to guide them as to the correct aspect of what they were trying to depict. When the ring was wide open extending beyond one pole and covering the other, they would tend to see the whole thing merged together as a great oval, with two small dark patches inside it representing the spaces of sky between the east and west limbs of the globe and the 'ansae' (i.e. handles) of the ring, so-called because they seemed to resemble the handles of a vase. When the ring was fairly well open, its near side passing in front of the globe would be hard to detect owing to lack of contrast, so the observers, thinking of handles and unaware of a surrounding ring, would be inclined to draw the entire globe in front of the ring.

Curious theories (Hevel, Roberval, Hodierna, Wren, Fabri)

In retrospect it is hard to realise how extraordinarily difficult seemed the problem of Saturn's appendages to the foremost scientists and mathematicians of the mid-seventeenth century: several strange theories were evolved. Hevel, after much observation and study, found that the phases of Saturn had a period of about 15 years, but mistakenly thought the line of the ansae was parallel to the orbit. He published in 1656 a treatise, with illustrative diagrams, entitled *Dissertatio de Nativâ Saturni Facie*, in which he merely classified the various forms of Saturn as mono- and tri-spherical, spherico- and elliptico-ansated, and spherico-cuspidated, and concluded that the planet was ellipsoidal in shape with two appendages attached to the surface. He therefore failed to get to the root of the matter. As Proctor has pointed out (*Saturn*, p. 52), though some of the appearances were consistent with the assumption of attached handle-like structures, it was incredible that the motion of the planet should so exactly synchronise with that of the Earth in its orbit that the diameter through the ansae should always be at right angles to the line of sight from the Earth. Yet the ansae, when visible at all, seemed always to extend to the same distance from the limbs of the globe.

The French mathematician Gilles Personne de Roberval, an original member of the Académie Royale des Sciences, supposed Saturn to be surrounded by a torrid zone giving off vapours, transparent if in small quantity, reflecting sunlight at the edges if of medium density, producing an elongated elliptical aspect if very thick.

A Sicilian priest and mathematician, Giovanni Hodierna, wrote a small pamphlet presenting the naïve theory that Saturn was a spheroid with two dark patches on it.

During the period 1653–56 Sir Christopher Wren, a scientist and astronomer before becoming an architect, was making researches concerned, among other things, with the Saturn problem, assisted by Sir Paul Neile, a well-to-do amateur. He made models in wax, cardboard and copper, and by 1658 he and Neile are said to have been evolving a theory that the planet had an elliptical corona, meeting the globe in two places and rotating with Saturn once in each sidereal period, the axis of rotation coinciding with the plane of revolution. But Wren never published his theory because, before he was ready to do so, news came of Huygens's solution. According to J. Summerson (*Sir Christopher Wren*, pub. Collins, 1953) Wren showed the disinterested love of truth characteristic of him; he was so delighted with the neatness of Huygens's hypothesis that he said: 'I loved the invention beyond my own'.

Very different was the attitude of Honoré Fabri, S.J., a French professor of philosophy and mathematics, who, as will be seen, engineered an attack on Huygens's theory in order to publicise his own. Fabri thought he could explain all the phases of Saturn by means of the movements of two hypothetical large, dark, unreflecting satellites close to the planet, and two large bright ones somewhat farther away. He was prepared to throw in, for good measure, the real little satellite (Titan) discovered by Huygens, though it was of no importance to his theory.

Huygens discovers Titan (1655) and solves the Saturn problem

Christiaan Huygens who, at the age of 26, found the true solution to the appendage puzzle, was one of the most remarkable figures of the seventeenth century, for, like William Herschel a century later, he was a skilful observer, a brilliant theorist and a master craftsman: he was also a great inventor. He achieved fame in other sciences related to astronomy: realising the importance to the astronomer of instruments that would record time accurately, he made discoveries in the theory of the pendulum and invented the pendulum clock, and also, it is said, the spring watch. Anxious to improve the optical parts of telescopes, he studied optics and lens-making, finding a new method of giving lenses a more accurate curvature, and inventing the compound eyepiece named after him the Huyghenian. His correspondence and scientific papers, with translations and copious notes, fill twenty-three large volumes: *Œuvres Complètes de Christiaan Huygens*, published by the Société

Hollandaise des Sciences, with much information about other scientists of the period also. Volume XV (pub. 1925) contains his Saturn observations and drawings, and the full Latin text with French translation of his three pamphlets on Saturn: *De Saturni lunâ observatio nova* (1656), *Systema Saturnium* (1659), and *Brevis Assertio systematis Saturnii sui* (1660), on pp. 172–177, 209–353 and 439–467 respectively. It also contains (pp. 403–437) the text of the opposing pamphlet *Brevis Annotatio in systema Saturnium* (1660).

In the spring of 1655, having with his brother Constantyn made a telescope 12 feet long, of 10½ feet focal length, small aperture and magnification of 50, Christiaan Huygens discovered Saturn's largest and brightest satellite, now known as Titan, and within the next few months solved the Saturn problem. Though he was not ready to publish his full solution, he was persuaded by his friend Jean Chapelain and by Gassendi and others, to make known the satellite discovery, so in March 1656 he did so in a short pamphlet entitled *De Saturni lunâ observatio nova*.

This stated that on 25 March 1655 he saw near Saturn to the west a little star about 3 minutes of arc distant and in line with the ansae. Doubting whether it was a planet he noted its position and that of Saturn in relation to another star at a similar distance on the other side of Saturn but not in line. Next day he found the western star at about the same situation and distance relative to Saturn, but the other star nearly twice as far away. Hence he inferred that the latter was a fixed star, while the former was accompanying Saturn in its then retrograde motion and so must be a satellite. Observations on the following days removed all doubt: the satellite moved gradually round to the other side of the planet, completing its period of revolution on the sixteenth day; 'its greatest digression was seen to be a little less than three minutes' (actually the apparent mean distance at Saturn's mean opposition distance is 3′ 17″3, and Huygens himself three years later amended his determination of greatest apparent distance to 3′ 16″).

Huygens stated in this pamphlet that the discovery of a moon of Saturn had 'opened the way' to explaining the ansae and why they sometimes disappeared, and to the prediction of future changes. He forecast that towards the end of the following month (April 1656) the ansae would reappear extending in a straight line on either side of Saturn, and concluded by inviting anyone who thought he had solved the problem to send him his solution, and stating his own solution in the form of an anagram, so that it could not be said afterwards that either had copied from the other.

Hevel, Roberval and Hodierna responded to the appeal, sending particulars of their theories, but Saturn itself seemed to have let Huygens down, for no sign could he see of the ansae either in April or at any time until the planet disappeared 'into the Sun' in June 1656. This was in spite of the fact that he was now using a 23-foot telescope with a compound eyepiece giving a magnification of 100, but the effective aperture was only 2⅓ inches, too small to show the ansae when they formed a thin line. The failure of his prediction, the distraction of inventing the pendulum clock, and the full explanation of his theory proving a harder and lengthier task than he had expected, all combined to delay the completion of his treatise until 1659.

However, on 13 October 1656, Huygens found that the ansae had reappeared more or less in the shape he had predicted for the previous April and much like they had been in March 1655 (see Figures 4a, 4b), and by 1659 he had many more observations to support his thesis.

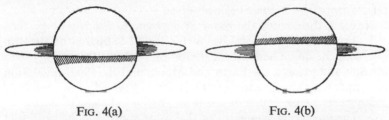

FIG. 4(a)   FIG. 4(b)

FIGURE 4. Huygens: Saturn (a) 25 March 1655; (b) 13 October 1656. Huygens used the same sketch inverted for the second date. (Redrawn after reproduction of his sketches in *Oeuvres Complètes*, vol. XV, from *Systema Saturnium*, and both inverted to show south at top).

Systema Saturnium, 1659 (Huygens)

Huygens began his celebrated treatise *Systema Saturnium*, published in July 1659, by saying that when the telescope, invented in the Netherlands, was first used by Galileo to survey the heavenly bodies, his most astonishing observations were those of Saturn. Neither Galileo nor any other astronomer since had been able to explain the causes of the planet's variable figure: various astonishing forms had been shown in drawings and many people considered them to be optical illusions. Huygens had therefore studied to improve lenses; he gave particulars of his telescopes and illustrations of their performance. He then gave a full account with drawings and diagrams of all his observations of Saturn and Titan, a table of Titan's motion and an explanation of how its place could be calculated.

Huygens then discussed drawings made by others during the

preceding forty years (see Plate I), stating that some were distortions through poor telescopes of forms shown in others, that no. IV suggested by Hevel was an impossible shape, and that no. X, by the excellent telescope-maker Divini, could be considered as showing the most rigorously correct form of Saturn except that the shading around the edges of globe and ansae seemed to be a personal addition. He expressed the opinions that Saturn, when without ansae, was round, not elliptical, and that the impression of some observers that there was a difference of size (and of distance from the globe) as between the two ansae was false.

Huygens then dealt in turn with the theories communicated to him by Hevel, Roberval and Hodierna, refuting them courteously but with most convincing arguments. He showed a touch of irony regarding Hodierna's theory, recommending him to make two dark marks on an egg and turn it slowly round; he would then see that it failed to show some of the observed forms of Saturn, and showed other forms that had not been observed.

Huygens then stated the bases of his hypothesis:

1. The discovery of Titan revolving round Saturn in an orbit in line with the ansae and with a period of 16 days. By analogy with the Sun and planets, the Earth and Moon, and the rapid revolution of Jupiter's moons, he inferred that Saturn, situated at the centre of Titan's orbit, must rotate on its axis in much less than 16 days.

2. The unvarying symmetry of the ansae and the globe suggested that Saturn must be surrounded by another symmetrical body, a ring around the middle, with plane perpendicular to the axis of rotation.

3. Observation showed him that the line of the ansae was inclined at an angle of more than 20° to the ecliptic: hence the same ring would appear sometimes as a wide ellipse, sometimes as a narrow one, sometimes even as a straight line.

Saturn's equator, he said, has a constant tilt with respect to the plane of its orbit, just as the Earth's equatorial plane has to the plane of the Earth's orbit.

4. The fact that the ring forms ansae showed that it is separated from Saturn on all sides by an equal interval.

He then gave the solution of his anagram as follows: *Annulo cingitur tenui, plano, nusquam cohaerente, ad eclipticam inclinato,* which may be translated: 'it is surrounded by a thin flat ring, nowhere touching, and inclined to the ecliptic.'

He estimated the outer diameter of the ring to be in the ratio of 9 : 4 to the globe's diameter (a good approximation), and the width of the ring from outer to inner edge to be about equal to the width of the space between it and the globe (too low an estimate of the width of the bright rings).

He then proposed to answer in advance two possible objections: (a) the attribution to a heavenly body of a form never found before and contrary to the accepted principle of the spherical shape; (b) the placing around Saturn of a solid and permanent ring, not attached to the globe in any way, but rotating and revolving rapidly with the planet while remaining at the same distance. He stated that his theory was not due to his imagination like the epicycles of the ancient astronomers which never appear in the sky, but was forced upon him by the evidence of his eyes; a ring-shaped body could rotate about its centre as easily as a spherical one; and, tending to approach the centre by equal force on all sides, such a body would remain in equilibrium at a distance from the centre everywhere the same (he did not mention centrifugal force).

He refuted the suggestion of Hevel that the line of the ansae was parallel to Saturn's orbit and inclined only 2½° to the ecliptic, and mentioned that an observation of Grimaldi in 1650, reported by Riccioli, showed that the ansae were then considerably inclined to the ecliptic.

Huygens then gave a very long and detailed explanation and proof, illustrated with diagrams, regarding the ring's inclination and the causes of its phases: two of the diagrams and some of the main points of the discussion will be found in Chapter 5 and Plate II (1). He predicted disappearances of the ring for July 1671, March 1685 and the end of 1700, and he had the pleasure of seeing the first two of these forecasts approximately realised.

The last part of the treatise was devoted to observations and calculations of the diameters of the planets in relation to that of the Sun. He made the ratio between the diameter of Saturn's globe and that of the Sun, assuming both at unit distance, to be 1 : 7·4. (The correct value is 1 : 11·5.)

Though all the rival theories would fit one or more of the observed forms of Saturn, no hypothesis save that of Huygens could account for all the phases and also explain the mechanism which produced them. He had made the greatest of all discoveries about Saturn and presented it with full observational evidence and the most convincing arguments and explanations. Why did his theory at first meet with such a poor reception? Probably because it was too bold and novel for many of his contemporaries.

His best friend, Chapelain, accepted the ring theory (as Wren

95

did) with enthusiasm, but his other great friend Ismael Boulliau (Bullialdus) had a serious objection: he could not see why, if there were a ring, it should ever vanish. (Boulliau's idea was largely right, but the disappearances of the ring were due to its thinness and the inadequacy of early telescopes, not to any unsoundness in the theory.)

Huygens was especially disappointed that Prince Leopoldo de' Medici, to whom his treatise was dedicated, failed to respond. Apparently Leopoldo had consulted Boulliau, and when he did reply after several months he said he was convinced by the arguments and observations of Huygens, but that he had hesitated at first because there did not seem to be anything like Saturn's ring in connection with any other heavenly body.

Hevel rightly insisted that Saturn was elliptical, whereas Huygens had drawn it round; Riccioli disliked the ring theory and thought of writing a pamphlet against it. Then Huygens was warned by letters that an attack on his theory was in preparation, ostensibly by Divini, but, as he was a poor latinist, really by Fabri, who seems to have written pamphlets on various subjects, sometimes under pseudonyms, sometimes in the names of other men.

None of the critics queried the rash assertion of Huygens in the dedication of the treatise that, with the discovery of Saturn's satellite, the perfect number of 12 planets and satellites (6 of each) was complete, and therefore no more would be found; this, apparently, was a common notion of the time. The wild surmise of the lensmaker, Friar Anton Schyrle (Rheita) that Saturn had six satellites and Jupiter nine, turned out eventually to be nearer the truth.

Ring theory attacked (Divini and Fabri)

In August 1660 the opposing pamphlet, *Brevis Annotatio in systema Saturnium* duly appeared, stating as its purposes:

1. To refute the accusation of fraud made against Divini. This referred to Huygens's criticism of the shading on his drawing, and the refutation was somewhat weakened by the admission that the shading had been added to give relief, i.e. to depict the planet as a sphere.

2. To show certain errors of Huygens in attributing to Saturn 'a garland formed of a resplendent ring' (an ironical reference to a sentence in Huygens's dedication).

3. To explain certain principles of Fabri's system, based on the hypothesis of a motionless Earth which orthodox astronomers 'have the duty of defending against heterodox Aristarchians'. (Aristarchus

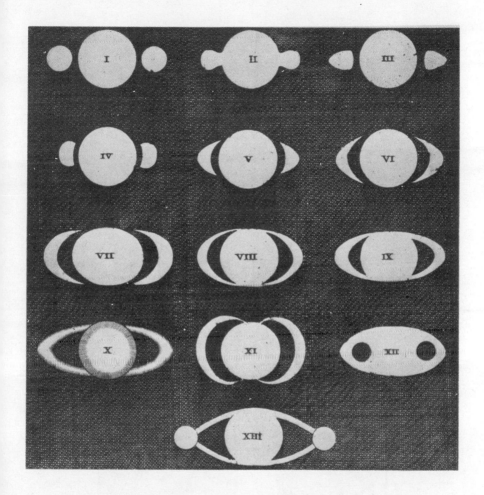

PLATE I.—Early drawings of Saturn (from *Systema Saturnium*, 1659)
I Galileo (1610); II Scheiner (1614); III Riccioli (1641 or 1643);
IV–VII Hevel (theoretical forms); VIII, IX Riccioli (1648–50);
X Divini (1646–48); XI Fontana (1636); XII Biancani (1616),
Gassendi (1638, –39); XIII Fontana and others at Rome (1644, –45).
Riccioli made a drawing in 1646 rather like XI but less distorted.

Plates I and II (Fig. 1) were photographed from the copy of
Systema Saturnium in the British Museum Library by courtesy of the
Director.

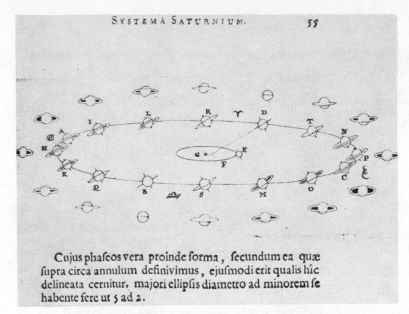

Cujus phaseos vera proinde forma, secundum ea quæ
supra circa annulum definivimus, ejusmodi erit qualis hîc
delineata cernitur, majori ellipsis diametro ad minorem se
habente fere ut 5 ad 2.

FIG. 1.—Huygens: Diagram of Ring Cycle (from *Systema Saturnium*).
For explanation see pp. 101–2 of text.

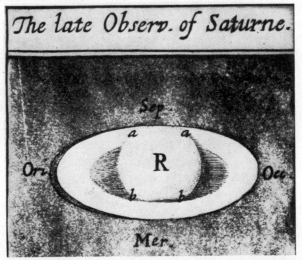

FIG. 2.—Hooke: Drawing of Saturn 29 June 1666 (from
Phil. Trans., vol. 1, plate). See text pp. 108–9. *bb* was later
thought to indicate crepe ring. Hooke's drawing shows
North at top, and should be inverted to show usual view
with inverting telescope for observers in Northern
Hemisphere.

PLATE II

anticipated Copernicus in suggesting a heliocentric system—see Chapter 2).

The pamphlet poured sarcasm on Huygens's telescopes, observations, and 'fantastic' theory; insisted that the ansae really appeared sometimes as spheres, and that the 'cavities of the ansae' were darker than the surrounding sky. Huygens's dusky band bisecting Saturn when without ansae was dismissed as 'a pure fiction': the band was shown by Huygens sometimes above, sometimes below the centre: if there were a ring Saturn could never be without ansae.

More reasonable criticisms were: that Huygens's theory was based on too few observations over too short a period during which Saturn had traversed only a fraction of its orbit (though of course he had in fact carefully studied the earlier observations of others); and that, though spots showed the Sun was rotating, Huygens had no proof that Saturn or any other planet rotated.

Agreement was expressed with Huygens on two points and his discovery of the satellite Titan was admitted. Then came the extraordinary statement that this alleged moon of Saturn was not really a moon because it did not circle round the planet; a similar remark was made about the four bright satellites of Jupiter. Fabri could not allow his large hypothetical satellites to cross in front of Saturn, since they would cause eclipses and other phenomena not observed: they therefore had to circle behind the planet, and he seems to have been afraid that Titan and Jupiter's moons might adversely affect his theory if it were conceded that they could cross the face of their respective planets.

Divini's pamphlet went on to say that Huygens departed from truth in claiming that the various phenomena of Saturn could only be explained by a ring—'other systems explain them easily and I like best that one worked out by Father Honoré Fabri'. A very detailed explanation of Fabri's theory with diagrams followed.

Prince Leopoldo, to whom this amazing pamphlet was also dedicated, complimented the alleged author, but said he would suspend judgment until the periods of the motions of the four hypothetical satellites had been found. 'That', as Huygens remarked, 'might take a long time.'

Counter-attack by Huygens

Amazed at the manner of the attack on his theory Huygens quickly produced a crushing but humorous reply. Important new evidence and his skill in turning his opponents' arguments against themselves made his own case even stronger. His *Brevis Assertio systematis Saturnii sui* (September 1660) began by remarking that two adversaries were ranged against him under one name; though Divini was named as author he had been aided by Father Fabri. Huygens continued: 'I had believed that there would be subtle objections unforeseen by me and drawn from profundities of astronomical science with that politeness and modesty befitting a man dedicated to liberal studies. But I was deceived—they attack my observations without solid arguments, accusing me openly of inventing them contrary to truth.'

By careful reasoning based on his observations and table of Titan's motion, Huygens showed that Divini's observations of the satellite were inaccurate, and that the suggestion that it did not circle round Saturn was nonsense. He also showed the absurdity of the assertion that the 'cavities' (spaces between ring and globe) were darker than the rest of the sky. He himself had seen the ansae looking like small globes, but only when he observed them with a small and inferior telescope.

Fabri's assertion that the dark band across the disk in the round phase could not be seen with the best telescopes merely showed that their telescopes must be inferior to those of Huygens. He then said: 'In case anyone should think that I have invented this phenomenon, it has also been seen in England.' Huygens had received a letter

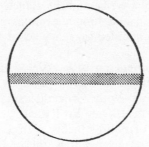

FIGURE 5. Huygens and William Ball: Saturn's 'round phase' 5 February to 2 July 1656. (Redrawn after reproduction of Huygens's sketch from *Brevis Assertio* in *Oeuvres Complètes*, vol. XV, inverted to show south at top).

drawing attention to this dark band as having been observed by William Ball* (of Mamhead, near Exeter) and asking Huygens whether he had seen it. Ball had noted the band diametrically across the globe February–July 1656, when Saturn was without ansae, and his drawing was just like Huygens's own (see Figure 5). Later, when the ansae returned and the band was very difficult to see, its

* Ball, Wren and Neile were all acquaintances of Samuel Pepys.

position was less correctly shown by Ball. In 1657 the English observers saw at the same period as Huygens the straight ansae (see Figure 4b); Ball's sketch differed from Huygens's only a little in the thickness of the arms. Ball's drawing of the wider open ring 1657–58 was in absolute agreement with his. Huygens pointed out that these English observations were independent testimony to the phases of Saturn, and as Ball had throughout been unaware of the ring hypothesis, he could not have been biased even unintentionally by it.

Fabri's theory seemed to Huygens so far removed from reason and reality as hardly to need refutation. That 'pretty fiction of four globules' reminded him of a conjuring trick with black and white balls alternately appearing and disappearing. Nevertheless he easily showed the flaws in the 'ridiculous system': to account for the various drawings the hypothetical satellites would have to vary in size; to fit Divini's drawing the dark ones would have to be smaller than the bright ones, contrary to the hypothesis, and in any case spherical bodies could not produce the appearance of elliptic arcs.

Divini and Fabri ought to be required to show the orbits and periods of their satellites, but Huygens thought no such period could exist, since Divini had claimed that Saturn had appeared to him as in drawing no. X (Plate I) for the space of three years.

With reference to his hypothesis and the Copernican system Huygens wrote: 'No one in my opinion could reasonably reproach me with having adapted my system to the Copernican . . . I could easily substitute for the Copernican system that of Tycho. For the phenomena in question it does not matter which of the two I use. All the same the truth of the matter cannot be explained otherwise than in following Copernicus; and furthermore our system of Saturn strongly corroborates his.'

Opposition to the ring theory collapses

In 1661 Divini and Fabri returned to the charge with another pamphlet *Pro sua Annotatione . . . adversus . . . Assertionem*, more moderate in tone but more fantastic in content, attempting to bolster up Fabri's theory by increasing the number of hypothetical satellites to six of different sizes, with parabolic instead of circular orbits. This amused Huygens and he soon dropped the idea of a further reply: he could well afford to wait. Hevel, perhaps the best observer[*] among the critics, was so greatly impressed by Huygens's *Brevis*

---

[*] Hevel, like Cassini, was an ardent observer who boasted that he never missed a clear night.

Assertio that he abandoned his own theory and accepted that of the ring. Then, early in 1665, Huygens was delighted to hear from two different sources that even Fabri, convinced by the appearance of Saturn which he saw in a good telescope of Guiseppe Campani, had at last been forced to retract. He did so in his *Dialogi physici* (see *Œuvres Complètes* vol. XV, p. 401 and footnote), nevertheless clutching at one doubtful straw, for Fabri seems to have supposed (incorrectly) that Huygens agreed with him about the ring-plane of Saturn being parallel to the plane of the Earth's equator. Hence, after referring to the ingenuity of that 'most distinguished and learned man' Christiaan Huygens in discovering the ring of Saturn, Fabri went on to draw the strange conclusion that the parallelism of the ring with the Earth's equator supported his own hypothesis of a motionless Earth and dealt a final blow to the Copernican system!

CHAPTER 5

HUYGENS'S RING PHASES AND RECENT EXAMPLES

Ring's inclination and cycle (Huygens)

In this chapter it is proposed to explain the working of the ring theory of Huygens. This can best be done by reproducing two of his excellent diagrams, mentioning a few only of the points made in his detailed discussion and proof, and then making use of modern examples with more accurate data than he was able to give.

One of the bases of his theory was the discovery by observation that the line of the ansae is inclined at a large angle to the ecliptic. His diagram (Figure 6) shows the successive positions *L*, *M*, *N*, *O* of Saturn observed by him on 9, 10, 11, 12 April 1655 in relation to a fixed star C, *RQ* being a line parallel to the ecliptic. It is at once evident that, while Saturn's apparent path *LO* is only slightly inclined to *RQ*, yet the line of the ansae, always parallel to *PQ*, makes a large angle of 20° or more with both orbits. Huygens considered the line of the ansae to be more nearly (though not exactly) parallel to the plane of the Earth's equator, and hence that Saturn is considerably tilted, as the Earth is, with respect to the ecliptic.

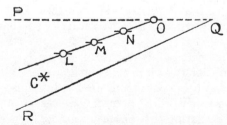

FIGURE 6. Huygens: diagram showing inclination of Saturn's ring to Saturn's apparent path and the ecliptic. (Redrawn after the reproduction in *Oeuvres Complètes*, vol. XV, of the diagram, based on observations 9–12 April 1655, in *Systema Saturnium*).

He used the diagram shown here as Plate II(1) for the detailed explanation of his theory. The orbits of the Earth and Saturn he assumed to be coplanar, *ABCD* representing Saturn's orbit, *FE* the Earth's, and G the Sun. The symbols by *DR*, *AH*, *BS* and *CP* show respectively the directions of the zodiacal constellations Aries, Cancer, Libra and Capricornus. A and C are the two positions of maximum opening of the ring as seen from Sun and Earth; B and D the two positions where the ring-plane passes through the Sun, the ring at such epochs being invisible for several months in the mid-seventeenth century telescopes. This 'round phase', as he called it, had been observed by Galileo in 1612, by Gassendi in 1642–43, and by Huygens and others in 1656. He said that I, H, O, P are positions of rather wide opening; L, K, M, N narrower; R, Q, S, T very narrow, suggesting the illusion of three spheres in small imperfect telescopes. The outer ring of sketches on the diagram shows how the planet actually appears to the observer at each phase.

Plate II(1) is a facsimile of part of a page in a copy of Huygens's book preserved in the British Museum. The Latin sentence under the diagram is part of the text referring to phases A and C (widest opening) and may be translated: 'Thus the true form of this phase, as we have already shown in regard to the ring, will be such as is depicted here (i.e. in the outer sketches), having the greatest diameter of the ellipse in the ratio of almost 5 : 2 compared with the least diameter'. (Actually the proportion at widest opening is more like 9 : 4—see Frontispiece—and the ring then covers most of one hemisphere including the polar limb.)

Huygens pointed out that the same phase would always be repeated when Saturn again reached the same place in its orbit; the planet's rotation on its axis, however rapid, would make no change in the situation (i.e. apparent inclination) of the ring with respect to the Earth. He stated that Saturn would be much more brilliant, even to the naked eye, when the ring was wide open than when it was invisible.

In the earlier part of his discussion of his hypothesis Huygens gave the impression (which probably misled Fabri) that he regarded the ring-plane as being parallel to the Earth's equator, and he said that to an observer it would naturally appear to be so. But he then qualified this by saying that another consideration showed that the parallelism was not exact. If the two planes had been parallel the ring ellipse would have been narrowest when Saturn was in either of the directions now known as ecliptic longitudes 180° and 360°, whereas he had found in fact that the narrowest phase occurred when Saturn was (in modern parlance) at ecliptic longitudes $170\frac{1}{2}°$ and $350\frac{1}{2}°$. He deduced an inclination between the two planes of just over 4°, though the true value in his time seems to have been at least twice as great.

He explained the persistent invisibility of the ring for several months by the fact that when Sun and Earth were on opposite sides of the ring-plane, observers would be looking at the unillumined face of the ring. He considered the question: 'why is it that when the ring-plane is directed towards us, the edge which is illumined by the Sun is invisible? Is this due to the ring's extreme thinness? Not at all . . . the very dark band . . . seen during the round phase . . . shows indubitably . . . that the ring . . . has a certain thickness, but that the edge reflects the light of the Sun either not at all or very feebly'. (Here Huygens thought of the true explanation—the ring's extreme thinness—but rejected it; he did not realise that the dark band was the projection on the globe of the unilluminated surface of the ring, not its edge). From the fact that the ring's ansae could not be seen under a low Sun, he inferred that the surface was flat, not mountainous like that of the Moon.

Modern particulars of ring cycle

The inclinations of the Earth and Saturn have been calculated by M. B. B. Heath from *N.A.* data as follows for 1960 January 1·5d. U.T.:

Inclination of

Earth's equator to ecliptic	$23°4445$
Earth's equator to Saturn's ring-plane	$6°6462$
Saturn's equator to Saturn's orbit-plane	$26°7343$
Saturn's equator to ecliptic	$28°0666$
Saturn's orbit-plane to ecliptic	$2°4899$

The ring-plane is, subject to slight oscillations, almost exactly in the plane of Saturn's equator, and remains in an almost fixed position with reference to the stars, but just as the direction of the Earth's axis of rotation has a slow precessional movement with a period of nearly 26,000 years, so Saturn's axis of rotation has an even slower precession, the period being upwards of 412,000 years (Proctor: *Saturn*, p. 89).

From the fixity of the ring-plane with reference to the stars it follows that from the viewpoints of Sun and Earth the tilt of Saturn's ring appears to be continually changing, as shown in Huygens's diagram. At alternate intervals of about 13 years 9 months and 15 years 9 months the ring-plane passes through the Sun, and within a few months before and after each such occasion the ring-plane will pass either once or three times through the Earth. A full explanation of this phenomenon with diagrams and modern examples will be found in Chapter 27.

During the interval of 13¾ years between one edgewise position of the ring and the next, the south pole of Saturn is tilted towards Sun and Earth and the southern face of the ring is exposed to view, the ring gradually covering and then uncovering the northern hemisphere of Saturn. During this interval Saturn passes through perihelion, and so is traversing the smaller part of its orbit at a greater mean velocity. On the other hand, in the period when the north pole of Saturn is tilted towards Sun and Earth and the northern face of the ring is exposed to view, covering and then uncovering the southern hemisphere, Saturn has to traverse the larger part of its orbit and at lower mean speed, passing through aphelion. This explains why the two intervals between successive edgewise positions are so unequal. The third column of the table of oppositions in Chapter 1 gives an example of the general change in ring aspect during an entire ring cycle. The fine series of Lowell Observatory photographs 1909–21 in Plate XI covers most of a period when the southern hemisphere and the southern face of the ring were presented to the Earth.

It may not be realised that from the viewpoint of the Sun there is much similarity between the cycles of the Earth and Saturn. Both have the equatorial plane considerably tilted with respect to the plane of the orbit; the main differences are that Saturn's revolution period and cycle are $29\frac{1}{2}$ times as long as the Earth's and that the Earth has no ring. The solstices of the Earth in a way correspond to the widest open phases of Saturn's ring, the equinoxes to the edgewise ring positions of Saturn.

In theory, and according to earlier writers such as Webb, there are three possible circumstances in which the ring can disappear: (a) when the Earth is in its plane, since the ring is too thin to be seen when presented edgewise; (b) when the Sun is in the ring-plane, so that neither surface is directly illuminated; (c) when the Earth and Sun are at small elevations to the ring-plane but on opposite sides of it, so that the dark side of the ring is being viewed by observers. It has recently been pointed out (B.A.A.J., vol. 70, p. 52) that condition (a) is the only one that ever causes complete disappearance in large modern instruments. Even in the largest telescopes the bright illuminated edge is too thin ever to be seen; the thin line of light that has been observed just before or after the time of the Earth's passage through the ring-plane is not really the edge but an extremely foreshortened view of the surface. Examples of the experiences of observers illustrating these points will be found in later chapters, e.g. Chapters 14 and 26. Owing to the inclination of $2\frac{1}{2}°$ between Saturn's orbit-plane and the ecliptic, the Sun and Earth are usually at somewhat different elevations on the same or on opposite sides of the ring-plane, sometimes one being at a greater elevation, sometimes the other.

Short-term variations in opening and tilt

From the point of view of an imaginary observer situated at the centre of the Sun there would appear to be a steady opening out of Saturn's ring from the edgewise to a widest open position, then a steady closing up until the next edgewise phase was reached, and so on. Though this is also true in general for observers on the Earth, in detail the changes are much more complicated owing to the Earth's rapid motion in an orbit slightly inclined to Saturn's. Proctor (in his chapter 4) gave an example of this for the synodic period from Saturn's conjunction of 14 October 1864 to that of 26 October 1865, showing that the ring not only appeared to open, then close, then open again during parts of the period but also that the speed of apparent opening and closing varied. The only dates when the extent of tilt was similar for Earth and Sun were at opposition and conjunction. The general result was that, although the ring had appeared to be closing during five months of the period, it had by the end opened out by the same amount to an observer on the Earth as to an observer supposed to be placed at the Sun's centre.

A more recent example (for 1947 and 1948) is shown by the following figures (to two places of decimals) selected from the *N.A.* of those years:

Date	Tilt of Ring (Earth)		Tilt of Ring (Sun)
1947 Jan. 1	−18°63		−19°63
Jan. 26	−19°38	opening	−19°33
(Opposition)			
Apr. 3	−20°69		−18°52
Aug. 5	−16°90	closing	−16°95
(Conjunction)			
Nov. 29	−12°88		−15°40
Dec. 31	−13°24	opening	−14°96
1948 Feb. 9	−14°52		−14°41
(Opposition)			
Apr. 17	−15°99		−13°45
Aug. 19	−11°54	closing	−11°65
(Conjunction)			
Dec. 13	− 7°27	opening	− 9°93
Dec. 31	− 7°41		− 9°66

The minus sign shows that the south pole of Saturn and the southern face of the ring were on view. The steady closing for the Sun is in marked contrast with the erratic alternation as presented to the Earth. This difference is also illustrated by Hepburn's diagram, Figure 20 in Chapter 32. The slight difference in tilt for Sun and Earth at opposition and at conjunction is due to the small inclination of the plane of the Earth's orbit to that of Saturn. For the same reason observers on the Earth are occasionally able to see the ring slightly more open than it would ever appear from the Sun's centre. The maximum opening of the ring for an observer on the Earth can slightly exceed 27° but can never be as much as 28°.

Tilt at oppositions 1943–73 (Heath)

To supplement the list of opposition places 1943–73 given in Chapter 1, Mr. M. B. B. Heath has also kindly supplied the following list of ring tilts at the same oppositions, calculated to the nearest half degree. The error, in most cases of much less than half a degree, makes no appreciable difference to Saturn's appearance except when the ring is almost edgewise. When the tilt is positive the planet's north pole and the northern face of the ring are visible. These figures show the broad progress of the cycle as observed from the Earth and ignore the detailed fluctuations between each opposition and the next.

Year	Oppn. Ring Tilt	Year	Oppn. Ring Tilt	Year	Oppn. Ring Tilt
1943	−26½°	1953	+13°	1963	+14½°
1944	−25½°	1954	+17½°	1964	+10°
1945	No opposition	1955	+21½°	1965	+ 4½°
1946	−23°	1956	+24°	1966	− 1°
1947	−19½°	1957	+26°	1967	− 6½°
1948	−14½°	1958	+26½°	1968	−12°
1949	− 9°	1959	+26°	1969	−17°
1950	− 3°	1960	+25°	1970	−21½°
1951	+ 2½°	1961	+22°	1971	−24½°
1952	+ 8°	1962	+18½°	1972	−26°
				1973	−26½°

The list shows that the change in tilt is very slight from one opposition to the next when the ring is wide open, and most marked when it is closing to or opening from an edgewise position.

Stellar magnitude and ring tilt

The *N.A.* values of Saturn's stellar magnitude at the oppositions from 1943 to 1955 (from *B.A.A. Hdbks.*) are given below; those for oppositions from 1956 onwards are shown in Appendix I.

Year	Oppn. Magnitude	Year	Oppn. Magnitude	Year	Oppn. Magnitude
1943	−0·3	1948	+0·1	1952	+0·6
1944	−0·3	1949	+0·4	1953	+0·5
1945	No opposition	1950	+0·7	1954	+0·4
1946	−0·2	1951	+0·8	1955	+0·3
1947	+0·0				

Stellar magnitude at opposition agrees well on the whole with ring tilt, because the planet is very much brighter with the ring wide open than partly open, and faintest with the ring edgewise, being then usually +0·8. But Heath's paper in Appendix I points out that the distances of Saturn from Sun and Earth also affect the apparent magnitude at opposition. Hence when the ring is wide open on the northern face, as in 1958 and 1988, the opposition magnitude is only +0·2, because the planet, near aphelion, is very far away. On the other hand, in 1943 and 1973, when the southern face of the ring is fully exposed to view and Saturn is near perihelion, the stellar magnitude is −0·3. The apparent brightness can, however, vary considerably within a year, e.g. for 1948 the *N.A.* gave the value at opposition (February 9) as +0·1 but showed it declining to +0·9 in the following autumn after conjunction owing to the greatly increased distance between Saturn and the Earth.

CHAPTER 6

SATURN 1660–1725: CASSINI'S DISCOVERIES OF THE RING DIVISION AND FOUR SATELLITES

The study of Saturn through the telescope had been founded by Galileo in the early seventeenth century and enriched by the genius of Huygens in the middle years. During the later seventeenth century the dominant personality in the observation of the planet was, without any question, Jean Dominique Cassini. To continue the history of the observations from the point where it was left at the end of Chapter 4, however, it is desirable to mention several observations of the decade preceding Cassini's great discoveries.

Observations of the 1660s (Campani, Hooke)

According to Antoniadi (*L'Astronomie*, vol. 44, pp. 158–60) the first to notice, shortly after 1660, that the ring is less bright in its exterior part (now known as ring A) than in the interior part (now ring B) were the telescope-makers Divini and Campani. He also stated that about 1662 the shadow of the globe on the ring was discovered by Campani and about the same time by the French mathematician Adrien Auzout. The shadow of ring on globe had already been detected by Huygens (see Figures 4a, b).

On 29 June 1666 Robert Hooke* reported (*Phil. Trans.*, vol. 1, p. 246) a rather interesting observation accompanied by a sketch (see Plate II, Fig. 2), which showed the ring rather wide open with a narrow black line (bb) where it crossed the globe, and a narrow black line (a) by each limb of the globe where it crossed the far part of the ring.

As to the black lines Hooke remarked: 'whether shadows or not, I dispute not.' Probably the observation was made near opposition, when sometimes the optical illusion of a shadow on the ring is seen by one limb of the globe when the true shadow is by the other limb. His dark line at the edge of the ring across the globe was considered two centuries later to have been an early glimpse of part of the crepe ring. Hooke himself remarked that the ring seemed somewhat

---

* Hooke was a very versatile experimental physicist and inventor, known to Pepys; curator of experiments, Roy. Soc., 1662.

brighter than the globe, and that he could plainly see that 'the southern part of the ring was on this side of the body and the northern part behind'. Huygens's ring hypothesis was still recent enough for Hooke to feel satisfaction that he could see that the ring was truly a ring; he was of course observing its northern face, so what he meant was that the ring crossed in front of the globe in the southern hemisphere and behind its north polar region. Hooke was using a 60-foot long telescope. (Saturn looked very similar in 1960 with 3-in. aperture).

'Round phase', 1671 (Hevel, Huygens, Flamsteed)

The latter part of a letter of Hevel from Danzig (*Phil. Trans.*, vol. 5, p. 2087) gave particulars of Saturn's position as observed on 26 August 1670 with a 50-foot long telescope, and stated that the ring was narrower and more oblique than it had been in 1666 and 1668.

Huygens had predicted that the ring would vanish and the 'round phase' of Saturn would be seen for a year from July or August 1671. Part of another letter in Latin from Hevel (*Phil. Trans.*, vol. 6, p. 3032) gave Saturn's position on 11 and 12 September 1671 as Pisces 17°,* and said that as bad air had prevented effective observation from 1 June to 11 September it was scarcely possible to tell whether the 'round phase' could be seen during that summer. He added: 'The arms appear extremely narrow even in 60- or 70-foot tubes but I think not altogether vanished'.

Hevel's misgivings about Saturn's failure to fulfil the forecast of Huygens were experienced even more acutely by Huygens himself, who wrote (in French) the following interesting account of what took place (*Œuvres Complètes*, vol. XV, p. 501):

'When I invented the hypothesis of Saturn and found out that there was a flat circle surrounding its globe, I had only observed it for a year, and the observations which I found made before by others with less good glasses hampered me more by their falseness than helped me in this research. This is why I do not think that people have found strange during the past year the prediction I had made 13 years before concerning the return of the "round" shape of Saturn if they found . . . the prediction a little different from observations.

'I had said that Saturn would be seen round, without ansae, in 1671, and it was so in fact 2 months earlier than I had expected, viz. the end of May.

* Equivalent ecliptic longitude, 1950, about 351°, near 11 Psc and Pisces–Aquarius border.

'Then a little interruption of the round shape happened which I had not foreseen, but . . . as soon as I had news of the return of the arms on 14 August I predicted that they would soon be lost again— which was also found to be true for from 4 November I observed Saturn with arms so nearly effaced that I was doubtful whether they were still visible, although M. Cassini assures me that he could still see them on 13 November.

'In this loss of arms what was remarkable was that they remained always broad enough to be seen, but that they lost little by little their distinctness which shows that it was the obliquity of the rays of the Sun on the surface of the ring which marred the view this time instead of the former round appearance from the end of May to 14 August. It was the obliquity of the rays of our view on the same surface of the ring which prevented us from seeing it.

'After this time the round shape lasted till Saturn was hidden near the Sun, which having left now and having returned visible, its arms appear again, conforming also to my prediction . . . and M. Cassini saw them for the first time 5 June 1672 but they were so clear that doubtless they had come back some time before.'

Editorial notes in vol. XV of *Œuvres Complètes* explain that when the ring-plane began to cross the Earth's orbit in 1671, the Earth was ahead of it and owing to greater orbital speed drew away from the plane for a time, so that the ring appeared to grow larger but dimmer. Cassini was able to see the ansae until 13 December 1671; the ring-plane passed through the Sun on 5 January 1672, and finally the Earth met and passed through the plane on 23 February 1672. The passages of Earth and Sun therefore both occurred near Saturn's conjunction, so that the circumstances were rather similar to those of 1950 (see Chapter 27 and Figure 17b).

The other forecasts of Huygens were somewhat late in fulfilment: in 1685 the 'round phase' came in July–November, not in March; the one he predicted for the end of 1700 came in June–August 1701.

The Rev. John Flamsteed, who five years later was appointed the first Astronomer Royal, was also observing Saturn towards the end of 1671 and gave the following particulars in two letters from Derby (*Phil. Trans.*, vol. 6, p. 3034): 'Octob. 12 last past, at my first viewing Saturn with my less Tube, I thought I saw something on each side of him, amidst the Colours of my Glass and the spurious rays of his body. Directing my longer Tube (of 14 feet) to him, I could see his Anses somewhat more distinctly, but very slender, and to one that thought not of them, scarce discernable.' In the second letter he wrote: 'Last Thursday night, Novemb. 30. (the only clear night we have had of late) I observed Saturn with my 14 foot Telescope, the aperture being 1½ inch, and its Eyeglass drawing 2 inches. He appeared perfectly round, free from rays and colours, and no Ansae to be seen. My worthy friend Mr. Townly in his last to me, bearing the date Novemb. 20. 1671, desireth me to continue the observations of Saturn, telling me, that he looked at him one night, and could hardly distinguish his Line of the Ansulae, but plainly saw a dark line through him near his upper part. A week or two after . . . Octob. 12 . . . I had frequently the same appearance, though in a wider aperture than I use at present. . . .'

Cassini discovers Iapetus, 1671

Jean Dominique Cassini in his earlier career had been known as Gian Domenico Cassini, for, like Galileo, he was an Italian, and for nineteen years, from the age of twenty-five, professor of astronomy at Bologna, where he distinguished himself greatly both as an observer and as a computer. He discovered, for example, the rotation of Mars and of Jupiter and some of their surface features, and computed the motions of Jupiter's bright satellites. In consequence of his renown he was invited to Paris in 1669 by King Louis XIV to supervise the new observatory, becoming its first director and a naturalised French citizen. Cassini, unlike Huygens, did not make his own object glasses but used and tested lenses obtained from the best instrument-makers of the period, such as Campani and Divini. In 1671, using one of Campani's telescopes, Cassini made his first discovery of a satellite of Saturn.

The following is part of a translation made at the time ('English 't out of French') of what would appear to be Cassini's own report of this memorable discovery (*Phil. Trans.*, vol. 8, p. 5178):

'About the end of October 1671 Saturn pass'd close by Four small Fix't Stars, visible only by a Telescope, within the sinus of the water of Aquarius, which Rheita once took for new Satellits of Jupiter . . . but which Hevelius . . . shew'd to be some of the Common Fix't Stars. . . .'

(Old drawings of the constellation figure show the water poured from the pot held by Aquarius making its first considerable curve near the stars ψ^1, ψ^2 Aquarii, some three degrees south of the ecliptic, so the four telescopic stars are probably near these.)

'This passage of Saturn gave us occasion to discover in the same place, within the space of 10 minuts, by a Telescope of 17 ft., made by Campani, Eleven other smaller Stars, one of which, by its particular motion, shew'd it self to be a true Planet: which we found by comparing it not only to Saturn, his Ordinary Satellit, discovered 1655 by Mr. Hugens, but also to other Fix't Stars. . . .

'These Observations shew a motion of this New Planet that is very manifest in respect of the Fix't Stars but less sensible in respect of Saturn. Yet it appears, that from Octob. 25 unto Novemb. 1, his distance from Saturn increased Westward, and from that time unto Novemb. 6 it diminished; so that his greatest digression from Saturn hapned in the beginning of Novemb. and was found to be of 8 minuts or of 10½ diameters of Saturn's Ring, whence it was consequent, that, if this Planet were a Satellit of Saturn . . . his revolution about Saturn was of a long duration, since for 12 days together he not only remain'd on the same Occidental side of Saturn, but there was also little change of apparent distance between him and Saturn. The greatest digression of this Planet was treble that of the ordinary Satellit, and this enabled us to judge the Time of his revolution to be quintuple, applying to the Satellits that proportion, which Kepler hath noted in the Principal Planets, between the periodical Times and their Distances.'

(These estimations of Cassini were quite good, because the average elongation of the orbit of Iapetus from Saturn is of the order of 500 seconds of arc, and the synodic revolution periods of Iapetus and Titan are respectively a little less than 80 days and slightly under 16 days. The remaining part of Cassini's report which follows shows that he also noted the pronounced inclination of the plane of Iapetus's orbit to that of Saturn's equator.)

'But there was one circumstance, which made us doubt, whether it were a Satellit or a Principal Planet, which was; That in the last observations we took notice, that he had a little Southern latitude in respect to the Line of the wings of Saturn, which we had not observed in the first, when he was nearest to Saturn; which happens not to the other Satellit, which hath always the more latitude, the nearer he is to Saturn; yet it might well be, that the Circle of this Planet might have some declination from the Circle of the other Satellit, as it comes to pass in the Principal Planets, the Circles of which are inclined to one another. However this difficulty made us suspend our Judgment until we could make such a number of observations as might suffice for a more precise determination.'

Cassini discovers Rhea, 1672

Towards the end of 1672, a year after the finding of Iapetus, Cassini discovered a third satellite, now known as Rhea, much nearer to Saturn than Titan and Iapetus are. A report which he made

(*Phil. Trans.*, vol. 12, p. 831) gave the following particulars: 'greatest digression' from the centre of Saturn $1\frac{2}{3}$ times the diameter of the ring; period of revolution 4^d, 12^h, 27^m (a very accurate determination). He also stated that when the ring was wide open the satellite did not touch the ring or Saturn at conjunction, but (presumably owing to the glare of the planet) Cassini was unable to detect Rhea at conjunctions, and so he said it was ordinarily seen only every third day and rarely for two days together. He added the remark: 'The apparent magnitude of these Planets is so little, that posterity will have cause to wonder, that their discovery was begun by a Glass of 17 foot.'

Periodical vanishing of Iapetus

In the same report Cassini drew attention to a remarkable and puzzling peculiarity concerning the satellite now known as Iapetus: 'One of these two Planets, which is distant from the centre of Saturn 10 diameters and a half of his Ring, maketh his revolution about Saturn in 80 days. He was discover'd at the Parisian Observatory, A. 1671 about the end of Oct: and in the beginning of Nov: in his greatest Occidental digression, and after many cloudy days he ceased to appear, for a reason which was then unknown, but hath been discover'd since. For, after that many revolutions of this small Planet had been observ'd, he was found to have a period of apparent Augmentation and Diminution, by which period he becomes visible in his greatest Occidental digression, and invisible in his greatest Oriental digression. . . .'

Cassini then argued that this could not be due to varying distance from the Earth and Sun, and consequent varying exposure to the Sun's light, because, as he pointed out, in one revolution the distance of Iapetus from the Sun does not vary a hundredth part, and in fact the greatest diminution of the light of the satellite occurred when it was approaching the Sun and Earth. He continued:

'But it seems, that one part of his surface is not so capable of reflecting to us the light of the Sun which maketh it visible, as the other part is. . . .'

Cassini compared the different parts of Iapetus in reflecting quality to the land and sea areas of the Earth, and suggested that opposite hemispheres of Iapetus would be seen in turn from the Earth if the satellite revolved with the same hemisphere always facing Saturn, as the Moon does with respect to the Earth. He stated that after its first discovery in 1671 the satellite could not be found until the middle of December 1672, and then disappeared till the beginning of

February 1673, when it had been observed for thirteen successive days, so that its motion had been determined. Since then it had always been seen in the western part of its orbit and at conjunctions with Saturn, but 'he could never be seen in his Oriental digressions, where he remains invisible in every revolution of 80 days for a whole month together. . . .'

He added that the satellite begins to appear two or three days before inferior conjunction and to disappear two or three days after superior conjunction, and concluded with the comment: '. . . And sometimes after he hath begun to disappear in a Telescope of 32 foot, he hath been sought for with a Telescope of 45 foot, but in vain.'

Cassini held to his excellent explanation until 1705, but in September 1705 Iapetus began to be visible also in the eastern half of the orbit, so in 1707 Cassini dropped his explanation as being made too hastily (*Mémoires de l'Académie des Sciences*, 1705, p. 121; 1707, p. 96): a century later, however, William Herschel considered that Cassini's explanation was the only possible one.

Cassini's division, 1675

Cassini found the first breach in the supposedly solid, rigid and opaque ring of Saturn by discovering in 1675 that the breadth of the ring was divided into two parts by a dark line, now known as 'Cassini's Division'. In most textbooks he has been quoted as expressing the incorrect opinion that the two parts of the divided ring are of equal width, but although this may have been his first impression there is reason to suppose that he soon amended it. Herschel stated (*Sci. Papers*, vol. 1, p. 431) that Cassini saw the ring divided into 'two equal parts' according to Lalande, but 'two almost equal parts' according to Laplace. Cassini's own letter of August 1676 (*Phil. Trans.*, vol. 11, p. 689) seems to be quite non-committal on this point: '. . . we have discerned on the globe of Saturn a dusky zone (*zona subobscura*), a little farther south than the centre, similar to the zones of Jupiter.'

(This seems to be the earliest recorded observation of the south equatorial belt; the letter continued:)

'Also the breadth of the ring was divided into two parts (*dividebatur bifarium*) by a dark line, apparently elliptical but in reality circular, as if into two concentric rings, the inner of which was brighter than the outer one. This aspect I saw immediately after Saturn's emersion from the Sun's rays and through the whole year till immersion. . . .' He added that he had used a 35-foot long telescope, and afterwards a smaller one of 20-foot length.

Cassini's sketch of Saturn in 1676 (reproduced as Figure 7) which is said to be the earliest sketch showing the division and the south equatorial belt ever to be published, certainly suggests that by 1676 at any rate he considered the inner part of the ring (now called ring B) to be wider than the outer (now ring A). In the following years he is said to have detected other belts also, noticing that they sometimes seemed inclined to the plane of the rings; he rightly regarded them as atmospheric, not surface, features.

FIGURE 7. J. D. Cassini: sketch of Saturn 1676, showing Cassini's division in the ring; said to be the first published drawing of the division. (From the reproduction in Hutchinson's *Splendour of the Heavens*, 1923, p. 363).

About eighty years ago there was considerable discussion as to whether Cassini's discovery of the dark division had been anticipated ten years earlier by William Ball of Mamhead, and some of the nineteenth-century writers, such as Webb and Proctor, even went so far as to claim the priority for Ball and to refer to the gap in the ring system as 'Ball's division'. This idea was, however, completely wrong, as was shown by W. T. Lynn (in the *Observatory*, October 1882) and by Professor J. C. Adams (in *M.N.*, vol. 43, p. 92, January 1883). Adams made it clear that, while Ball's observations of Saturn were much esteemed by Huygens (as mentioned in Chapter 4), there was no real evidence that Ball, or Huygens either for that matter, had noticed any indication of a division in the ring: apparently the idea had originated, not from any claim by Ball himself, but from a later misunderstanding of a comment made by an anonymous correspondent on one of Ball's observations. The comment was based, not on any marking on the ring in Ball's drawing, but on the depressed shape given to the outer edge of the ring at the minor axis. It is therefore clear beyond doubt that the sole credit for the discovery of the great division in the ring system belongs to Cassini.

Cassini also discovers Dione and Tethys, 1684

In March 1684, using unwieldy aerial telescopes with object glasses of 100 and 136 feet focal length made by Campani, Cassini found two more satellites, both fainter and nearer to the planet than Rhea is. They are nearer to Saturn than the Moon is to the Earth and have revolution periods of less than three days; they are now known as Dione and Tethys. According to Proctor, Cassini found that the orbits of all five satellites discovered up to that time are in accord with Kepler's laws, that those of the inner four are in planes very nearly coinciding with the plane of the rings, and that that of Iapetus is in a plane inclined at about 15° to the ring-plane. Further details of all the satellites and their orbits will be found in later chapters, especially Chapter 23.

Just as Galileo had named the four bright moons of Jupiter 'Medicean' satellites in honour of his patron, the Grand Duke of Tuscany, so Cassini wished to call the four Saturnian satellites he had found after Louis XIV, but their present names, derived from classical mythology, were suggested by Sir John Herschel early in the nineteenth century.

Occultations and Conjunctions (Hevel, Boulliau, Flamsteed, Cassini)

In a letter of Hevel (*Phil. Trans.*, vol. 6, p. 3027) the observation of an occultation of Saturn by the Moon on 1 June 1671 was reported and he added that he had seen Saturn occulted by the Moon only twice before in 41 years, viz: on 29 June 1630 in the Danish Sound near Hveen, and on 3 August 1661 at Danzig. It seems quite a testimony to Hevel's diligence as an observer that he should have managed to witness this phenomenon as many as three times. Boulliau (*Phil. Trans.*, vol. 12, p. 969) observed an occultation of Saturn by the Moon in 1677, giving details of the position and discussing discrepancies found in the tables. In 1692 Cassini is said to have observed the unusual phenomenon of the occultation of a star by Titan.

Flamsteed and Hevel (*Phil. Trans.*, vol. 13, pp. 244, 325) recorded observations of conjunctions of Saturn and Jupiter in October 1682 and February and May 1683. Flamsteed's report gave many details of the positions and distance measurements.

According to Proctor, about 1707 or 1708, Dr. Samuel Clarke and the Rev. William Whiston, using a telescope of 17 feet focal length, are said to have observed a star in one of the dark spaces between rings and globe, a rare observation, of which unfortunately there seems to be no reliable first-hand evidence.

Huygens continued to observe Saturn (though intermittently) until 1693, using larger and larger tubeless aerial telescopes made by his brother Constantyn, the longest being of 210 feet focal length with an object glass of $8\frac{1}{2}$ inches diameter, but he made no further discoveries. He observed Titan many times, and also (after Cassini had discovered them) Rhea and Iapetus, but he seems never to have been able to detect Dione and Tethys, the fainter satellites found by Cassini.

Titan's period (Halley, Huygens)

'That Ingenious Astronomer' Edmond Halley (eventually to become second Astronomer Royal) corrected Titan's period to $15^d 22^h 41^m 6^s$ (*Phil. Trans.*, vol. 13, p. 82, 1683), and Huygens's ultimate assessment of the sidereal period was $15^d 22^h 41^m 11^s$. Both are close to H. Struve's value (1907) of $15^d 22^h 41^m 23 \cdot 165^s$, the best modern approximation.

Early traces of the crepe ring (Picard, Hadley)

In the mid-nineteenth century when the faint inner ring (ring C) had been discovered, a search was made of early drawings and records to find evidence of its visibility in former times. As a result it appeared that indications of the faint ring across the globe, but not in the ansae, had been detected by Campani in 1664, Hooke in 1666, Picard in 1673 and Hadley in 1720. None of these had guessed what it was and had evidently assumed it to be a belt. J. R. Hind, (in *M.N.*, vol. 15, p. 32) quoted the following remark of Picard under date 15 June 1673: 'Saturne étoit sorti des raïons du soleil: il y avoit deux barres noires qui marquoient les deux bords intérieurs et extérieurs de l'anneau.'

Presumably in this case the dark line at the outer edge of ring A was the shadow of the ring across the globe, and that at the inner edge of ring B was the projection of the crepe ring across the globe. Jean Picard, who also made measures of the diameter of Saturn's ring, was an assistant to Cassini at Paris Observatory and introduced improvements in the instruments and methods of observing.

Investigations made in 1851 by Otto Struve into early evidences of the existence of ring C (*M.N.*, vol. 13, p. 22) showed that former observers had often referred to a dark 'equatorial belt', which crossed the globe adjoining the inner edge of ring B. In Struve's opinion the observations of Hadley in the 1720s seemed to establish beyond doubt the identity of that belt with the obscure ring, the full discovery of which did not occur till 1850. John Hadley, best known as the maker of the first really satisfactory reflecting telescope, gave an interesting account of his observations (*Phil. Trans.*, vol. 32, p. 385). He stated that in the spring of 1720 when Saturn was about 15 days past opposition, he saw the planet's shadow on the ring and plainly saw the ring 'distinguished into two parts by a dark line concentric to the circumference': the outer part of the ring seemed the narrower, and the dark line which separates them 'stronger next the body and fainter towards the upper edge of the ring'. He also discerned two belts on the globe: one of these crossed Saturn close to the ring's inner edge and 'seemed like the shade of the ring on Saturn, but when I considered this with respect to the situation of the Sun, I found that belt could not arise from such a cause.'

Continuing his report in August 1723, Hadley stated that the dark line on the ring parallel to the circumference (evidently Cassini's division) was chiefly visible in the ansae but he had several times been able to trace it 'very near if not quite round': especially in May 1722 he could discern it outside the north limb of Saturn 'in that part of the ring that appeared beyond the globe of the planet'. The globe, at least towards the limb, seemed to him to reflect less light than the inner part of the ring, and he said that he sometimes distinguished it from the ring by the difference of colour. The dusky line which in 1720 he had seen by the inner edge of the ring across the disk, 'continues close to the same, though the breadth of the ellipse has much increased since then'.

Hadley's comparison of the observations of 1720 and 1723 therefore showed that the dusky line adjoining the ring had not kept a fixed position on the globe but had followed the changing position of the ring.

Vanishing ansae, 1714 (Maraldi)

Meanwhile in October 1714 when the Earth was nearly in the ring-plane, a few days before the ring's disappearance, Jacques Maraldi, nephew of J. D. Cassini and assistant at the Observatory, noticed that the narrowing of the ansae seemed unequal in extent. At his first observation the eastern ansa seemed the larger, but two nights later it had disappeared, while the western one was visible, though only as a faint line of light. Maraldi deduced that the ring is not uniform in thickness, and that it revolves round Saturn. This second inference is very important as being the first evidence of the rotation of the ring system. Strangely enough Maraldi based this correct conclusion on the assumptions that the ring system is not only solid, but also a rigid solid, assumptions both of which were subsequently proved to be wrong.

Solidity of the rings questioned (the Cassinis, Wright)

Two years before Maraldi's observation the great Cassini had died, being succeeded in the supervision of Paris Observatory but not in the official title of director by his son Jacques, who in 1717 published a table of the distances, mean motions and orbits of Saturn's satellites, which he had carefully investigated.

It would seem that neither Cassini shared Maraldi's view of the solidity of the rings: Antoniadi (*L'Astronomie*, vol. 44, p. 54) quoted the following sentences to show that J. D. Cassini had recognised the true nature of the ring system: 'Cet anneau pourroit être formé comme d'un essaim (swarm) de petits Satellites . . . l'apparence de l'anneau est causée par un amas de très petits Satellites de différents mouvements, qu'on ne voit point séparément' (*Mém. de l'Acad. Roy. des Sciences*, 1705, p. 18). Antoniadi considered that the latter sentence proves also that Cassini visualised the ring satellites nearest to Saturn moving with a greater angular velocity than those further away. Curiously enough, it seems that Jacques Cassini did not mention his father's opinion when, in 1715, he announced: 'L'anneau de Saturne est formé d'une infinité de petites Planètes fort près l'une de l'autre' (*Mém. de l'Acad. Roy. des Sciences*, 1715, p. 48).

This revolutionary theory of the rings consisting of swarms of small particles was also suggested independently in 1750 by Thomas Wright of Durham; at that period it was purely speculative, being unsupported by mathematical proof, and received scant attention until proved correct in the mid-nineteenth century (see Chapter 13).

Ring measures (Bradley)

Among many planetary observations the Rev. James Bradley made a few of Saturn, the most important being a series of micrometer measures April to June 1719, and the only known record of the ring's reappearance in 1730. His measures closest to modern values were on 7 May 1719: ring's diameter, outer $43''7$, inner $29''2$; Saturn's diameter $19''0$, but when he had taken means corrected for mean distance from the Sun the final results were too small: ring's outer diameter $41''25$, Saturn's $17''75$. He also made Titan's greatest elongation $2' 56''75$ and Iapetus's $8' 34''$.

On 25 August 1730 he noted: 'I looked at Saturn through my reflector, and found that his ring was become visible again, it having disappeared when I last saw him, which was about a month ago. It now appears like a line, but somewhat brighter and broader towards the ends than near the body.' He saw satellites 1 and 2 (Tethys and Dione) nearly in the line of the ansae to the west and 'the Hugenian satellite' (Titan) to the east also in line. (Bradley's *Miscellaneous Works*, 1832, pp. 340–64).

CHAPTER 7

LATER EIGHTEENTH CENTURY: HYPOTHESES OF LAPLACE, AND OBSERVATIONS OTHER THAN HERSCHEL'S

The 'great inequality' (Laplace)

The celebrated French mathematician, Pierre Simon de Laplace, during the later eighteenth century investigated mathematically the whole mechanism of the solar system on the basis of Newton's law of gravitation, solving a number of problems of the planetary movements. Astronomers had been puzzled by an abnormal variation found in the orbital velocities of Jupiter and Saturn which in the course of 2,000 years would produce a difference of 3° 49' in the celestial longitude of Jupiter and 9° 15' in that of Saturn, amounts respectively equivalent to nearly eight and over eighteen times the Moon's apparent diameter. In 1773 Johann Lambert pointed out that the motion of Saturn from being retarded had become accelerated. In 1784 Laplace, after a careful investigation, explained the 'great inequality' as being due to the fact that five of Jupiter's revolutions round the Sun take 59·3 years, nearly equivalent to the 58·9 years required for two revolutions of Saturn; consequently there is a great reciprocal perturbation between the two planets. Jupiter, moving faster in an interior orbit, tends to pull Saturn slightly inwards towards the Sun and, though sometimes retarding Saturn's motion, tends on balance to accelerate it. Saturn has the opposite, though smaller, effects on Jupiter's motion.

The period of the 'great inequality' originally given by Laplace as $929\frac{1}{2}$ years is somewhat too long. The figure depends on the values adopted for the mean annual motions of Saturn (S) and Jupiter (J). Good modern values are: $S = 43996''204$ (per annum)

$$J = 109256''640 \text{ (per annum)}$$
$$\text{Whence } 5S - 2J = 1467''740$$

To complete one revolution at $1467''740$ per annum would take 883 years, but Dr. J. G. Porter, who derived this figure, recommends 890 years, as given by W. M. Smart's *Celestial Mechanics*, since the actual value is not a constant.

Laplace's investigation cleared up one of the factors that had for centuries helped to produce discordances between tables and observations, and gave a further verification of the law of gravitation.

Nebular hypothesis (Kant and Laplace)

A nebular theory of the origin of the solar system had been suggested by Immanuel Kant in his work on the Nature and Theory of the Universe (1755). Laplace's similar nebular hypothesis was first put forward by him in a note in his *Exposition du système du monde* (1796) and afterwards fully elaborated. He had noted that the Sun and planets are relatively close together compared with the much greater distances of the stars, and that there are numerous regularities in the solar system which are not necessary consequences of the law of gravitation, facts which seemed to indicate very strongly a common origin for the system. The regularities he noted are as follows: the seven planets and fourteen satellites known at the time all revolve round the Sun in the same direction and they and the Sun all rotate in the same direction; the revolutions all occur in planes with only small inclinations to each other, and the eccentricities of the orbits are all small. The chance of such an arrangement being fortuitous is so extremely slight that even the subsequent discovery of members of the system that do not conform to these regularities does not vitiate the argument, and in Laplace's time the evidence for common origin was overwhelming. His theory was that the Sun and planets had been formed out of a primeval nebulous mass which through cooling, concentration under gravity, and increasingly rapid rotation broke up into successive nebulous rings, which in turn breaking up would reunite to form planets, those nearer the central mass having progressively shorter periods of revolution. The same process of rotation, contraction and ring formation would take place on a smaller scale in the planets, resulting in some cases in the formation of satellites. In the case of Saturn this process had not been completely carried out: hence the continued existence of Saturn's rings.

Of course objections were subsequently found to Laplace's nebular hypothesis, the chief ones being that such rings would condense into many bodies rather than single planets, and that it failed to explain the fact that ninety-eight per cent of the angular momentum of the system is possessed by the outer planets, less than two per cent is attributable to the Sun's rotation, and a very minute proportion to the revolution of the planets (Mercury, Venus, Earth and Mars) which are nearer to the Sun than Jupiter is. Because of these objections the nebular hypothesis was out of favour in the early twentieth century, and cosmogonists proposed quite different theories of the origin of the solar system. In recent years, however, evidence that stars are being formed in gaseous nebulae, such as the Great Nebula of Orion, and other reasons have led to a revival of the nebular hypothesis in a modified form (see Chapter 40).

Stability of Saturn's rings (Laplace)

In 1785 a memoir by Laplace, *Théorie des attractions des sphéroids et de la figure des planètes*, investigated the problem of the stability of Saturn's rings, and might well have shattered the eighteenth century belief in the solidity of the ring system. His first conclusion was that such a ring must be in rotation round the planet; otherwise it must collapse and fall on the planet. Secondly, the ring or rings must be very narrow; otherwise they would be disrupted by the strains to which they are subjected. Even assuming that Cassini's division separates the ring into two (an assumption by no means generally accepted at that time), these two rings would be much too broad for the requirements of Laplace's theory. He therefore considered that the two rings must each consist of many narrower rings, separated by gaps of a similar nature to Cassini's division but much narrower and therefore beyond the power of then existing telescopes to detect. He went on to suggest that rotating rings, even as narrow as he had envisaged, would be broken up unless they were eccentrically situated with respect to the planet and also weighted unequally in different parts of their circumference.

The arguments which led to these conclusions were summarised by Proctor as follows. The chief forces operating on every part of the rings would be the huge attraction of Saturn's mass, tending to drag the rings towards the centre of gravity of the system, and the attractive force of all the other parts of the rings, tending to weaken them by dragging the inner parts outward and the outer parts inward: nothing but rotation of the rings about the globe could prevent them from crumbling and falling on to the planet's equator. He then considered at what rate the rings should revolve and decided that, to remove all strain, each particle of the rings must revolve at the rate appropriate to a satellite at the distance from Saturn of that particle, or each ring should revolve at the rate appropriate to a satellite at the mean distance of the particles of that ring from the centre of the planet. But even then the outer parts of the ring would have too great a velocity and would tend to seek a larger orbit, while the inner parts would have too small a velocity and tend to seek a smaller orbit, unless prevented by the ring's cohesion. The cohesion of a flat ring of the dimensions of Saturn's ring system would be quite insufficient to resist these tendencies, but the strains would be much less in a narrow ring and its cohesion would be sufficient to resist them. Hence Laplace deduced that the ring system must be divided into several concentric rings, and he calculated the rate of revolution appropriate to each part of such a system. He next proved that a perfectly uniform solid ring could only go on rotating round a perfectly uniform planet if subject to no disturbing influences. If once disturbed, even slightly, it could never recover its equilibrium. But the rings are subject to the attractions of the revolving satellites, the other planets, the Sun, and even the stars. Hence the rings, since they continue to exist, cannot, if solid, be uniform. Laplace considered, but did not give such a thorough mathematical investigation to, the case of a solid, non-uniform ring. He argued that the destruction of such rings could only be avoided by the removal of the centre of gravity to a sufficient distance from the centre of figure. This could not be accomplished merely by eccentricity of position and by the breadth of each ring differing in different parts of its circumference; there must also be variations in the density and thickness of each ring or irregularities on its surface. The disturbing influences would tend to draw the rings out of the plane of Saturn's equator, but he considered that the attraction of the planet's equatorial bulge would be sufficient to overcome these influences, and to force all the rings to move in a single plane which would very nearly coincide with the plane of the planet's equator.

Laplace's theory, based on mathematical and physical considerations, postulates a ring system which seems much too artificial to have a real and continuing existence, and one cannot help being surprised that it should have held the field for half a century. From such a theory to the total abandonment of the idea of solid rings seems but a short step to take. William Herschel was aware of the theory and was no doubt perturbed by it, for he was, to start with, a firm believer in a single, solid ring, and was not even prepared to admit that Cassini's division was a true gap through the rings until he had been able to test this properly by observation. He was even more sceptical of the existence of other smaller divisions, although he himself had glimpsed one such marking on the inner ring four times in 1780. Very probably Laplace's theory gave Herschel added encouragement to investigate thoroughly by observation the structure of the ring system, to ascertain whether the rings rotate about Saturn, and if possible to measure their speed of rotation. His progress with these investigations will be related in the next chapter.

Belts, positions, elements (Messier, Bradley, Hornsby, Lalande)

Before discussing Herschel's comprehensive investigations of Saturn it seems desirable to see whether any observations were made between Bradley's of 1730 and Herschel's earliest view of the planet in 1774, and also what observing of Saturn was done by Herschel's contemporaries. It seems that until the mid-eighteenth century neglect of the planet was almost complete. It is true that Gottfried Heinsius, a German professor of astronomy at St. Petersburg, made a prediction* of the disappearance of the ring for 1743–44 (*Phil. Trans.*, vol. 42, p. 602), but a record of the observation of that disappearance has yet to be found. The earliest disk observation after 1750 seems to be that made by Charles Messier on 28 March 1766 (*Phil. Trans.*, vol. 59, p. 459) with an achromatic reflector of 10 foot 7 inches focal length. Messier, best known as a comet seeker and for his catalogue of nebulae, wrote: 'I perceived on Saturn's globe two darkish belts; they were indeed extremely faint, and difficult to be discerned, directed, however, in a right line parallel to the longest diameter of Saturn's ring.' As in 1766 the ring must have been rather wide open covering much of the northern hemisphere, the belts he saw must have been the south equatorial belt and perhaps the south temperate one.

A further observation by Messier in 1776 (*Phil. Trans.*, vol. 66, p. 543) stated: 'I have observed since the 14th of May, a belt of a fainter light on the body of Saturn, opposite to the part of the ring behind the planet. It is pretty broad, and almost as distinct as those of Jupiter.' He made the observation with a 3½-foot focus achromatic by Dollond, and claimed no originality for the discovery, saying that the Cassinis had seen it at the end of the previous century. As the ring would (1776) have been narrowly open on the northern face, the belt referred to was probably the north equatorial one. Although Messier has been credited with the observation of a spot on Saturn in 1776 he made no mention of that.

When Astronomer Royal, Dr. Bradley with three successive assistants determined Saturn's place on 146 dates, 1751–62, some in each year, thus tracing its path from Ophiuchus to Pisces. About 24 of the observations were Bradley's own, and 39 of 1757–60 were made by Charles Mason (A. Auwers: *Bradley'schen Beobachtungen*,

* Kindly verified by Dr. A. Armitage as only a forecast.

124

vol. 1, p. 605). Mason with another English astronomer, Jeremiah Dixon, surveyed 1763–67 the southern boundary of Pennsylvania, thenceforth known as the Mason–Dixon Line.

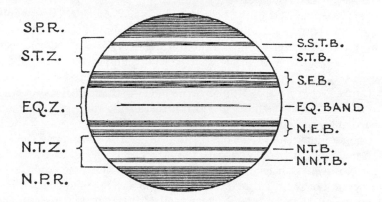

FIGURE 8. Diagram showing nomenclature of Saturn's belts and zones. The parts of the temperate zones adjoining the north and south equatorial belts are sometimes called (as for Jupiter) the north and south tropical zones respectively; throughout this diagram T. means temperate; P.R. means polar region. The positions of the belts are somewhat variable, and usually few are visible; also part of one or other hemisphere is usually covered by the rings, which have been omitted from this diagram.

The Rev. Thomas Hornsby, first Radcliffe observer at Oxford, used the transit instrument there to make two series of very accurate determinations of Saturn's position, recording the R.A. and Decl. on 80 dates 1774–79 and on 121 dates 1790–97 (*M.N.*, vol. 96, p. 148). A comparison based on calculations from Newcomb's tables showed maximum differences of 8″6 in Decl. and 0·36 sec. in R.A., the differences in most cases being much less.

Among the few contemporaries of Herschel to observe Saturn was Joseph de Lalande, professor of astronomy at the Collège de France and later in charge of the Paris Observatory. He corresponded regularly with and visited Herschel, who referred to him as 'one of our most eminent modern astronomers'. Lalande published a memoir on the elements of Saturn's orbit, and Herschel used in his calculations Lalande's estimate of 3 minutes of arc for Titan's mean distance from the centre of Saturn (modern value 3′ 17″3), and his estimate of the breadth of the ring as nearly one-third of that of the globe.

'Round phase', 1773–74 (Varelaz)

In the autumn of 1773, the year before Herschel began observing Saturn, Joseph Varelaz at Cadiz recorded his impressions of the planet just before and after the ring's disappearance. The translation of his letter of 12 October 1773 (*Phil. Trans.*, vol. 64, p. 112), probably the sole record of that disappearance, seems to have been overlooked by previous writers about Saturn. Varelaz reported: 'It has been my luck to observe the celebrated phaenomenon of the round phase of Saturn . . . recommended to astronomers in the Gazette of France of July 23. From the 24th of September to the 4th of October, I saw, clearly and distinctly the two ansae of the ring; but with this particular circumstance, that the occidental ansa appeared more strongly illuminated than the oriental. The atmosphere was thicker on the 5th, and I could only see the occidental. The 6th, I thought I could discern some faint remains of the ring; but that might be a deception of my sight, because the atmosphere remained very thick, and the planet could not be seen well terminated. On the 7th, the atmosphere being more transparent, and the heavens clearer than I have ever seen them, I observed the total disparition of the ring; and, having repeated the same observation the following day, I was convinced that this famous phaenomenon took place the 6th day of the month. . . . The most striking circumstances of this phaenomenon were the following: 1. The occidental ansa constantly appeared more bright than the oriental. 2. On the disc of the planet one could clearly distinguish the line of the shadow projected from the thickness of the ring. 3. On the extremities of this, some luminous points were perceived, which reflected the light more strongly than the others. 4. I did not observe a sensible variation in the apparent diameter of the ring. . . .' He concluded by stating that these observations had been made with a 5-foot long reflector of Short, a 4-foot one of Naime, and 'an excellent achromatic telescope' of Dollond, 'to which I applied one of the strongest and brightest magnifiers'.

CHAPTER 8

WILLIAM HERSCHEL'S DISCOVERY OF TWO SATELLITES AND INVESTIGATIONS OF THE SATELLITES AND RING (1774–94)

Herschel's Papers on Saturn

William Herschel, the outstanding astronomer of the eighteenth century and one of the greatest of all time, was a Hanoverian who settled in England as a young man, earning his living as an organist, music teacher and orchestra conductor. Taking up astronomy in 1773, at first as a spare-time hobby, he soon decided to make his own telescopes, choosing the reflecting type invented by Gregory and Newton in the seventeenth century and later improved by Hadley and others. Herschel became very expert at casting, grinding and polishing speculum metal mirrors, and was so industrious that by 1795 he had made no fewer than 430, many of which he used in telescopes which he made and sold. All this activity left little time for observing, and his planetary observations of the 1770s, though carefully recorded, were few compared with his prodigious output of later years. But he had already started to carry out a grandiose scheme of surveying all stars down to certain magnitudes to find out more about star distribution and the shape of the universe. While engaged on this work in 1781 he discovered, almost by accident, the major planet Uranus. This led King George III to award him an annual salary and other generous grants, enabling him not only to devote thenceforth his whole time to astronomy, but also to carry out his plan of building a giant telescope, 40 feet long with a mirror 48 inches in diameter. Unfortunately the seeing conditions were rarely suitable for the great telescope, it was extremely cumbersome, and condensation of moisture injured the speculum: though he used it occasionally, nearly all his observing continued to be done with his 20-foot reflector of 18·7 inches aperture and other smaller telescopes.

His life and remarkable contributions to many branches of astronomy have been well recounted in the late J. B. Sidgwick's *William Herschel* (Faber and Faber, 1953) and in Dr. J. L. E.

Dreyer's introduction to *The Scientific Papers of Sir William Herschel* (pub. by Royal Soc. and Royal Astronomical Soc., 1912). This latter work, in two volumes, contains 71 of his papers and other data; his Saturn observations, contained in seven of the papers and part of an eighth, are the source of almost the whole of this chapter and the next. The following list gives volume and page references to the *Scientific Papers* and the *Philosophical Transactions*:

	S.P.	P.T.
1. Discovery of Sixth and Seventh Satellites and Construction of the Ring (1789)	1, p. 370	80, p. 1
2. The Satellites and Rotation of Ring (1790)	1, p. 382	80, p. 427
3. The Ring, and Rotation of Fifth Satellite (1791)	1, p. 426	82, p. 1
4. Quintuple Belt (1793)	1, p. 452	84, p. 28
5. Rotation of Saturn (1794)	1, p. 458	84, p. 48
6. Singular Figure of Saturn (1805)	2, p. 332	95, p. 272
7. Figure, Climate and Atmosphere of Saturn and its Ring (1806)	2, p. 360	96, p. 455
(8.) ... New irregularity in Saturn's Figure (1808)	2, p. 403	98, p. 145

Herschel's first paper gave a very impressive survey of Saturn and its problems, including: discovery and preliminary data of two additional satellites, shadow transit of Titan; the solidity, brightness, thinness and flatness of the ring, the dark line on it, its appearance when nearly edgewise, and the probable shape of the edge; also (to be related in the next chapter) the belts, probability of an atmosphere, situation of axis of rotation, and polar flattening.

Herschel discovers Enceladus and Mimas, 1789

On 19 August 1787 Herschel saw with his 20-foot telescope an object which he suspected to be a sixth and previously undiscovered satellite, but owing to his discovery of two satellites of Uranus and other preoccupations was prevented from following up the suspicion until 28 August 1789, when he was able to use his newly completed 40-foot reflector with a power of 189. His original record of this observation (*Sci. Papers*, vol. I, p. lii) ran as follows: 'Saturn with 5 stars in a line, very beautiful. The nearest of these five is probably a satellite, which has hitherto escaped observation. It is less bright than the others. What makes me take it immediately for a satellite is its exactly ranging with the other four and with the ring. The ring is very bright but extremely slender . . . The new satellite is nearer to the end of the ring than the length of the projecting part of the ring is beyond the body of Saturn. . . .' (see Figure 9a). In his list of observations he stated that the new satellite and the second (Dione) were on the preceding side, the others on the following one.

In his first paper he wrote: . . . 'the very first moment I saw the

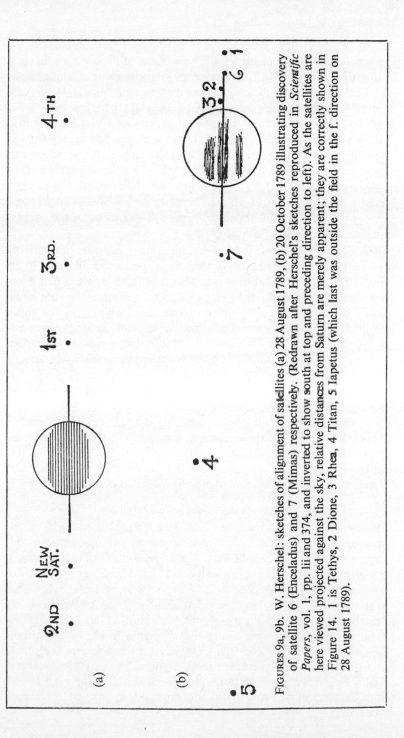

FIGURES 9a, 9b. W. Herschel: sketches of alignment of satellites (a) 28 August 1789, (b) 20 October 1789 illustrating discovery of satellite 6 (Enceladus) and 7 (Mimas) respectively. (Redrawn after Herschel's sketches reproduced in *Scientific Papers*, vol. 1, pp. lii and 374, and inverted to show south at top and preceding direction to left). As the satellites are here viewed projected against the sky, relative distances from Saturn are merely apparent; they are correctly shown in Figure 14. 1 is Tethys, 2 Dione, 3 Rhea, 4 Titan, 5 Iapetus (which last was outside the field in the f. direction on 28 August 1789).

planet. . . . I was presented with a view of six of its satellites, in such a situation, and so bright, as rendered it impossible to mistake them. . . . The retrograde motion of Saturn amounted to nearly $4\frac{1}{2}$ minutes per day, which made it very easy to ascertain whether the stars I took to be satellites really were so; and, in about two hours and a half, I had the pleasure of finding, that the planet had visibly carried them all away from their places.' This following up for $2\frac{1}{2}$ hours was done with the 20-foot telescope, with which on September 8 and 14 1789, he detected a still fainter object which he suspected to be a seventh satellite. He was able to confirm its discovery on 17 September with the 40-foot reflector. Figure 9b seems to be the earliest satellite diagram with Mimas (7).

His discovery of these very faint innermost satellites was mainly due to his skill as an observer and the superior quality and light-grasp of his telescopes, but it was also facilitated by their being less swamped than usual by the glare of Saturn, owing to the ring being nearly edgewise, and this circumstance also resulted in the inner satellites, whose orbits are almost in the plane of the ring, appearing in alignment with the ring on either side of the planet, instead of scattered all round Saturn as they would appear when the ring is seen tilted.

William Herschel always referred to the satellites by numbers: those previously known being numbered by him 1 to 5 in order of mean distance from the planet, and his own discoveries 6 and 7, though 7 (Mimas) has the innermost orbit, and that of 6 (Enceladus) is between those of Mimas and 1 (Tethys). It was to avoid this confusion that early in the following century his son, Sir John Herschel, gave them all the names by which they are now known.

Observations of the seven satellites

Between July and December 1789 Herschel made some hundreds of observations (listed in his second paper) of the positions and appearance of the seven satellites, and derived tables of their motions. He worked out the revolution periods of Mimas and Enceladus respectively as 22^h 37^m $22 \cdot 9^s$ and 1^d 8^h 53^m $8 \cdot 9^s$, which are very close to the modern values. He concluded that the orbits of these two are exactly, or very nearly, in the plane of the ring, as he repeatedly saw them apparently running along the very thin ansae of the almost edgewise ring. He noticed that Tethys, even when nearer the planet than Enceladus, and therefore more dimmed by the glare, 'was still visibly brighter than the latter'. He found Mimas to be even fainter than Enceladus; he wrote: 'It is incomparably smaller . . . and, even in my forty-feet reflector, appears no bigger than a very small lucid point. I see it, however, also very well in the twenty-feet reflector; to which the exquisite figure of the speculum not a little contributes. . . .' He found particular difficulty in determining the period of Mimas, because its orbit is so small that it was hard to obtain enough observations: often it was in front of or behind the planet or the ring, and at other times so near them that very good seeing conditions were needed to fix its position accurately. Another thing that bothered him in these satellite observations was the appearance of 'lucid spots' on the almost edgewise ring: for a time he even suspected that there might be an eighth moon even nearer to Saturn than Mimas, but he eventually concluded against this, and that those bright spots unaccounted for by satellites were some phenomenon of the ring itself.

Shadow transit of Titan

While engaged on these observations he was able to witness a phenomenon less common on Saturn than on Jupiter because it can only be observed during the four years or so (in every period of about fifteen years) when the ring system appears nearly edgewise or at a small tilt to the Earth: the transit of a satellite's shadow across Saturn. As Titan emerged from transit across the planet at the south preceding limb, he saw the round black shadow of Titan on the following limb, and watched it crossing slightly north of Saturn's equator until, almost 2^h 9^m later, the shadow reached the centre of the disk.

Light-change of Iapetus

In Chapter 6 it was related how Cassini had found a very notable variation in the brightness of Iapetus, and had suggested that the light-change might be due to the satellite's rotating on its axis in the same period that it takes to revolve round Saturn. When Herschel saw that Iapetus always showed the same brightness in the same part of its orbit, and that the change in brightness was regular and periodical, he thought that the cause must be a rotation on its axis and decided to find out a way of determining its rotation period. He therefore (as stated in his third paper) made a long series of observations, noting all changes of apparent brightness. He found that in about one-third of its orbit around western elongation Iapetus maintained a brightness not more than one magnitude short of that of Titan and outshone all the other satellites of Saturn. On the other hand, throughout the eastern half of its orbit, Iapetus was fainter than Rhea and hardly, if at all, brighter than Dione or Tethys. He estimated the drop in brightness at about three magnitudes; nowadays it is considered to be about two magnitudes, which means that Iapetus is over six times as bright round western elongation as it appears in the other half of its orbit. Having seen the satellite lose and regain its light regularly during many revolutions round the planet, Herschel felt sure that its rotation period could not differ much from the period of its revolution round Saturn, and was greatly strengthened in this opinion by the fact that Cassini had said that 'it disappears regularly for about one-half of its revolution, when it is to the east of Saturn'. Herschel had traced the regular change through more than ten revolutions (a period of over two years); Cassini had seen the same change in 1705 in the same part of the orbit some 397 revolutions (about 86 years) earlier. To overcome the difficulty of the lack of observations between Cassini's and his own, Herschel argued that if Iapetus had made a single rotation on its axis more or less than the number of its revolutions round Saturn since Cassini's time, the difference would amount to nearly one degree per revolution, or ten degrees during the period of his own observations, a quantity he would probably have noticed. Moreover, he was able to fill a little of the gap by taking account of the observations of a French astronomer, Bernard, in 1787, which agreed with his and Cassini's and extended Herschel's period of observation from ten to twenty revolutions. He therefore considered that he was justified in extending the period back to all the 397 revolutions since Cassini's time, and in concluding that the rotation period of Iapetus was equal to its revolution period of 79 days 7 hours 47 minutes.

He made three further deductions from the appearance of Iapetus: that by far the largest part of its surface reflects much less light than the rest; that neither the darkest nor the brightest side of the satellite faces the planet, but part of each and probably rather less of the bright side; and that the great regularity in the change of brightness suggested that Iapetus resembled the Moon in possessing little or no atmosphere, as well as in keeping the same face always turned towards its primary.

'Round phase', 1773–74

There must have been three passages of the Earth through the ring-plane in 1773–74, since Varelaz had recorded a disappearance of the ring in October 1773, and Herschel made the following statement about the commencement of his Saturn observations early in 1774: 'The planet Saturn is, perhaps, one of the most engaging objects that astronomy offers to our view. As such it drew my attention so early as the year 1774; when, on the 17th of March, with a 5½-feet reflector, I saw its ring reduced to a very minute line. . . . On the 3ᵈ of April, in the same year, I found the planet as it were stripped of its noble ornament, and dressed in the plain simplicity of Mars. . . .'

Black 'zone' on ring

By 1778 the ring was rather wide open and the dark marking now known as Cassini's division was prominent. Herschel commented: '. . . the black disk, or belt, upon the ring of Saturn is not in the middle of its breadth; nor is the ring subdivided by many such lines, as has been represented in divers treatises of astronomy; but . . . there is one single, dark, considerably broad line, belt, or zone, upon the ring, which I have always permanently found in the place where my figure represents it. I give this, however, only as a view of the northern plane of the ring, as the situation of the planet has hitherto not afforded me any other. . . .' He stated that he hoped shortly to examine the markings (if any) on the southern face which, in 1789, was coming into view. Meanwhile, he noted that, unlike the belts of Saturn and Jupiter, the black marking was not variable in colour and shape, and so was probably due to 'some permanent construction of the surface of the ring'. But it could not be the shadow of a mountain chain because it was visible all round the ring; such a shadow should be absent at the ends of the ansae when Saturn was in opposition; for similar reasons it could not be a cavity or groove in the surface. In 1780 when the ring was wide open he could see it all round the ring, but gradually decreasing in breadth towards the middle: hence its boundaries must be two concentric circles.

Evidently referring to Laplace's theory, he expressed the opinion that a division even into two rings was too artificial for acceptance unless the phenomenon absolutely demanded it, while subdivision into 'narrow slips of rings' would deprive them of the only dimension that could keep them from falling on to the planet. He admitted, however, that if they were rotating about Saturn at different velocities, that rotation might keep them up. He added that even if the southern face should be found to have a dark zone in the same situation and of equal breadth to the one he had seen on the northern, he would still feel that judgment should be suspended until a star, occulted by Saturn, could be seen between the ring openings as well as between the ring and Saturn. (This final proof was not obtained until long after Herschel's time.)

Ring's solidity and brightness

He considered the ring to be as solid and substantial as the planet, among the reasons being the shadow it cast on Saturn, and the irregularities in the movements of the satellites which he attributed to the ring's attraction, though he admitted that part of these irregularities must be due to the pull of the planet's equatorial bulge.

He next drew attention to the brightness of the ring, and gave instances of observations in 1777, 1780 and 1781, at each of which the ring was brighter than the globe; in the observation of March 11, 1780, he had made the comparison using successively higher magnifying powers with the same result: the light of the ring was more intense than that of the globe, and with the high powers the ring continued to look white, while the planet seemed to assume a yellowish colour.

Ring's thinness

He then emphasised one of the most remarkable features of the ring structure: 'its extreme thinness', having regard of course to its enormous extent and surface area. As during the late summer and autumn of 1789 the Earth had been nearly in the ring's plane, the conditions had been favourable for investigating this peculiarity, and he felt that his observations had been so complete that there could be no doubt of the ring's thinness. He had repeatedly watched Rhea, Dione, Tethys and Enceladus passing in front or behind along the line of the almost edgewise ring, so that they could be used as micrometers to estimate its thickness. He gave in his paper a number of examples of such observations, of which the most striking were his estimate on July 23 of the line of the ring appearing to be less than half the thickness of Dione, and that of August 29 when the thickness of the line of the ring seemed to be scarcely a quarter, and certainly less than a third, of the diameter of Rhea. He pointed out that such estimates might be rendered unreliable if there were a rarified atmosphere on the two faces of the ring which, by refraction, might cause a satellite whose diameter was really equal to the thickness of the ring to appear to project above and below the ring-plane. He had no means of making an accurate estimate of the diameters of the satellites, and thought that the smallest of them might have a diameter of a thousand miles. But modern estimates put the diameter of Dione at only about 600 miles and that of Rhea at about 850 miles. Consequently Herschel's two estimates mentioned above had (without his being aware of the figures they produced) set an upper limit to the thickness of the ring system of less than 300 miles and possibly little more than 200 miles. In later years, as will be seen, the figure was still more drastically cut down by other astronomers, but Herschel has the distinction of being the first to attempt to estimate the thickness of Saturn's ring, and to draw attention to its limitation.

It may seem curious that he did not subscribe to the argument that the ring system must be extraordinarily thin to become for a short time invisible when the Earth passes through its plane. His objection to this argument was that astronomers who used it assumed and drew a square edge to the ring, whereas he considered the shape of the edge must be spherical or spheroidal. He then made this curious statement reminiscent of the opinion of Huygens: 'Nay, I may venture to say that the ring cannot possibly disappear on account of its thinness; since, either from the edge or the sides, even if it were square on the corners, it must always expose to our sight some part which is illuminated by the rays of the sun: and that this is plainly the case, we may conclude from its being visible in my telescopes during the time when others of less light had lost it, and when evidently we were turned towards the unenlightened side, so that we must either see the rounding part of the enlightened edge, or else the reflection of the light of Saturn upon the side of the darkened ring, as we see the reflected light of the earth on the dark part of the new moon. I will, however, not decide which of the two may be the case; especially as there are other very strong reasons to induce us to think, that the edge of the ring is of such a nature as not to reflect much light.'

It seems that in this matter Herschel was somewhat led astray, chiefly by his assumption that the ring system was solid, and partly by the marked superiority of his own telescopes over those of his contemporaries and predecessors.

As will be seen (in Chapter 28) Dr. Louis Bell in 1919 expressed the opinion, now universally accepted, that it is the extreme thinness of the ring system which causes it, when fairly edge-on to the Earth, to disappear in all but the very largest telescopes and to be of doubtful visibility in those.

Luminous points on ring

Luminous points on the almost edgewise ring, seen by Herschel in 1789, and called 'condensations' or 'knots' by nineteenth- and twentieth-century astronomers, gave him much trouble and made him change his opinion more than once as to their nature. He said that, like several other astronomers, he had formerly supposed that

135

there was a roughness or irregularity in the surfaces of the ring; this idea arose: '. . . from seeing luminous parts on its extent, which were supposed to be projecting points, like the moon's mountains; or from seeing one arm brighter or longer than another; or even from seeing one arm when the other was invisible'.

He evidently had in mind such observations as those of Maraldi in 1714 and Varelaz in 1773. But Herschel during 1789 changed his opinion when '. . . one of these supposed luminous points was kind enough to venture off the edge of the ring, and appeared in the shape of a satellite'. He then compared the positions of suspected luminous points with the calculated positions of the various satellites, and found that they appeared all to be accounted for by satellites in front of or behind the almost edgewise ring. He therefore concluded in his paper of November 1789, that he had no evidence that the ring was not of uniform thickness. Satellites seemed a much better explanation than ring projections, which would have to be impossibly large to show up at Saturn's great distance. He soon found however, that the supposed solution was too simple and that Saturn was beguiling him.

As he explained in his second paper, when he had corrected the tables of all the satellites, taking into account observations to the end of 1789, although the great majority of the bright points observed were completely accounted for by calculated places of satellites, there was a large residue (nearly fifty) chiefly in October–December 1789, which would not fit the position of any satellite. He was tempted for a time to assume the existence of at least one undiscovered moon, but though some of the bright points seemed to be moving, none was obliging enough to come off the ring, and also he could not get the majority of the positions of the bright points to fit with a revolution period longer than about $15\frac{1}{4}$ hours: if they had done so he might have tentatively assumed the existence of one more satellite exterior to the ring system.

Ring's rotation period

Herschel tried to turn the impasse to advantage by using the 'lucid points' to prove a rotation of the ring and calculate its period. He sagely remarked that a bright and apparently projecting point did not necessarily imply any great irregularity or distortion of the ring surface such as mountains upon it, since a vivid light could appear to project greatly beyond a surface on which it was situated, and hence the luminous places on the ring might be merely very bright reflecting regions or of a more fluctuating nature, such as violent fires. He found by comparing several observations that he could derive a revolution period of 10^h 32^m $15 \cdot 4^s$ for the brightest and best observed spot. He then calculated its apparent distance from the centre of Saturn as $17\rlap{.}''227$ which would place it on the ring system. Unless the latter had a groove or division in which a satellite could revolve, this result must imply a revolution of the ring itself. He then discussed and showed the objections to the idea that the ring was of a fluid consistency which would allow a satellite to revolve in the midst of it, and the difficulties in the way of assuming that such a body was revolving in a groove in the ring.

Finally he drew up a table for the revolution of the ring system for the whole year 1789, based on the period he had calculated, and applied it to all the observed positions of bright spots not accounted for by satellites. He found that to reconcile all the observations he had to assume five bright moving spots on the ring, and admitted that with this number of spots it would be possible to find other periods with which some of the observations would agree. But as about 20 of the observations agreed with the calculated positions of one particular spot (α), this alone seemed to him to prove both the rotation of the ring and the rotation period he had worked out.

In 1791, when he had verified that Cassini's division was a true gap, dividing the ring into two 'entirely detached from each other', he had to consider to which ring his calculated rotation period applied. He felt satisfied that (α) and two of the others were on or near the outer edge of the outer ring, and that therefore the period applied to that ring. He thought that the remaining two spots were probably on the inner part of the same ring, but even if they were on the inner ring he had too few observations of them to work out a period. He considered that there was probably a difference, though a small one between the periods of the two rings. Though there was one other occasion while he was still observing—about 1803—when the ring system was presented almost edgewise, he did not revert to the subject of the ring's rotation in his later papers.

It will be seen (in Chapter 10) that during the following half-century many attempts by other astronomers to detect movement by bright spots on the almost edgewise ring failed. P. H. Hepburn, Director of the B.A.A. Saturn Section, writing in 1926 (in *Splendour of the Heavens*, p. 368) said that what Herschel saw was an optical phenomenon often since observed and not attributable to the rotation of the ring. The fact of rotation and also that of differential rates had been much later (1895) confirmed by the spectroscope (see Chapter 15): Herschel's period of about 10^h 32^m corresponds to that of a particle near the outer edge of the inner ring (B). It seems therefore that Herschel had managed to obtain a correct result, though it was based on illusory data.

Cassini's division a true gap

During 1791 Herschel had been able to get a good enough view of the ring's southern face to determine whether the appearance and position of the dark marking corresponded with his observations of it on the northern face. He did not get the chance of applying the test of an occulted star, but he was able to make use of the fact that he had (to his own satisfaction) proved that the ring was rotating once in approximately every $10\frac{1}{2}$ hours. He inferred from this that during the course of any evening's observation, the part of the dark marking in front of the globe which appeared foreshortened, and the part hidden behind the globe, must have duly passed to the ansae. If there had been any difference of breadth at any part of the marking's circumference, he must have noticed it as a variation in breadth at the ansae for which he had been watching, but he had found none. This appeared to be an absolute confirmation for the ten years during which he had observed it on the northern face, that the dark marking was uniform in breadth, colour and sharpness of outline all the way round. These facts made a strong prima facie case for regarding the marking as a true gap through the ring system; if it should present the same aspect and occupy the same position on the southern face, that would be decisive.

In his third paper he then listed several observations of August–October 1791, made with various telescopes, including the 40-foot reflector, and found in every case that the black division on the southern plane was in the same place, of the same breadth and at the same distance from the outer edge as he had seen it in earlier years on the northern plane. He also found that it nearly always seemed to be as dark as the background sky. He concluded: '. . . I think myself authorised now to say, that the planet Saturn has two concentric rings, of unequal dimensions and breadth, situated in one plane, which is probably not much inclined to the equator of the planet. These rings are at a considerable distance from each other, the smallest being much less in diameter at the outside, than the largest is at the inside.'

Dimensions of ring system

Herschel then gave estimated proportionate breadths of the rings based on measures of his drawings, as in column (1) of the following table. Assuming the total width of the ring system was equal to about one-third the globe's diameter, he calculated the width of Cassini's division to be 2,513 miles: column (2) shows what his other figures would be in miles on this basis: column (3) modern estimates from the *N.A.*, as given in *B.A.A. Hdbks.*:

| | Herschel's Estimates | | B.A.A. Hdbk (*from N.A.*) |
| | (1) | (2) | (3) |
	Parts	Miles	Miles
Ring A:			
Outside diameter	8,300	181,375	169,300
Inside diameter	7,740	169,135	149,000
Breadth of A	280	6,120	10,150
Ring B:			
Outside diameter	7,510	164,110	145,500
Inside diameter	5,900	128,930	112,600
Breadth of B	805	17,590	16,450
Breadth of Cassini's Division	115	2,513	1,750

On comparing these figures it will be noticed that Herschel's estimates of the total span of the rings (outside diameter of ring A), and of the width of ring B, work out at only about 7 per cent larger than the modern values, and therefore agree reasonably well with them, while his estimate of the width of Cassini's division is nearly 50 per cent above the *N.A.* figure. It is therefore all the more surprising that his assessment of the width of ring A is so much too small. He stated that his proportions were 'merely taken from very accurate representations', as he had not at that time been able to use for the purpose a micrometer attached to his great telescope. From his drawings of Saturn (see Plates III, IV(1)) it is evident that he consistently drew the outer ring too narrow in proportion to the inner one, as compared with modern drawings and photographs. Possibly the brightness of the outer part of ring B caused him to place the division too far out, or the duskier outer part of ring A may have been more indistinct in his time than it is now.

Further on in his paper of December 1791 he gave several micrometer measures of the total span of the rings, taken in October and November of that year, some with the 20-foot, others with the 40-foot reflector. He worked out the mean of all his measures (for the mean distance of Saturn) at 46″677, which is less than 9 per cent above the *N.A.* figure of 43″96, but the values he computed from the mean micrometric measures for the span of the rings and the breadth of Cassini's division in miles are much too large: over 204,883 miles and 2,839 miles respectively.

Other alleged divisions

Though he had now satisfied himself by observation that there were two rings, separated by an open space, he did not accept the theory of a multiplicity of rings. Referring to the 'excellent theories' in Laplace's memoir recounting observations of many ring divisions, Herschel considered whether the ring system could be subject to continual and frequent changes, breaking up into a number of narrow rings and then reuniting into two. He said: '. . . the mind seems to revolt, even at first sight, against an idea of the chaotic state in which so large a mass as the ring of Saturn must needs be, if phenomena like these can be admitted. Nor ought we to indulge a suspicion of this being a reality, unless repeated and well-confirmed observations had proved, beyond a doubt, that this ring was actually in so fluctuating a condition.'

He then reviewed the observational evidence, starting with his own records, covering the period from 1774 to 1791. He could only find four, made with his 7-foot and small 20-foot reflectors on June 19, 20, 21 and 26, 1780, when he had detected, in addition to the main division, a narrower dark line which he called a second 'black list', close to the inner edge of ring B on the preceding side (see Plate III, Fig. 1). In June 1780 the rings had been wide open and therefore well placed for such an observation, but he had not been able to detect the marking on the following ansa, he had failed to see it in a clear view of Saturn on June 29, 1780, and had never seen it or any other minor dark line on either face of the rings on any other occasion. He therefore felt that these four observations should be 'set aside as wanting more confirmation'. Turning to the observations of others, the only evidence he could find for possible changes in the ring structure was Cassini's statement that the division he had observed divided the ring into two equal or almost equal parts (which did not agree with the position of the main division as Herschel saw it), and an assertion by Short, merely in conversation with Lalande, that he had seen many divisions on the ring with a 12-foot telescope. Herschel therefore concluded that the observational evidence up to that time was insufficient to show that Saturn's ring system was of a very changeable nature.

Brightness of rings compared

In a paper read in January 1794 Herschel made a few further remarks on the appearance of the rings which entirely agree with modern observations and photographs: 'The outer ring is less bright than the inner ring. The inner ring is very bright close to the dividing space; and, at about half its breadth, it begins to change colour, gradually growing fainter; and just upon the inner edge, it is almost of the colour of the dark part of the quintuple belt.'

His observations of the 'quintuple belt', which consisted of three dusky belts, separated by two narrow light zones, on the southern hemisphere of the globe of Saturn, will be dealt with in the next chapter.

CHAPTER 9

WILLIAM HERSCHEL'S OBSERVATIONS OF THE SHAPE, FEATURES AND ROTATION OF THE GLOBE (1775–1808)

Variable belts

Apart from the satellites and ring, Herschel, in his first paper on Saturn, discussed the globe also, giving a list of over two dozen observations of dusky belts and bright zones made during the years 1775–80, a time when the ring system was opening more and more, exposing its northern face and the northern hemisphere to view, but covering part of the southern hemisphere. The features he noted seem to have all been in the equatorial region, including the equatorial zone, the brightness of which he remarked a few times, but most of his notes were of a dusky belt and often of two, generally rather dark but sometimes faint, usually parallel to but at times inclined to the equator. The belt or belts (probably the north equatorial belt, sometimes seen in two components) appeared to vary in width and occasionally one of them was not uniformly dark. In June 1780 he observed a large, diffuse, dusky spot on the belts (see Plate III, Fig. 1); the spot seemed to have moved from the centre to the limb from one day to the next, suggesting a rotation of the globe on its axis, similar to that of Jupiter.

Atmosphere

His first deduction from these observations and other similar ones from 1780 to 1789 was that the variations in the aspect of the belts indicated that, like those of Jupiter, they were probably atmospheric, and not surface features. This sign of a considerable atmosphere was, in his opinion, confirmed by other observations, particularly of the occultations of satellites, as he had 'found them to hang to the disk a long time before they would vanish'. He admitted that it might be partly an optical effect, causing a satellite to appear to reach the disk sooner than it actually does, but he considered that a satellite could hardly remain in view so long after apparent contact except by refraction caused by an atmosphere. As examples, he stated that he had seen Mimas, in spite of its swift motion, appear to hang on the limb for 20 minutes, and that Enceladus had remained in view

142

for 14 or 15 minutes longer than it should have done. He concluded 'that very probably Saturn has an atmosphere of a considerable density'.

Axis of rotation

He next pointed out that, during the fourteen years of his observations, the belts had, subject to minor variations, always followed the direction of the ring. As the ring opened, and the north pole tilted more and more towards the Earth, they seemed to move southward and show a curvature corresponding to what the equator should have. When the ring closed again, the belts appeared to move northward, and in 1789 when the ring was central, he saw them one on each side of it with the bright equatorial zone between. The belts he had then observed were therefore the ones now known as the north and south equatorial belts. He inferred that the planet rotates on an axis perpendicular to the equator and ring system, just as Jupiter rotates on an axis perpendicular to the plane of its equator. He considered that his three observations of June 19, 20 and 21 1780, of a dusky spot in three different positions relative to the centre of the disk, clinched the argument, but he was able to bring forward still another in favour of the rotation, namely, that he had observed Saturn to be, like Jupiter, Mars and the Earth, flattened at the poles.

Polar compression

He cited three observations prior to 1789 at each of which he had noted that Saturn's equatorial diameter appeared to be greater than its polar diameter. On 14 September 1789, with the ring almost edgewise, he seized the opportunity to make several micrometric measures of each diameter, using his 20-foot reflector and a good parallel-wire micrometer. The means of his measures were: equatorial diameter 22″81, polar diameter 20″61. He therefore stated that Saturn has a considerable polar flattening, the ratio of the equatorial diameter to the polar one being nearly as 11 to 10, and that the planet's axis is perpendicular to the ring-plane. Modern measures derived from the *N.A.* and given in the *B.A.A. Hdbk.* for 1960 would make the diameters at mean opposition distance respectively 19″52 and 17″46, the ratio being slightly more than 11 to 10, so that Herschel's value for the degree of flattening is reasonably good.

Features of southern hemisphere

His next paper dealing with any globe features was the fourth, read in December 1793, where he quoted half a dozen observations of November–December of that year recording the appearance of the belts and light zones in Saturn's southern hemisphere. As he had seen a number of these, all running parallel to the planet's equator, he presumed by analogy with Jupiter, that Saturn's period of rotation was also probably a short one, and 'that every conclusion on the atmosphere and rotation of the one, drawn from the appearance of its belts, will apply equally to the other'. The most interesting of the observations recorded in the short fourth paper was that of November 11 1793 (see Plate III, Fig. 2), made with one of his smaller telescopes, a 7-foot long reflector, and a magnifying power of 287. He thus described the features of the southern hemisphere, starting from the line where the inner ring crossed the globe:

1. 'Close to the ring of Saturn, where it passes across the body of the planet, is the shadow of the ring; very narrow, and black.' From this description and Herschel's drawing it seems almost certain that this narrow very dark band adjoining the inner edge of the bright ring consisted not only of the ring shadow but also of the dusky inner ring known later as the 'Crepe Ring', of whose existence Herschel was, however, not aware.

2. 'Immediately south of the shadow is a bright, uniform, and broad belt.' This is now known as the equatorial zone, and is usually the brightest part of the globe; light areas are now called *zones*, not belts.

3. 'Close to this bright belt is a broad, darker belt; which is divided by two narrow, white streaks; so that by this means it becomes to be five belts; namely, three dark and two bright ones; the colour of the dark belt is yellowish.' This is the feature which Herschel named, 'the quintuple belt'; the three northern sections of it presumably correspond to the south equatorial belt, which is often seen as two dusky components separated by a narrow light zone. To the south of this would be the light south tropical zone, not usually so narrow and the remaining part of the five-fold belt was presumably the south temperate belt.

4. 'The space from the quintuple belt towards the south pole of the planet which is in view, is of a pale, whitish colour; less bright than the white equatorial belt, and much less so than the ring.'

5. 'The globular form of Saturn is very visible, so that it has by no means the appearance of a flat disk.'

Rotation period of Saturn

Herschel's next paper, read in January 1794, was almost entirely devoted to the investigation of the rotation period of Saturn. He quoted a large number of observations of the belts made during

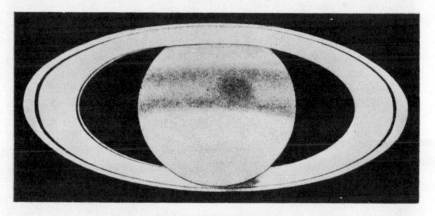

FIG. 1.—19 June 1780, shows linear marking by inner edge of ring B, p. ansa, called by Herschel a 'black list'.

FIG. 2.—11 November 1793, shows feature called by Herschel the 'quintuple belt', and crepe ring across globe involved with ring shadow.

PLATE III.—W. Herschel: Drawings of Saturn (from *Sci. Papers*, vol. I, plate XII, fig. 1, and plate XIII, fig. 1).

FIG. 1.—W. Herschel: 'square-shouldered' figure of Saturn, 18 April 1805 (from *Sci. Papers*, vol. II, p. 333). By North (lower) edge of ring across globe is undoubtedly the crepe ring, assumed by Herschel to be a belt.

FIG. 2.—H. Kater: Saturn, 17 December 1825, 6¼-inch refractor (from fig. 1 of plate in *Mem. R.A.S.*, vol. 4)—minor 'divisions' on ring A.

PLATE IV

November and December 1793 and January 1794, taking particular note of the dates and times at which he had recorded any slight variation of the width, darkness or other feature of any of the belts. He then selected pairs of very similar and strongly marked observations, as far apart in date as possible, assumed that the likeness between each pair of observations meant that the same parts of the belts had been on the central meridian on both occasions, and calculated a tentative rotation period that would give an exact number of rotations from the earlier observation to the later one. For example, he selected an observation on December 4 and one on January 7, at both of which the belts had appeared perfectly uniform, and he found that the total time interval between the two observations would allow for 79 rotations of Saturn if each rotation took 10^h 15^m 40^s. In seeking a rotation period he had one thing to guide him: his earlier deduction that the ring rotated about Saturn in roughly ten hours and a half; he probably used this as an upper limit, and sought a period of the same order but somewhat shorter. He next worked out in the same way a tentative rotation period for two observations, of December 6 and January 4, at each of which the northernmost of the belts was much darker and broader than the most southerly one; the period he derived was 10^h 16^m 51^s. As a check on the first result, he took two intermediate dates when the belts had appeared uniform, and worked out two more periods for December 4–18 and December 4–January 2; these results were both less than 10^h 18^m, which showed, he thought, sufficiently good agreement. He therefore decided to adopt provisionally the mean of the first two results (10^h 16^m $15\frac{1}{2}^s$) as an approximate rotation period for Saturn, and then followed his usual procedure of constructing tables of longitudes based on this period, and applying them to all his observations of the belts. He considered that if the calculated and observed appearances were found to agree, the solution of the problem would be achieved. After a few trials, some minor corrections and a further good observation, affording an interval of 100 rotations by which to check the assumed rotation period, he finally computed a rotation period of 10^h 16^m 0^s4, with which nearly all the observations closely agreed, and he was able to conclude that the period he had then found was correct to within 2 minutes either way.

To show how good this result was, it may be compared with those mentioned in later chapters (e.g. Chapters 15 and 34).

Herschel's method of arriving at a rotation period for Saturn's globe was similar to the procedure he had successfully used to determine the revolution periods of the satellites and the rotation period of the ring, but in those investigations, as in the later ones of Asaph Hall and of the B.A.A. Saturn Section, there had been definite bright or dark spots to observe. It must seem extraordinary, then, that at the first attempt ever made to find the rotation period of Saturn, Herschel should have even tried, let alone succeeded, by the use of rather slight observed differences of intensity as between ill-defined stretches of belts. It is an arresting example of his power to compel vague, unpromising indications to yield a definite and accurate numerical answer to the problem he had set himself to solve.

Shape of shadows

Although his paper of January 1794 is almost wholly concerned with observations of the 'quintuple belt' and the process of deriving a rotation period therefrom, Herschel quoted a few remarks he had recorded on the appearance of the shadows. For example, on 13 December 1793, he noted: 'On the south following part of the ring, close to the body of the planet, is the shadow of the body. The shadow of the ring upon the body of the planet close to the ring, is not parallel to the ring at the two extremes, but a little broader there, than in the middle; the ends turning towards the south.'

This widening of the shadow at the ends on the limbs of the globe has often since been observed and drawn, and is clearly shown on photographs of Saturn. Herschel, in a later paper, considered that it was 'partly a consequence of the curvature of the ring which in the middle of its passage across the body hides more of the shadow in that place than at the sides'. He might well have cited also the considerable curvature of the globe.

In the paper of June 1806 he also mentioned the peculiar shape of the globe's shadow across the rings: a little broader at the north than at the south edge of the ring, and narrower in the centre, making the edge of the shadow appear concave. Peculiarities in the shape of the shadow crossing the rings have been observed by many astronomers since Herschel's time and have given rise to much discussion, but later chapters (e.g. Chapter 29) show it to be an optical phenomenon, the apparent narrowness of the shadow at the brightest parts of the rings being caused by irradiation.

Colour of polar regions

Herschel also discussed in his paper of June 1806 periodical changes which he had observed in the colour of the polar regions of Saturn. In some observations of 1793–96 the south pole, which had then been for a long time turned towards the Sun, seemed to have lost a good deal of its former whiteness and become duller in tone. He next quoted some of 1806, when the north pole was turned towards the Sun; in those the north polar regions seemed to have lost much of their brightness, while the south polar regions had become very bright and white. Although he had been struck by the analogy to the changes in the polar caps of Mars, and had observed these changes on Saturn over a period of nearly half a Saturnian year, he wisely suggested that more confirmation was needed, and that the alterations in the appearance of Saturn's poles could be attributed to Saturnian atmospheric rather than surface causes. He quoted one more observation, of June 1806, as follows: 'The brightness which remains on the north polar regions, is not uniform, but is here and there tinged with large dusky looking spaces of a cloudy atmospheric appearance.'

He considered that this was further evidence of the existence of an atmosphere on Saturn. As regards the changes he had observed in the brightness of the polar regions he suggested: 'The regularity of the alternate changes at the poles ought however to be observed for at least two or three of the Saturnian years, and this, on account of their extraordinary length, can only be expected from the successive attention of astronomers.'

'Square-shouldered' figure of globe

In 1805 Herschel had been much concerned with a very strange appearance in the shape or figure of Saturn, and to this subject he devoted a paper read in June of that year. It seemed to him that the polar flattening was not as gradual as on Jupiter, but, beginning at a higher latitude, was much more sudden (see Plate IV, Fig. 1). His description of the observation on 18 April 1805 when he made this drawing is as follows: . . .'10-feet reflector, power 300. The air is very favourable, and I see the planet extremely well defined. The shadow of the ring is very black in its extent over the disk south of the ring, where I see it all the way with great distinctness. The usual belts are on the body of Saturn; they cover a much larger zone than the belts on Jupiter usually take up. . . .' Hence he assumed that the shading along the northern edge of the ring across the globe, which was certainly due mainly if not entirely to the crepe ring, was a belt.

He took micrometric measures at various dates from April to June 1805 to try to determine as accurately as possible the latitude of the points of greatest curvature of the globe. He thought at first that this latitude was about 45° but in the end adopted a mean value of the measures he made which was 43° 20'. He also measured carefully with a micrometer on more than one occasion the equatorial and polar diameters and the diameters from 43° 20' N. to 43° 20' S., and decided that the diagonal diameter was the greatest, being in the proportion to the equatorial and polar diameters respectively as is 36 : 35 : 32. During the period he examined Saturn most carefully with various telescopes and magnifying powers and compared its shape repeatedly with that of Jupiter which was also in the sky. He concluded that Saturn was slightly flattened at the equator as well as having a very considerable polar flattening, and that its disk was shaped rather like a rectangle with rounded corners. He even made an accurate drawing based on the measures he had taken and found that it matched exactly with the image of the planet as seen through his telescope.

A good deal of his next paper, of June 1806, was also devoted to this subject of Saturn's apparently singular shape.

In an observation made in 1788, Saturn appeared 'a very little flattened at the equator' as well as much flattened at the poles. He then detailed his observations of April to June 1806. He found that the ring, being more widely open than in the previous year (1805), obstructed the view of the curvature more in higher latitudes, but afforded a better view of the equator, which appeared more curved than it did the year before. But he still considered that, in general, the shape of Saturn was similar to that of 1805, and that the diagonal diameter was the greatest, but the ratio was reduced to 36 : 35·4 : 32. He also still found a difference in the apparent shape of Jupiter from that of Saturn, though both planets were rather low in the sky.

In a paper of April 1808 he reported how both he and his son John had, in June 1807, noted that the south pole of Saturn appeared much more curved than the northern polar region, and that he had seen the same phenomenon on two more occasions in the same month and several more times afterwards. Nevertheless he concluded that this was an optical illusion due to the position of the ring.

Herschel's observations of 1805 and 1806 of what became known as the 'square-shouldered figure' of Saturn presented a problem which caused much perplexity and discussion among astronomers of the earlier decades of the nineteenth century. His outstanding quality as an observer, and the fact that he had investigated, tested and measured the peculiarity in every possible way with different telescopes and magnifying powers on numerous occasions, made it very difficult to dismiss his conclusions out of hand as an optical illusion. The next chapter will show how this mystery was cleared up. He seems to have been the victim of a very subtle optical trick by Saturn to which the ring position, the polar shading and the planet's low altitude all probably contributed.

Though Herschel made one or two mistakes about Saturn his contribution to the knowledge of the planet was immense: he made important discoveries, he verified those of Cassini, he used ingenious methods to solve its problems, he investigated the planet in a most comprehensive way, rescuing it from comparative neglect and pointing the way to many lines of observation and research that have been followed up by astronomers during the century and a half since his lifetime.

As an expression of Sir William Herschel's real enthusiasm for the planet Saturn, as a summary of his more important discoveries, and as a fascinating picture for his successors, what could be better than the opening paragraph of his paper of 1805:

'There is not perhaps another object in the heavens that presents us with such a variety of extraordinary phenomena as the planet Saturn: a magnificent globe, encompassed by a stupendous double ring: attended by seven satellites: ornamented with equatorial belts: compressed at the poles: turning upon its axis: mutually eclipsing its ring and satellites, and eclipsed by them: the most distant of the rings also turning upon its axis, and the same taking place with the farthest of the satellites: all the parts of the system of Saturn occasionally reflecting light to each other: the rings and moons illuminating the nights of the Saturnian: the globe and satellites enlightening the dark parts of the rings: and the planet and rings throwing back the sun's beams upon the moons, when they are deprived of them at the time of their conjunctions.'

CHAPTER 10

AFTER HERSCHEL: VARIOUS OBSERVATIONS IN THE EARLY AND MID NINETEENTH CENTURY

Sir William Herschel, knighted towards the end of his life, had aroused much interest in the peculiarities and problems of Saturn, and in the 1790s while Herschel was still observing the planet, Johann Schröter had already begun to follow up his work. Schröter, who had bought one of Herschel's telescopes and had a private observatory at Lilienthal, near Bremen, is chiefly renowned for his lunar work, including the discovery of the clefts on the Moon. Assisting him for some years was Karl Harding, who discovered the minor planet Juno and afterwards became professor of astronomy at Göttingen.

Most of the observations to be recounted in this chapter can be regarded as continuations of Herschel's work, and in one or two respects, notably the 'lucid spots' on the almost edgewise ring and the 'square-shouldered' figure of Saturn, the early nineteenth century observers succeeded in clearing up the problems he had left them.

Shadows (Schröter, Lassell, Coolidge)

Herschel had drawn attention to slight irregularities in the shape of the ring shadow across the globe and in that of the globe's shadow athwart the rings: his followers, starting with Schröter in 1792 (*M.N.* vol. 21, p. 261), got into worse difficulties by finding these shadows notched and jagged, a striking example being Lassell's impression in 1849 (*M.N.* vol. 21, p. 236). Most of the leading observers of the mid nineteenth century devoted much attention to the observation and drawing of these anomalous shadow appearances, a very striking set of examples being provided by the observations and drawings of Sidney Coolidge, using the Harvard 15-inch aperture refractor in 1854–57 (Harvard C.O.A. vol. 2, pt. 1, pp. 80–92 and 125–34, and figs. 84–120)—in those years the rings were widely opened with the far part visible beyond the south pole of Saturn, and the appearance of the globe's shadow on the rings was very peculiar.

As it was firmly believed during the early nineteenth century that the rings were solid, it is not surprising that the observers should have supposed the irregular shadows to indicate irregularities in the surface of the rings. Even in the later nineteenth century when their solidity had been disproved, the anomalous shadows continued to puzzle astronomers, and it was not until early in the present century that these appearances were proved to be optical, due to atmospheric tremors, irradiation and diffraction* (see Chapter 29).

Spaces between globe and rings (Schröter, Schwabe, Harding)

According to the Rev. T. W. Webb (*Celestial Objects for Common Telescopes*, p. 195), Schröter in 1796, with a 19-inch aperture reflector, particularly examined the space on each side of Saturn's globe, between it and the interior edge of ring B, and found those spaces uniformly dark. This negative observation is of interest in view of the discovery of the crepe ring half a century later, and the fact that neither Herschel nor Schröter with their powerful telescopes could detect any lack of uniform darkness in the spaces on either side of the globe, has suggested to astronomers that the crepe ring may have become more visible during the ensuing half century. Webb also stated that in 1826 Wilhelm Struve, Director first of Dorpat and later of Pulkovo Observatory, measured Saturn repeatedly with a 9·6-inch aperture achromatic refractor but saw no sign of a faint inner ring.

The impression that the globe is not set quite centrally with respect to the ring system had, according to Webb, been noted by Gallet as far back as 1684, but the first observer to draw general attention to the matter was Heinrich Schwabe, famous as the discoverer of the sunspot cycle. Schwabe's statement about the position of Saturn's globe was made as the result of observations in 1827, and in the following year Professor Harding of Göttingen confirmed the appearance. Though unable to explain it as an optical deception, Harding could not consider it in any other light, and felt that actual measurement alone could decide, so at his request John Herschel and James South made a series of micrometric measures. Although the results they obtained were rather contradictory, John Herschel, after careful examination, was convinced that the following (eastern) space appeared the larger (*M.N.* vol. 2, p. 74). This verdict was apparently the reverse of Gallet's impression. Observers at the Rome Observatory also seem to have considered, in 1842 and 1843, that

---

* *Diffraction* is the reinforcement or cancelling out of light through the interference of light of different phase. It is caused by the nature of light, its wave-motion enabling it to bend round very small objects placed in its path. A point-source of light cannot give a point image in any optical instrument, but only a *disk* of light. For *irradiation* see footnote on p. 240.

the slight eccentricity of the globe's position was confirmed, but that it was rapidly variable. The knowledge that Laplace's theory would require the ring system to be eccentrically placed with respect to the planet must have made some at least of the astronomers of the early nineteenth century more prone to accept the results of these observations. The question has been canvassed and tested by measurement on subsequent occasions, but it seems likely that the true explanation lies in one of those optical effects for which Saturn is notorious (see also Chapters 33 and 35).

Spots (Schröter, Dawes, Lassell)

Sir William Herschel had shown the usefulness of spots for determining rotation periods, and it was natural that further attempts should be made to decide by this means the rotation speed of Saturn. Schröter and Karl Harding had detected some spots in 1796 and 1797 and deduced from the observations rotation periods ranging from 11^h $40\frac{1}{2}^m$ to over 12 hours. These results were not satisfactory because, even if the spots were in a temperate zone of Saturn where rotation is slower than in the equatorial region, the periods derived are too long. This is clear from a paper written more than a century later by the Rev. T. E. R. Phillips (*B.A.A.J.*, vol. 44, p. 29, Nov. 1933), stating that, while the difference between rotation periods of the various parts of the globe of Saturn is very much greater than anything observed on Jupiter, yet the range on Saturn is from rather less than $10\frac{1}{4}$ hours at the equator, through 10^h 20^m in the equatorial belts, to 10^h 37^m or 38^m in about latitude 35° to 40° both north and south.

Webb said that spots and streaks on Saturn were uncommon and the instances easily enumerated: for the period of eighty years following Schröter and Harding's detection of spots in 1797, the list given by Webb was quite brief. Yet the latter part of that period was a time when the planet was under frequent and careful survey by able astronomers in several different countries. It included, for example, the very long series of observations of Saturn carried out at Harvard College Observatory from 1847 to 1856 under the leadership of William C. Bond, the watchmaker and amateur astronomer whose zeal and ability caused him to be made its first Director, and who became, with his son, George P. Bond, the pioneers among the great American observers of Saturn. Spots on the planet must have been rare indeed throughout that period because even in his short list Webb seems to have included items that were not, properly speaking, spots on the globe at all, but vague mottlings, or bits of shadow or bright points on the rings. The only indications of dark globe spots in the Harvard observations seem to be on 1 December 1852 and 8 January 1853 (a dark spot on the preceding limb), which were not mentioned by Webb, and his list would seem to reduce to a spot seen by Schwabe on the edge of a belt in 1847, a round bright spot near the south limb observed by Busch and Luther in 1848, and the objects detected by Dawes and Lassell in 1858 recounted below (from *M.N.* vol. 18, pp. 72, 231):

The Rev. W. R. Dawes wrote: 'I have lately observed a well-marked light spot on the southern hemisphere of Saturn, which I estimated to be at about 40° or 45° of south latitude . . .' He stated that he had seen it on 11 and 14 January 1858, on both occasions a little past the central meridian, and remarked that 'distinctly visible phenomena of that kind on Saturn are not of very frequent occurrence'. In the course of an account of observations of the planet on 17 April 1858 Lassell wrote: '. . . . Near the preceding limb . . . and a little south of the equatorial dark belt, was a brighter portion, rather too large to be called a spot, yet sufficiently defined or marked out to be useful in determining the rotation of the planet, if one might be favoured with a more prolonged recurrence of so fine a state of atmosphere . . .'

Lassell there indicated the difficulty: a persistence of the spot and of good weather is needed to observe a number of transits over the central meridian during a few weeks so as to derive a reliable rotation period.

Of the two famous English amateur planetary observers, the Rev. William Rutter Dawes and his friend William Lassell, much will be said in this and the next few chapters, for they, along with the Bonds in America, were rightly regarded as the leading experts on Saturn during the middle of the nineteenth century, much as Huygens and Cassini had been two centuries earlier. Dawes was a nonconformist minister, who had formerly qualified and practised as a doctor; his exceptional skill as an observer compensated for the fact that he used mainly refractors of smallish aperture. Lassell was a brewer who, like Sir William Herschel, developed remarkable skill in the construction of telescopes and in the grinding and polishing of large speculum metal mirrors; he therefore observed mainly with large aperture reflectors, and earned a high reputation as a careful and critical observer.

They strove for absolute exactness in observing and description, not overlooking or failing to record any detail however slight. Though both (and especially Dawes) were shrewd interpreters of their own observations, the outstanding value of some of the details they noted was fully realised only by later astronomers.

'Lucid spots' on ring (Bond)

Another attempt to follow up the work of Sir William Herschel consisted in the observation of bright spots on the edgewise ring. Schröter and Harding watched some in 1802 and 1803, Schwabe in 1833 and 1848, W. and G. Bond in 1848, and Angelo Secchi, S.J. (Director of the Collegio Romano Observatory) in 1862. All these observers found the 'lucid spots' immovable, and therefore useless for determining a rotation period for the ring. W. Bond was very definite about this (*Harvard C.O.A.*, vol. 2, p. 115)—he stated that the bright prominences on the ring in 1848 so much resembled satellites that they were often to be distinguished only by their fixed position relative to the globe. 'Thus, on July 11th, for an interval of more than two hours the motions of the satellites "were very evident, but the bright spots on the ring did not change their positions".' He gave similar examples from observations on July 18 and 21 and August 29, and added that the same phenomena were noticed when the unilluminated northern face of the ring was turned earthwards from September 1848 to January 1849. He summed it up by stating: 'The evidence of their stability, not only from hour to hour, but for months in succession, was most conclusive.'

Satellite studies (J. Herschel; Beer and Mädler; G. Bond; Lassell)

Early in the nineteenth century a good deal of progress was made, especially in the investigation of satellite orbits and revolution periods, and much interesting information about this was given by J. R. Hind (*The Solar System*, 1851, pp. 111–16).

Schröter made estimates of the diameters of some of the satellites, and while his estimates for Rhea and Iapetus were much too large, his figures of 500 miles for the diameters of Dione and Tethys and 2,850 miles for Titan are reasonably near modern values. Sir John Herschel, who made satellite observations at the Cape 1835–37, was unable to detect Mimas with his large reflector, but from his father's observations and his own made at the Cape he worked out the revolution period of Enceladus at $1^d 8^h 53^m 6^s8$, which agrees to the nearest minute with the modern estimate. Sir John Herschel also noticed the interesting and important relationships between the revolution periods of the inner satellites: that the time of revolution of Mimas is very close to half that of Tethys, and that of Enceladus approximately half that of Dione (*M.N.*, vol. 7, p. 24).

Bessel, by observations with a heliometer, computed the period of sidereal revolution of Titan to be 15^d 22^h 41^m 24^s86, which is practically the value used at present. John Lamont, a Scot who became Director of Munich Observatory, found in 1836 that the eccentricity of the orbit of Tethys is very small, and that its inclination to the plane of the planet's equator is less than 2°. Wilhelm Beer (a banker and brother of the composer Meyerbeer) became the pupil and partner in astronomy of a teacher, Johann Mädler, and their partnership produced much notable astronomical work, including, in 1837, a famous book and map of the Moon. Beer and Mädler, from Sir William Herschel's observations, found a value for the revolution period of Mimas of 22^h 36^m 17^s705, which is within one minute of that now adopted. They also deduced that Mimas has an elliptical orbit (it is more elliptical than those of the other four inner satellites), and that its distance from Saturn was 118,000 miles—the modern value is 115,000 miles. They considered that the orbit of Enceladus is circular and in the plane of the rings; modern figures for the eccentricity and tilt are very small. They estimated the distance of Enceladus from Saturn at 152,000 miles, only 4,000 miles more than the present-day estimate.

With regard to the visibility of the innermost satellites, Lassell said that Enceladus was instantly seen with his large reflector when within 40° or 50° of greatest elongation; on the other hand he found Mimas incomparably fainter than Enceladus and an object of extreme difficulty under all but the most favourable conditions (*M.N.*, vol. 8, p. 43).

Francesco De Vico, S.J., who preceded Secchi as Director of the Collegio Romano Observatory from 1839 and is well known for his work on comets, observed Mimas and Enceladus to redetermine their periods of revolution. He tried to overcome the handicap of using a small telescope by introducing a small metal disk into the field of the eyepiece to cut off the glare of the planet.

The *Harvard Observatory Annals* (vol. 2, part 1) record a considerable number of micrometric measures of satellite positions and comparisons of their brightness made with the 15-inch aperture Merz refractor from 1848 to 1852, the observer in most cases being George P. Bond. Lassell was very active in satellite observing, especially at Malta, with his 24-inch aperture reflector in 1852–53 and again in 1861–64 assisted by Albert Marth and using a 48-inch aperture reflector which he had also constructed. A very good series of satellite observations was carried out at the East India Company's Madras Observatory, under the direction of Capt. W. S. Jacob in

1856–57 (*M.N.*, vol. 18, p. 1 gives an abstract of the results). The telescope used had an aperture of only 6·3 inches, so although Jacob was able to derive the elements of the orbits of Iapetus, Titan, Rhea, Dione and Tethys, he was dissatisfied with his observations of the fainter moons, Enceladus and Mimas, considering the margin of error in longitude too large.

Shape of globe (Bessel; Main)

Many observers during the earlier part of the nineteenth century made measures to determine the polar flattening of Saturn, and a great deal of attention was paid with varying results to the question whether the globe had a regular elliptical shape or had the distorted 'square-shouldered' appearance repeatedly recorded by Sir William Herschel in his later observations. According to Proctor, Schröter found the planet not perfectly spheroidal in August 1803, and Dr. William Kitchiner (whose main hobby seems to have been gastronomy) confirmed Herschel's opinion with two different achromatics in the autumn of 1818, when the ring was too narrow to account for the appearance. George Airy, the Astronomer Royal, is said to have had a similar view on one occasion, but on another thought the planet seemed flattened at latitude 45°, about the position where Herschel had considered it to have the longest diameter. Moreover, Airy showed (according to Hind and *M.N.*, vol. 10, p. 91), that the Herschelian shape could not be accounted for by the attraction of the ring.

While it is true that in 1848 G. Bond noted on July 11, and W. Bond on August 29, that Saturn's north pole seemed flattened, the *Harvard Annals* do not appear to support Proctor's suggestion (*Saturn*, p. 66) that the Bonds found other irregularities and distortions on other dates: those peculiarities were all recorded by S. Coolidge, an assistant at the observatory; the Bonds seem to have been observing mostly the rings, belts and satellites, not the shape of the globe except when they took measures of it in 1848.

It was at the beginning of 1855 and once towards the end of that year that Coolidge obtained the impression that the equatorial diameter was not the longest; it was chiefly in November–December 1855 that the south polar region seemed to him to project farthest on to the rings sometimes to the left, occasionally to the right of the true pole; he also a few times thought that the limb seemed flattened on one side or the other of the south pole. All these impressions were obtained during a period when the rings were very widely open, and a peculiar-shaped shadow of the globe on the rings was contiguous to the south polar region and rather to one side, a position of affairs that might easily produce optical deceptions. By the spring of 1857, although the opening of the rings was still nearly as wide, Coolidge noted that there was no distortion and in a detailed view on March 16 in good definition he found the outline perfectly regular. Moreover, in October–December 1848, when the ring was practically edgewise, W. Bond on three dates, and G. Bond on one date, made micrometric measures of the equatorial and polar diameters and of diameters inclined at 45°: on each occasion they found the equatorial diameter the longest; on October 9 the value for the inclined diameter was less than that of the polar one. Perhaps the most decisive evidence was provided by the measures made by Bessel and Main.

Friedrich Wilhelm Bessel, who had once been a merchant's clerk and then an assistant to Schröter, was for 36 years Director of Königsberg Observatory, produced a catalogue of over 3,000 stars, and was the first astronomer to succeed in measuring the parallax of a star: this was in 1838 and the star was 61 Cygni. To determine the general form of Saturn's globe as well as the correct value of the ellipticity he made a long and excellent series of measures of the disk with the Königsberg heliometer in the years 1830–33. Although none of his measures was made in the periods of total disappearance of the ring, some at least were effected when the ring was extremely narrow, and at this time, on three nights in February and March 1832, Bessel measured the planet in five directions and convinced himself of its elliptical shape. Altogether he took 70 measures of the equatorial diameter and 68 of the polar diameter of Saturn, and the value he derived for the polar compression was 1/10·2. Bessel published his results (in *A.N.*, vol. 12, p. 169), but Main stated (*M.N.*, vol. 13, p. 152) that he had been ignorant of their existence until after he had completed his similar investigations in 1848 and published his findings, which were therefore completely independent.

The Rev. Robert Main was chief assistant at Greenwich Observatory, and like Bessel, was highly skilled and experienced in micrometer work. He laid down as a first principle that 'all measures made with a double-image apparatus (the type of micrometer he used) capable of producing a definitive result for the ellipticity of the planet must be made during the time of the invisibility of the ring'. All his measures of 1848 were therefore made under that condition, and with the same objects as Bessel had had, namely, to determine the general shape of Saturn and the correct value of the polar flattening. Main was equally convinced by his results of the planet's regular shape, and his mean value for the ellipticity was 1/9·2. He pointed out that if those measures made by Bessel when the ring had nearly disappeared were selected, fifteen in number, of February and March 1832 and 1833, and the mean taken, that mean value would be 1/9·06, in close agreement with Main's own result. His figure for the polar compression also accords reasonably well with more recent opinion among astronomers; for example, the globe dimensions of Saturn used by the *N.A.* and given in *B.A.A. Hdbk.* 1960 would make the polar compression 1/9·475.

The Council of the R.A.S. stated in 1850 (*M.N.*, vol. 10, p. 91) that they attached 'a high value to the micrometrical measures of different diameters of the planet, made at the Royal Observatory by the Rev. R. Main, first assistant in that establishment. Judging by the eye only, Sir W. Herschel conceived that the form of the planet deviated from the elliptic spheroid by considerable protuberance at middle latitudes. . . . Mr. Main's measurements have shown beyond doubt that the form of the planet does *not* differ from that of the elliptic spheroid; and the Council have the very great satisfaction of stating that Sir John Herschel, with his usual frankness, has communicated to Mr. Main his avowal that Mr. Main's observations are perfectly decisive on this point, and that there cannot remain the smallest reason for suspecting the existence of this anomalous form.'

Belts (Lassell)

Letters from Lassell (in *M.N.*, vol. 13, pp. 11, 15) describing his observations of Saturn with his large reflector at Valetta, Malta, in the autumn of 1852, gave the following description of the belts of the southern hemisphere of the planet during late October: '. . . The two principal equatoreal belts are of a ruddy brown colour, changing pretty suddenly at the southern edge of the most southern of these belts into a dusky bluish-green colour, very much deepened around the south pole, the exact place of which seems to be marked out by a circular lighter shade. This dusky portion is much variegated by minute dark stripes and lines. . . .'

In early November he was able to record a wealth of detail on the southern hemisphere of the globe such as is very rarely to be seen on that planet. South of the brilliant equatorial zone Lassell saw the chief belt, of a ruddy hue, reaching from about 7° to 28° south latitude; next a white zone about 5° wide, less brilliant, uniform and well-defined than the equatorial zone; then a yellowish-red belt terminating sharply at about 45°; next an extremely narrow dark blue or bluish-green belt. To the south of this was another light zone but of a deeper shade; then a narrow blue belt passing through the 60th degree; next a still deeper-shaded light zone; and to the south of that a sharply defined very dark blue belt, a little above the 70th degree and encircling the pole. About three-quarters of the circle of this belt could be seen, including within it the polar area, which appeared as a lighter coloured spot about 20° in diameter.

It is interesting that, whereas the Harvard observers had shown little detail on the globe in their drawings of Saturn before 1 December 1852, on that date G. P. Bond drew several very clearly defined belts, and a small round lightish polar cap with a dark marking at its edge. Again in January–February 1857, G. P. Bond drew and remarked on a succession of half a dozen darkish belts interspersed with lighter zones, and a very clearly marked small round lightish south polar cap. So many distinct belts on Saturn is a much less common sight than on Jupiter.

Occultations by the Moon

Saturn was occulted by the Moon on 8 May 1832 and on 8 May 1859. On each occasion there were numerous observers who timed the phenomenon, and some of them noted changes in the appearance of Saturn (*M.N.*, vol. 2, pp. 104, 111, 127, 133; vol. 19, pp. 239–41 and 267). In the 1832 observations some found Saturn 'small, dull, gauzy' on emersion and yellowish-green for three or four diameters of distance from the Moon.

At the 1859 occultation, while Professor Challis saw no change in Saturn's aspect, others described the planet on emersion as: 'of a dull earthy colour', 'dull, leaden blue', or very faint and 'ashy'. Dawes remarked: 'The very pale greenish hue of Saturn contrasted strikingly with the brilliant yellowish light of the Moon.'

The exiguous ring, 1848–49 (W. and G. Bond; Dawes)

From April until 3 September 1848, and again from 12–13 September until 19 January 1849, the Earth and Sun were on opposite sides of the ring-plane, so during those periods the ring was considered to have disappeared and to be theoretically invisible. A long and excellent series of observations by W. C. Bond and his son from 25 June 1848 to 23 January 1849, made with the 15-inch aperture Merz refractor at Harvard, recorded fully the varying aspect of the planet and ring during this interesting epoch (*Harvard C.O.A.*, vol. 2, pp. 3–42 and 108–20, and figures 2–55). On June 25 all that could be seen of the ring was a thread of light on either side of the planet and a very dark narrow band across the disk. During early July the ansae generally seemed broken into several fragments with bright spots (like satellites but brighter) here and there on the upper (southern) edge. By the end of August the ansae were exceedingly faint and not more than $0''1$ thick, the following one a fine thread, the preceding one broken in the outer part. On September 3, when the Sun had passed through the plane to the same side as the Earth, the ring looked quite bright, and the bright prominences and the dark band across the disk had disappeared. However, the Earth having passed through the ring-plane again on September 12–13, the ansae again became scarcely visible, presenting a 'beaded and broken' appearance, and the dark trace across the disk was once more in evidence. On October 7, W. Bond remarked that the ring in the ansae was 'like a fine thread of light broken into globules towards the extremities'. Later in October and in November they found that the dark trace across the disk was slightly south of the line of light in the ansae, was arched southward and had increased in thickness to $0''5$; on November 22 it was suspected at moments of being divided lengthwise. All this time the ansae appeared discontinuous and with points of light. On 19 January 1849 the Earth finally passed to the same side of the plane, so the ring 'reappeared', but to W. Bond it looked 'rough, almost resembling detached globules', and the dark trace across the disk was still visible. G. Bond noted on January 23: 'The ring is a very different object from what it was before its reappearance: not nearly as delicate, much brighter, and without breaks. Near the ball, the ring is not so bright as . . . at a little distance from it; presenting the appearance of two loops meeting near the limb . . .'.

The foregoing summary, brief though it is, includes the salient features noted by the Bonds during the observations of 1848–49. With regard to the increased difficulty of seeing the ring from the unilluminated side just before the reappearance, William Bond cited in confirmation the observations of the Rev. W. R. Dawes whom he described as 'a most careful and thoroughly practised observer'.

Bond gave, in the Harvard Annals, a very long and discursive commentary on the phenomena seen in 1848–49, but his much more concise statement (in *M.N.*, vol. 10, p. 16) said that the Harvard observers had, often, during the period of disappearance of the ring, noticed breaks or inequalities in it 'such as would arise from irregularities in its structure, were the matter . . . unequally distributed. . . .' He stated that, whereas something like this had often been seen before on the illuminated side of the ring, similar appearances on the unilluminated side were a new feature. He went on to say that, during the periods of disappearance, 'the light reflected from the edge of the ring (the only part then visible), instead of being uniformly distributed over a single line, was interrupted on each side of the planet by spaces of some seconds in breadth, where it was barely possible to trace the continuity of the edge. . . .' He then pointed out that the bright spots on the ring, so like satellites, were distinguishable by their fixity of position, like those Schröter had seen on the illuminated side. He concluded: 'The fact of these inequalities always retaining one unaltered position may be explained without precluding the possibility of a rapid rotation of the rings, by attributing them to the reflexion of the solar rays from their inner edges.'

Dawes gave an account of his observations (in *M.N.*, vol. 10, p. 46), made mostly with his 6½-inch aperture refractor at Cranbrook. Though this was a very small instrument compared with the one used by the Bonds, he was able to detect quite a number of interesting details. On 30 June 1848 he found the ansae not quite invisible, and of a 'deep coppery tinge'; he also glimpsed on each arm a bright point that he took at first to be a faint satellite, but after a further careful examination he thought each might be a small part of the ring which reflected more light than the rest; he estimated these bright points to be at about the outer edge of ring B. He had a similar view on July 15; on August 9 he noted: 'occasionally an exceedingly faint dusky red line (the arms of the ring) extends on each side of the shadow on the ball'. On August 20 and September 1, he found the trace on the globe very narrow and the arms invisible.

On September 2, the day before the reappearance of the ring was due, he noted: 'The ring is visible as an excessively narrow line of nearly the same colour as the planet. No shadow on the ball.' On the next day he found the ring bright, though rather duller than the planet, and added: 'There is a dusky line about the same degree of shade as the belts (which are strong to-night), crossing the ball precisely in the line of the ring. It is very different from the shadow of the ring'. On September 4 he had a similar view, though 'the dusky shade across the ball' was less distinct, and the following arm decidedly longer than the preceding one.

At the time of the next disappearance in mid-September, Dawes was on a visit to Lassell at Starfield, near Liverpool, and had the advantage of being able to observe with Lassell's 24-inch aperture reflector. On September 12 he remarked: 'The ring is the finest line imaginable, yet of a pretty clear bright planetary colour.' The Earth passed through the ring-plane at some time during September 12–13.

On September 13 with the large reflector Dawes could see no sign of the ring or its shadow; the next day the ansae were still invisible, though he saw 'a very narrow dusky line (on the ball) just south of the bright equatoreal region'. On September 19 the ring was 'just visible when the planet is best seen'.

After his return to Cranbrook, again observing with his small refractor, Dawes, on October 6, could not detect the ansae, but the line across the globe was 'very sharply defined and black, though I thought not quite uniformly so; the northern edge appearing blacker than the southern'. On October 11 he began to have glimpses of the ansae again, especially the following one. On October 26, and still more on the 30th, he had the impression several times 'of the dark line across the ball being divided by an exceedingly fine line of light'. He had this impression again on November 21, when, as on October 30, he also saw a bright point on each ansa of half the brightness of Enceladus.

On December 5 and 20 he found the ring easily visible but 'broken into portions, which look almost like small satellites'. On December 21 he saw the ring as a very fine line, occasionally but not always appearing broken. The Earth passed through the ring-plane to the same side as the Sun on 19 January 1849 but cloudy weather precluded observation until the 22nd, when Dawes noted: 'The ring is bright, though narrow. It is very much brighter in two places similarly situated on each arm; and they coincide with the extremities of the inner and outer ring'.

The brilliant deductions Dawes drew from his observations are of great interest, and they differ in some important respects from those of Bond. He remarked first that when the obscure side of the ring was turned towards the Earth, it was not invisible with moderate optical power, but that its visibility diminished as the Earth approached its plane. He said this was proved by the increased difficulty of seeing it on August 9 compared with July 15; by its invisibility on August 20 and September 1; by the increasing facility with which it was seen between mid-October and the end of November; and still more conclusively by its total invisibility even with Lassell's 24-inch aperture reflector on September 13, when both Earth and Sun were very nearly in the plane of the ring, and consequently the Sun's light might be expected to be reflected from its edge.

Hence Dawes inferred 'that the edge of the ring reflects extremely little light, and that the visibility of the ring, when its obscure side is turned towards the earth, does not arise from the sun's light reflected from the edge, but from a feeble illumination of the obscure surface;

and that it is this surface only which is seen, and not the edge'. This opinion was contrary to that of Bond, but was in accordance with that expressed by Barnard about sixty years later (see Chapter 26).

Dawes went on to say that his conclusion was supported by the appearance of stationary bright points on the dimly visible ring. If they were real inequalities on the edge, they should have been visible in the 24-inch aperture when the edge was directly turned towards him and fully illuminated. He therefore deduced that 'the appearance of bright points in the unilluminated ring arises from the greater reflective power of some portions of its surface'. He pointed out that the outer part of ring B is usually brighter in appearance than any other part of the rings. He did not consider that the fact of the brightness of the points not being always the same on each side, sometimes the eastern and at other times the western seeming the brighter, presented any serious difficulty: the ring at a given distance from its centre very likely would not have the same degree of reflectivity throughout its circuit, and the observed fact was in harmony with the ring's rotation, 'which the stationary appearance of irregularities on the edge can scarcely be'.

On the question of the source of the light rendering the obscure surface visible, he quoted Sir W. Herschel's opinion (see Chapter 8) that it seemed more likely to be reflected light from Saturn upon the side of the darkened ring than a reflection from the edge of the ring. Dawes, however, felt that the 'tarnished copper' colour he had seen, and the brightness of the bright points on the ring, could not be accounted for by reflected light from the globe of Saturn, with only one quarter of its illuminated surface turned towards that side of the ring. He considered it more likely that 'the illumination of the obscure surface of the ring arises from the refraction of the sun's light through an atmosphere surrounding each of the rings, and thus throwing a pretty strong twilight upon them'. This would have cleared up the matter rather nicely if the assumption that the rings had an atmosphere could have been maintained, but the abandonment less than ten years later of the idea of the solidity of the rings undermined Dawes's assumption, and so the problem of the bright points and the visibility of the dark side of the rings continued to worry astronomers long afterwards, as will be seen in Chapter 26. In that chapter, however, mention will be made of a twentieth century theory that the rings may be enveloped in a cloud of very fine dust, which seems like an attempt to revive Dawes's ring atmosphere theory in a modified form.

CHAPTER 11

ELUSIVE SUBDIVISIONS IN THE RINGS

Southern face, 1819–31 (Kater)

The astronomers of the early and mid nineteenth century seem to have been influenced, in their observations of Saturn, not only by the lines of investigation suggested by Sir William Herschel, but also by Laplace's theory that the stability of the ring system required it to be subdivided into a large number of narrow rings. As has been shown in Chapter 8, Herschel had been profoundly sceptical about alleged observations of ring subdivisions, and the discovery of such markings in the early nineteenth century seems, to start with at any rate, to have been accidental. Whenever such a discovery was made, however, the observer and others, fortified by the knowledge of Laplace's theory, usually followed it up by a deliberate search of the rings to see whether they could confirm the marking that had been seen and to look for more. A good summary of the earlier discoveries was given by J. R. Hind (*The Solar System*, pp. 107–9), and an important original source is the paper read in 1830 by Captain Henry Kater, Vice-President and Treasurer of the Royal Society (*R.A.S. Memoirs*, vol. 4, p. 383).

Apparently Kater had been in the habit of observing Saturn chiefly in order to test the performance of his two Newtonian reflectors, one by Dollond with a mirror of $6\frac{3}{4}$ inches diameter, the other by Watson with a mirror of $6\frac{1}{4}$ inches diameter and a focal length of 40 inches. He stated that although the performance of the latter was uncertain, he had never seen a telescope which, under favourable circumstances, gave a more perfect image. On 17 December 1825, the conditions were very favourable: no wind and a very slight fog. With the larger reflector he saw a very distinct image but nothing unusual, but with the $6\frac{1}{4}$-inch Watson and his most perfect single eyepiece (power about 280): '... I fancied that I saw the outer ring separated by numerous dark divisions, extremely close, one stronger than the rest dividing the ring about equally.... I have little doubt, from a most careful examination of some hours, that that which has been considered as the outermost ring of Saturn consists of several rings. ... The inner ring decidedly has no such appearance. ...'

Kater's drawing (see Plate IV, Fig. 2) shows three divisions around both ansae of ring A, the central one, at about the middle of the southern face of the ring, being the darkest and widest. On the same evening he requested two friends to examine the ring and make drawings: one showed six subdivisions on the ring, evenly spaced and all equally thin and faint; the other showed only one dark marking at about the middle of the ring.

He thought he glimpsed some markings on ring A twice in the following month with the 6¾-inch, but could not be certain, and in February 1826 and on a remarkably fine evening in January 1828 with that telescope he could see no sign of subdivisions in ring A, so he noted: 'I am, therefore, the more persuaded that they are not permanent.' He delayed submitting his paper until 1830 in the hope of getting another good view of the subdivisions in ring A, but did not succeed in doing so. He mentioned the matter to Professor Quetelet, Director of the Brussels Observatory, who told Kater that he had seen ring A divided into two in December 1823 at Paris, when observing with a 10-inch aperture achromatic. In 1826 both John Herschel and Wilhelm Struve tried to follow up Kater's observation but neither succeeded in detecting any subdivision in the rings, though observing with much larger telescopes.

Northern face 1833–48 (Encke, De Vico, Dawes, Lassell)

On 28 May 1837 Professor Johann Franz Encke, Director of Berlin Observatory, using a large telescope there, not only saw ring A on its northern face divided by a black line, but obtained micrometric measures of its position. His measures, as quoted by Hind, would place the line on the inner part of the ring at approximately one-third of the distance from its inner to its outer edge.

In the summer of 1838, Father De Vico, at the Collegio Romano Observatory in Rome, having heard of Encke's observations, scrutinised Saturn very carefully, and on May 29 saw very clearly and showed to others three dark lines, one being nearly in the middle of ring A and the other two on ring B. It is said that in observations on subsequent days, the number of zones in the rings seemed to vary with the seeing conditions, and that about the time of meridian passage as many as six rings were sometimes noticed, distinctly enough not to ascribe them to optical illusion.

Schwabe at Dessau in the course of observations of the rings made on thirty days in the summer of 1841 was able to glimpse Encke's division only on four occasions, on one or other ansa.

Dawes gave an interesting account (*M.N.*, vol. 6, p. 11) of the first view of a subdivision in ring A obtained by himself and Lassell on 7 September 1843 with the 9-inch aperture reflector at Lassell's observatory near Liverpool: . . . 'The sky was hazy and the stars dull . . . Mr. Lassell was electrified at the beautifully sharp and distinct view of the planet . . . examined it for a few minutes . . . and . . . requested me to examine carefully the extremities of the ring, and say if I observed anything remarkable. . . . I presently perceived the outer ring to be divided into two. This perfectly coincided with the impression Mr. Lassell had previously received . . . occasionally, for several seconds together, I had by far the finest view of Saturn that I was ever favoured with. The outline of the planet was very hard and sharply defined with power 450; and the primary (Cassini's) division . . . very black and steadily seen all round the southern side. When this was most satisfactorily observed, a dark line was pretty obvious on the outer ring. I was not only perfectly satisfied of its existence, but had time during the best views carefully to estimate its breadth, in comparison with that of the division ordinarily seen. The proportion appeared to me to be as one to three; but Mr. Lassell estimated it at scarcely one-third. It is certainly rather *outside* the middle of the outer ring, and is broadest at the major axis, being in this respect precisely similar to the primary division. *It was equally visible at both ends of the ring.*'

Dawes went on to say that other eyepieces were tried, and that although the secondary division was perceptible occasionally during best moments with a power of 400, no lower power showed it at all; he added: 'Neither Mr. Lassell nor myself obtained a single glimpse of any further subdivisions of the ring. The shading of the interior edge of the inner ring was very obvious, but no dark line was even suspected in that situation. . . .'

Professor James Challis, Director of the University Observatory at Cambridge, had some glimpses of this marking in ring A in the years 1842 and 1845, using the 11½-inch Northumberland telescope at Cambridge, at that time one of the largest refractors in the world and the one used by him in his fruitless search for Neptune in 1846. In 1845 also John R. Hind (a civil engineer who later became superintendent of the *Nautical Almanac* and discovered eleven minor planets, a nova and a nebula), saw what he considered to be Encke's division in the eastern ansa.

Southern face, 1849–61 (Dawes, Lassell, Tuttle, Jacob, Secchi, Coolidge)

An account by Dawes (*M.N.*, vol. 11, pp. 22–25) of Saturn

observations by Lassell and himself in November–December 1850 contains references to a division on the southern face of ring A. On 23 November Dawes had glimpses of a short and narrow line near the outer edge of that ring. The next day he received a letter from Lassell stating that, while observing the planet on November 21 with his 24-inch aperture reflector and powers of from 219 to 614: 'I had repeated impressions of a secondary division; and, if it be real, it is one-third of the breadth of the outer ring from its outer edge. The suspicion was the same for both ansae. . . .'

Encouraged by this independent confirmation from Lassell, Dawes continued the search, using a refractor with an object glass of very fine quality but an aperture of only $6\frac{1}{3}$ inches; he was, however, renowned for his skill in the detection of fine detail, and known for this reason as 'the eagle-eyed Dawes'. On November 25, using powers of 282 and 425, Dawes . . . 'was satisfied that, in finest moments, a very narrow and short line was discernible on the outer ring near its extremities'. . .' He noted on November 29: 'Having applied higher powers, and viewed the planet steadily for a considerable time, I obtained several satisfactory glimpses of a division near the extremity of the outer ring. It was occasionally seen with power 323; but far more certainly with 460. After a few seconds of uncommonly sharp vision, I involuntarily exclaimed, "Obvious". . .' Later the same evening Dawes made the following note: 'At times I am pretty sure of the lines on the extremities of the outer ring; but rather more so on the following side. The preceding extremity seems rather more dusky than the following one, and scarcely so distinct. . . .'

It is curious that at this period the Harvard observers were detecting subdivisions in ring B, but apparently saw no sign of one in ring A until two years later. On 10 October 1850 G. P. Bond suspected a subdivision near the inner edge of ring B, and on November 11 and again on January 7 1851, he was very confident of having seen it. W. Bond could see nothing of the kind on 5 September 1851, but one of his assistants, Charles W. Tuttle, had a remarkable view on 20 October 1851. A month earlier he felt almost certain that there was a subdivision in the ansae at about two-thirds of the distance from the inner to the outer edge of ring B. On October 20, Saturn looked steady, distinct and well-defined, and he found ring B to be 'minutely subdivided into a great number of narrow rings.' There was no uncertainty as they were distinctly seen by two other observers; the subdivisions commenced close to the inner edge and extended over about two-thirds of each ansa of ring B. They were quite apparent with a high power (861) but looked like a dark spot with a low one. Tuttle's later description, from memory, was: 'The divisions were not unlike a series of waves; the depressions corresponding to the spaces between the rings, while the summits represented the narrow bright rings themselves. The rings and the spaces between them were of equal breadth. . . . The exterior rings were not so well defined. . . .' His drawing showed the outer markings shorter and fainter, so that the whole thing gave the impression of the ends, in the ansae of ring B, of an elongated ellipse, continuing the line of the inner edge of the ring (*Harvard C.O.A.*, vol. 2, pp. 46–50, and fig. 65).

In early September 1851, with good views of Saturn through his large reflector, Lassell could see no sign of a division in ring A, but he noted: 'The . . . dusky surface of the exterior ring gave a striped impression, but no certain line', and also 'a faint but very evident shade' on the inner part of ring B, beginning at about the middle of that ring. On October 26, with his refractor, Dawes considered that a very narrow black line was steadily visible a little outside the middle of ring A, but that the most remarkable appearances were on ring B, which was clearly seen to be 'arranged in a series of narrow concentric bands, each of which was somewhat darker than the next exterior one. Four such were distinctly made out. They looked like steps leading down to the black chasm between the ring and the ball. The impression I received was that they were separate rings; but too close together for the divisions to be seen as black lines. The slightest undulation of the image was sufficient to confuse the different shadings, and to destroy the step-like appearance. . . .' (*M.N.*, vol. 12, pp. 9–13). It will be noticed that Dawes's observation of the appearance of ring B on 26 October 1851 confirms Tuttle's of six days earlier, except that the latter with a much larger telescope and higher magnification was able to separate the minute subdivisions: W. Bond called attention to this agreement in a note (*M.N.*, vol. 12, p. 155).

W. S. Jacob referred (in *M.N.*, vol. 13, p. 240) to observations of the faint division in ring A made with the 6·2-inch aperture equatorial refractor at Madras Observatory in 1852 with a power of 365, as follows: '. . . the fine outer division can, under favourable definition, be traced through more than one-half the circumference; its breadth cannot be estimated at more than one-third or one-fourth of the large division, say $0''1$. . . .'

He also remarked that Cassini's division appeared decidedly not black but of a slaty hue, and that while its inner edge (adjoining ring B) seemed pretty sharp, its outer edge (adjoining ring A) seemed rather shaded off.

In the autumn of 1852 Lassell had his 24-inch aperture reflector set up at Valetta, Malta, in order to take advantage of the lower latitude and clearer skies to make detailed studies of planets and their satellites. In a letter dated 1 November 1852, he described* the appearance of Saturn as viewed from Valetta (*M.N.*, vol. 13, p. 11) and had this to say about the marking on the outer ring: 'No division, properly so called, could at any time be traced or imagined on the ring A; but there was an evident *shade* in the middle of its breadth and occupying about one-third of it. This was seen night after night with different powers, from 297 to 650, so that I am perfectly satisfied of its existence. . . . Its depth of shade is rather deeper than the darkest shading of B close to its inner edge.'

With regard to ring B he wrote: '. . . The shading on the inner edge . . . extends clearly within one-third of its entire breadth of the outer edge, and during some rare glimpses appeared to me as if it were constructed in narrow concentric circles. . . .'

In 1854–55 the indefatigable Dawes was observing Saturn with a refractor of 7½ inches aperture, a full inch larger than that of his own telescope. This larger instrument had been loaned to him by the American, Alvan Clark (discoverer in 1862 of the dwarf companion of Sirius), who had made the object glass. With it Dawes observed and reported (in *M.N.*, vol. 15, p. 80) among many other details the following points regarding the southern face of the bright rings:

Ring A—'The interior edge of this ring is decidedly its brightest part: its light rapidly fades away towards the middle, where there is a very dark, narrow, well-defined line concentric with the ring, and about one-fifth of its breadth . . .' He went on to state that the line was always seen when the seeing was good and more readily than in the previous year. On 26 November 1854 he traced it more than half way round towards the globe, and saw it equally well in both ansae. He added that he had noticed this line during four apparitions, always at about the middle of the ring and of about the same breadth and darkness.

Ring B—He found this ring 'decidedly in stripes', as follows: about one-fifth of its breadth from the outer edge very bright; then a lightly shaded narrow stripe; then a lighter stripe; next a considerably darker stripe; then a much darker one extending nearly to the interior edge of the ring, where there is a very narrow bright line (cf. Lyot's diagram: Plate XVII, fig. 2).

Dawes also referred to observations made by Secchi at Rome with a 9-inch aperture refractor during the same winter: these agreed

* See Chapter 10.

closely with his own impressions in regard to the ring markings. Secchi had described the dark line on ring A as being like a pencil line drawn on it, and Dawes recalled that a similar line had been seen on the northern face of ring A by Encke in 1837 and by Lassell and himself in 1843 without knowledge of Encke's observation. Dawes commented: 'It may not be a division in the ring, as it was then supposed to be . . .', but he went on to point out that it was a remarkable circumstance that precisely the same appearance should exist on both surfaces of the ring. Secchi had also described 'the step-like concentric bands of shading on ring B . . .' exactly as Dawes had described them in 1851 and as he had occasionally seen them since.

In December 1852 G. Bond had suspected a subdivision in ring A, and in December 1853 both he and Tuttle strongly suspected a 'division, shading or seam' on that ring. In late December 1854 W. Bond noted that ring A was divided on both ansae at about two-fifths of the breadth from the outer edge, while Coolidge went so far as to state: 'There are two (and at times I suspect three) divisions' in ring A. From that time onwards the Harvard observers often recorded a subdivision in ring A and very occasionally some markings on ring B (*Harvard C.O.A.*, vol. 2, pp. 65–135); nearly all the observations of Saturn of 1855–57 seem to have been made by Coolidge. Some of the more striking of these were as follows:

On 9 January 1855 Coolidge saw very plainly 'the dusky seams in ring A' and also 'at the major axis of ring B close to its inner edge . . . three or four fine dark lines or divisions. They do not extend far round the ellipse. . . .' On January 25 G. Bond plainly saw a division near the middle of ring A, and Coolidge could trace it through nearly the entire circumference.

On 20 March 1856 Coolidge noted four or five narrow divisions on ring B 'giving its shaded part a wavy appearance'. A week later he found the 'seam' in ring A more distinct, narrower and sharper than usual, and traceable around most of the ring. On 12 February 1857 he remarked: 'There is a division in the outer ring (A) answering perfectly well to Secchi's pencil line'. It was a little outside the middle of the ring and one-fifth as broad as ring A at its major axis; there was a non-uniform shading over two-thirds of ring B. After a similar view on February 14, Coolidge found on March 4 that he could see no divisions in either ring, and G. Bond could barely see the 'seam' in ring A. Twice more in that month Coolidge managed with some difficulty to detect the 'pencil line' in ring A.

From all the observations of ring subdivisions cited in this chapter it seems evident that no true gaps in fixed positions are to be found in the ring system other than Cassini's division. The subdivisions seem more like surface ripples which come and go, not always in the same places. One near the middle of ring A, usually called Encke's division,* though by no means always visible, seems to have been more often observed and more nearly constant in location than any of the others. In 1857, as will be shown in Chapter 13, Laplace's theory of the stability of the ring system was superseded by that of Clerk Maxwell, so that the division of the rings into many narrow ones was no longer required by theory.

* For both faces of the ring, because originally thought to be a gap; 'Kater's' would have been a better name for the marking on ring A's *southern* face that corresponds to Encke's on the northern face.

CHAPTER 12

HYPERION AND THE CREPE RING

During the mid-nineteenth century careful and constant attention to Saturn and its satellites by numerous competent astronomers, some amateur, others professional, in various countries, brought about in quick succession two major discoveries: an eighth satellite and a third ring.

Hyperion discovered, 1848 (W. and G. Bond; Lassell)

William C. Bond, Director of Harvard College Observatory, reported (*M.N.*, vol. 9, p. 1) as follows the discovery of a new satellite: '... On the 16th September (1848), a point of light, resembling a star of the 17th magnitude, was noticed in the plane of Saturn's ring, between Titan and Japetus, by Mr. G. P. Bond, and entered by him in his diagram of the satellites and stars in the neighbourhood. On the 18th it was again seen similarly situated, and was recorded by us both, with a doubt expressed as to its character. ...' The report added that as nearly the same appearance recurred on the 19th, micrometer measures were made (in the direction of the plane of the ring) of the object's position with respect to Iapetus and a star, and '... the measures indicated that the suspected body partook of the retrograde motion of Saturn. ...'

The stars in the path of Saturn were mapped to avoid mistakes, and measures were made on several dates from September 21 onwards, chiefly of the object's position in relation to the centre of the planet, observations being 'continued long enough to identify the satellite by its motion.' By mid-October it was found that the satellite's motion among the stars was appreciable in three hours, and the discoverers were able to estimate the revolution period at 21 days and to decide that the orbit nearly coincided with the plane of the rings.

Before, however, the news reached England, the independent discovery of the same satellite had been announced by the celebrated amateur astronomer, William Lassell. The following is quoted from Lassell's report (*M.N.*, vol. 8, p. 195): '... On the 18th September (1848), while surveying the planet in the twenty-foot equatoreal, and looking out for Japetus, (which I expected to find following the planet and not far from the plane of his ring,) I remarked *two* stars exactly in the line of the interior satellites. Not being certain at the time which of these was Japetus (although the nearer of the two certainly seemed too faint), I made a careful diagram of their positions with respect to Saturn, and also to some neighbouring fixed stars.

'The next night, the 19th, . . . I was astonished to find that the *two* stars had both moved away from the fixed stars to which they had been referred, and were still accompanying Saturn; the more distant of the two had also gone northward, in conformity with the orbital motion of Japetus, while the nearer and fainter, remaining precisely in the line of the inferior satellites, appeared to have slightly approached the planet. . . .'

This suggested to Lassell that the more distant was Iapetus, and that the nearer and fainter must be a new satellite of Saturn, and he verified the suspicion by measuring the differences of right ascension between each and a fixed star. He went on to say: '. . . as the suspected new satellite was situated precisely in the line of the satellites interior to itself, I took micrometrical measurements of its situation at two epochs, four hours apart, and was satisfied that during that interval no perceptible change whatever took place in its position in the line of the satellites. As the motion of Saturn southwards in the same period amounted to 18", he must have left the suspected satellite obviously behind if it had been a fixed star. . . .'

He therefore concluded that he had found a satellite hitherto undetected. He obtained a good set of measures of its elongation on the 21st, but after that the weather was unfavourable, so from his data he was only able roughly to estimate its period at about 24 days.

From a series of observations extending over four months G. P. Bond computed the elements of the orbit, his values for the period (21·18 days) and the eccentricity (0·115) showing that the orbit lay between those of Titan and Iapetus and was the most eccentric among those of the eight satellites then known. Lassell made thirty observations of the satellite with his 24-inch aperture reflector at Malta from November 1852 to January 1853, and from these and five observations he had made in 1850 deduced a period of 21·297 days (*M.N.*, vol. 13, p. 181). He made several more observations with the 24-inch and one with his 48-inch aperture reflector at Malta in April 1860, and deduced from them a period of revolution of 21·284 days (*M.N.*, vol. 20, p. 292). These results compare quite well with the latest *N.A.* values (*B.A.A. Hdbk.* 1960) which are 21·276665 days for the sidereal period and 0·104 for the eccentricity.

It was also Lassell who chose the name Hyperion for the newly discovered satellite, in conformity with the names Mimas, Enceladus, etc., which had been bestowed on the others by Sir John Herschel.* Lassell also rightly inferred that Hyperion is intrinsically fainter than Mimas, although he found the former easier to see because of its far greater distance from the planet, and he suspected Hyperion of being variable in brightness.

Hind pointed out that it was on the very same evening, 19 September 1848, that the true nature of Hyperion was determined both in England and America, so that the discovery of the satellite by Bond and Lassell was not only independent but practically simultaneous. Hyperion was Bond's only satellite discovery, whereas Lassell two years earlier had found the chief moon of Neptune, and in 1851 detected two additional satellites of Uranus.

Faint inner part of ring B (W. Struve, Galle)

In Chapters 6, 9 and 10 it has been related how various seventeenth and eighteenth century astronomers observed and sketched the projection of a dusky inner ring across the globe of Saturn, but without realising what it was or detecting it in the ansae, and how even special attention to the globe spaces failed to arouse any suspicion of its presence in the minds of William Herschel and Schröter. Wilhelm Struve, in 1826 when measuring Saturn, does not seem to have suspected it either, though, according to Kater (*Mem. R.A.S.*, vol. 4, p. 388), Struve did make the significant comment in a paper that ring B 'towards the planet, seems less distinctly limited, and to grow fainter, so that I am inclined to think that the inner edge is less regular than the others'.

Webb (*Celestial Objects*, p. 196) related the following story: 'In 1828 a fine Cauchoix achromatic of 6⅓-inch aperture having been placed in the Roman observatory, it (ring C) was seen, as an old assistant informed Secchi, both in the ansae and across the ball, yet, strange to say, and little to the credit of Roman science at that time, *no notice whatever was taken of it*'. It is hard to tell how much credence should be given to this story; the reference to Secchi is an obvious error, since he was only ten years old at the time (1828), and the Director of the Collegio Romano Observatory was then the French Jesuit, Domenico Dumouchel, who had found the Observatory in a very miserable state in 1824, and had obtained the Cauchoix

---

* In Greek mythology Hyperion, Iapetus, Dione, Rhea, Tethys, and also Phoebe and Themis (see Chapter 22) were Titans, the first two being brothers, the rest sisters, of Saturn; Mimas was a giant, and Enceladus either a giant or a Titan.

Iapetus used to be spelt 'Japetus'.

telescope in 1825. The brief history of the Observatory, *La Specola Vaticana* (published 1952), mentions Dumouchel's observations of Halley's comet and De Vico's observations of Mimas and Enceladus, but contains no reference to the alleged discovery of Saturn's inner ring. More remarkable still, a letter from Father Secchi of 15 January 1851 (in *M.N.*, vol. 11, p. 52) merely confirmed that he had seen in the ansae 'in a confused manner' in November 1850 the crepe ring whose discovery had been already announced by Lassell and Dawes. Had this ring been detected in his own observatory 22 years before, he would hardly have refrained from mentioning and claiming priority for the discovery.

There can be no doubt that Johann Galle, well known as the finder of Neptune in 1846, had detected the crepe ring, though in the ansae only, in 1838. His observations, made with the 9-inch refractor of Berlin Observatory were discussed in an article by Dawes (*M.N.*, vol. 11, p. 184). It seems that they were published in a paper by Encke (in the *Transactions of the Berlin Academy of Sciences* for 1838) without attracting any attention until 1851. The more important ones are as follows: 'May 25 (1838). The dark space between Saturn and his ring seemed to M. Galle to consist, as far as its middle, of the gradual extension of the inner edge of the ring into the darkness, so that the fading of this inner ring has considerable breadth.

'June 10. The inner edges of the first ring fade away gradually into the dark interval between the ring and the ball. It seemed, if no illusion exists, that the ring, from the beginning of the shading inclusive, extends over nearly half the space towards the ball of Saturn.'

In Encke's paper no reference was made to any measurements of this appearance; Galle did obtain six days' micrometrical measurements at that time but did not publish them apparently until after the announcement in 1850 of the discovery of the dusky ring by Bond and Dawes. Galle's measurements (Dawes wrote) 'show that he saw the obscure portion of the ring encroaching on the black interval between the ring and the ball to almost precisely the same extent as it appeared to myself at the last apparition of the planet, namely, about 2″.' This would make the width of the dusky ring about 8,000 miles; the modern value is about $2\frac{1}{2}″$ or some 10,000 miles. With regard to the projection of the dusky ring across the globe, Dawes commented: 'It is remarkable that no intimation is given by Dr. Galle, that this projection was seen either by himself or any other observer; nor does it appear to have occurred either to Encke, Galle, or Mädler . . . that so dull an appendage would be visible as a dark line upon the ball. But, since the ring presented at that time a very broad ellipse, such a dark line would be much broader and more visible than it was last winter (1850–51). Why then . . . was it not observed . . . at the Berlin observatory? I think we may be justified in replying, that *it actually was seen*, but that its nature was not apprehended. . . .'

Dawes went on to point out that Encke's drawings of April 1837 and March 1838 showed a broad dark shadow at the inner edge of the rings across the globe, although on those occasions the elevations of Earth and Sun above the ring-plane were so similar that in the first case the shadow would have been very narrow and in the second invisible, so that what Encke had really drawn was the projection of the shadow of the dusky ring across the globe.

Discovery of inner dusky ring, 1850 (the Bonds and Tuttle; Dawes)

The full discovery of Saturn's third ring took place in America and England in November 1850. W. C. Bond gave a short account of the observations at Harvard College Observatory entitled *On the New Ring of Saturn* (*A.J.*, vol. 2, nos. 1 and 2), the main points being as follows: 'November 11th . . . We notice to-night, with full certainty, the filling up of light inside the inner edge of the inner ring of Saturn; also . . . where the ring crosses the ball . . . below the edge, there is a dark band, no doubt the shadow of the ring. But there is also a dark line . . . above the ring, very plainly to be seen, . . . where the ring crosses the ball. . . .

'November 15th, 7^h 30^m. Examined the new ring of Saturn with different powers, best definition with 400. New ring sharply defined edge next the ball. W. C. Bond thinks he sees the new ring clear of connection with the old, but the side next the old ring is not so definite as next the planet, so that it is not certain whether the new is connected with the old ring or not. Where the dusky ring crosses Saturn, it appears a little wider at the outside of the ball than in the middle. . . . (and) . . . not so dark as the shadow of the ring below on the body of the planet.

'8 p.m. The best definition of Saturn's ring we have ever had. G. P. Bond examined with powers 140 and 400. Cannot be sure that the new ring is divided from the old one, but there can be no doubt that it exists; its inner edge is sharply defined. I did once or twice fancy, with the higher powers, that there was a division between the old and new rings. . . .'

Bond then gave the following measures of the rings, the first two figures being in close agreement with those from the *N.A.* as given in *B.A.A. Hdbk.* 1960:

	Bond	N.A. (B.A.A. Hdbk)
Outer diameter of outer ring (A)	43″9	43″96
Inner diameter of inner ring (B)	29·3	29·24
Breadth of outer ring (A)	2·3	2·64
Inner diameter of dusky ring (C)	26·3	24·12
Distance of its inner edge from old ring	1·5	2·56

To the Bonds, therefore, the dusky ring at first seemed to be much narrower in the ansae than it did to Galle and (as will be seen) to Dawes, or to present-day observers—see Bond's drawing (Plate V, Fig. 1).

Tuttle, however, on 15 November 1850 estimated the width of the dark ring as about the same as that of ring A or perhaps a little less, and the final results of the measures made by the Harvard observers seem to have agreed with this.

A point of considerable importance that appears to have been hitherto overlooked is that it was Charles W. Tuttle, an assistant at the Harvard Observatory, who first suggested that what they were observing was a dusky inner ring. This is made clear by W. Bond's footnote (*Harvard C.O.A.*, vol. 2, p. 48) as follows: 'On the evening of the 15th (November 1850), the idea was first suggested by Mr. Tuttle of explaining the penumbral light bordering the interior edge of the bright ring outside of the ball, as well as the dusky line crossing the disc on the side of the ring opposite to that where its shadow was projected on the ball, by referring both phenomena to the existence of an interior dusky ring, now first recognised as forming part of the system of Saturn. This explanation needed only to be proposed to insure its immediate acceptance as the true and only satisfactory solution of the singular appearances . . . which we had previously been unable to account for'.

The telescope with which the Bonds made their discoveries was the Harvard Merz 15-inch aperture refractor: only about a fortnight after they first saw the dusky ring it was discovered independently in England by Dawes with a refractor of less than half the aperture. Unfortunately the story of Dawes's discovery in late November 1850, and Lassell's confirmation of it on December 3, as related in *M.N.*, vol. 11, pp. 21–27 and 49–51, is rather confused and requires some sorting out. Dawes's report should, chronologically, have preceded Lassell's; Dawes added further notes and emendations to his original report; worst of all, the editor and printer produced a garbled version of the original record of Lassell's observation of December 3 introducing the words 'Bond's ring', whereas it appears that the news of Bond's discovery did not reach the English observers until December 4.

Bad weather prevented Dawes from observing Saturn until the last week of November, and at that time, as related in Chapter 11, both he and Lassell were hot on the trail of a faint division on the outer ring, and in addition Dawes was looking for 'a suspected close satellite', so neither was giving particular attention to the inner part of the ring. On November 29, however, Dawes made the following notes in his journal:

'8ʰ 40ᵐ . . . There is an exceedingly narrow black line on the ball at the *southern* edge of the ring where it crosses the planet; and it is slightly broader at the east and west edges of the ball than near its middle. It is perhaps one-third the breadth of the shadow of the ring on the ball to the *north* of the ring. What can it be? It looks like a shadow; but how can the shadow of the ring be visible both on the north and south sides of it?

'8ʰ 55ᵐ. The interior portion of the inner ring is rather *suddenly* shaded off, and towards the inner *edge* scarcely reflects sufficient light to be always sure of its outline. It has struck me, that the dark line which I see at the southern edge of the ring where it crosses the ball, is nothing else but this shaded, or rather *unreflective*, portion of the ring, which at this part is projected into a very narrow line. . . .'

From the foregoing statement it will be seen that Dawes had grasped the essentials of the matter, connecting the dusky appearance in the ansae with the dark line across the globe, although he did not then, as Bond had done, suggest that it was a third ring because he had not seen a separation between it and ring B. He also stated that he had, in fact, seen the dusky appearance in the ansae very well on November 25, but had not been able to record it at that time owing to having visitors at his observatory. His amended very careful and accurate description of the appearance runs as follows: 'A dull, slightly reflective, zone, appearing as if illuminated by twilight, or like a very dull, unreflective portion of the moon, when the sun shines upon it *very* obliquely; extending towards the ball from a dark shaded boundary at the inner edge of the bright ring, to a distance about equal to the inner portion (or rather more than two-thirds of the breadth) of the outer ring; and seen projected upon the ball as a narrow dark line of elliptic form. This zone I considered to be a continuation of the breadth of the inner ring; the dark boundary between them not suggesting the idea of a complete *separation*.'

Lassell names it 'crepe ring'

It was Lassell who, on first seeing it through Dawes's telescope on December 3, summed it up in a neat, vivid phrase and gave the dusky ring a name that has endured. This is extracted from Lassell's description, contained in a letter to the Astronomer Royal: '1850, Dec. 3. Being on a visit of a few days at Wateringbury, the residence of my friend the Rev. W. R. Dawes, the clouds . . . suddenly cleared off about 9^h0, when we . . . turned the telescope, by Merz, of $6\frac{1}{4}$ ins. aperture . . . on the planet Saturn. The eyepiece principally applied was an excellent single lens, magnifying 298 times.

'. . . After surveying the planet for some time, I was struck with a remarkable phenomenon which . . . I shall proceed, as well as I can, to describe.

'It appeared as if something like a *crape veil* covered a part of the sky within the inner ring. . . . This extended about half way between what I should have formerly considered the inner edge of the inner ring and the limb of the planet, while there was a darker, ill-defined boundary line separating this crape-like appearance from the solid body of the inner ring. There was an exceedingly thin line, or shadow, running along the southern edge of the northern portion of the ring where it crossed the planet, and this line seemed somewhat broader at each end, where it touched the limbs of the planet. Mr. Dawes had previously drawn my attention to the appearance of this line. . . .'

Lassell's eye estimate of the crepe ring extending about half way from the inner edge of ring B to the limb of the globe is in good agreement with modern measures, which are 2″56 and 2″3 respectively for the widths of the crepe ring and space.

It was Otto Struve, son of the Wilhelm Struve previously mentioned and his successor as Director of Pulkovo Observatory, who suggested the convenient designations A, B and C for the three rings of Saturn, C being the crepe ring.

Suspected division in ring C (Dawes, O. Struve)

By January 1851 Dawes thought he could distinguish two parts of the crepe ring, the inner third appearing dimmer than the outer two-thirds, and that he could detect a dark boundary between the two parts of the dusky ring, and a rather wider dark boundary or division between rings B and C; he also made estimates of breadth, finding the projection of ring C at its minor axis very narrow compared with its breadth at its major axis (*M.N.*, vol. 11, p. 52). The importance of this last observation was later stressed by E. M.

Antoniadi (see Chapter 21), but the different degree of dimness and the dark boundaries were contested by Lassell who, observing on 9 September 1851 with his much larger telescope, stated (*M.N.*, vol. 12, p. 9): '. . . No difference in depth of shade of the "veil" could be suspected, and certainly no dark division separating it into two, neither could any darker division be made out between the "veil" and the bright ring. Once, for an instant, I suspected a darker boundary, but the impression did not return. Certainly the planet is seen to-night to such advantage, that, if such divisions exist, I think I ought to see them. . . .'

It would not appear that the Harvard observers ever saw a division in the midst of ring C, and as regards a division between rings B and C, their evidence was mainly negative, though they seem to have occasionally suspected one. Otto Struve, however, examining Saturn with the large refractor at Pulkovo in the autumn of 1851, found the crepe ring at the extremities of the ansae divided into two parts by a dark line, which he believed to extend throughout the entire circuit of the ring. He noticed that the gradation of light from ring B to ring C was very abrupt and striking, but he could not discover the slightest trace of a division between the exterior edge of C and the interior edge of ring B (*M.N.*, vol. 13, p. 22).

In a letter from Valetta, Malta, referring to observations of Saturn made in late October 1852 with his 24-inch aperture reflector, Lassell wrote: '. . . Notwithstanding the exquisite views I have had of the planet since my arrival here . . . I have never obtained a single glimpse of the division in the ring C, *seen and measured micrometrically* by M. Otto Struve; nor of the ring being concentrally divided into two, as seen by Mr. Dawes. On every occasion, and when most sharply defined, it appears of one uniform texture and depth of shade, constantly conveying . . . the idea of network or a crape veil. . . . Moreover, when best seen, the part crossing the ball seemed of uniform density, the northern edge not appearing sensibly darker than the southern. . . .' (*M.N.*, vol. 13, p. 12.)

Lassell sent Struve particulars of his observations of Saturn's rings, and Struve, commenting on these in a letter to the Astronomer Royal, wrote (*M.N.*, vol. 13, p. 75): 'Besides the transparency of the obscure ring, the most interesting observation (of Lassell) appears to be the disappearance of the division in the obscure ring, which is confirmed by my own observations of the present winter (1852–3). This division being easily seen last year, and with perfect surety, there is no doubt now, that the obscure ring undergoes rapid changes in very short intervals.'

Translucency of crepe ring (Jacob; Lassell) and of Cassini's division (Lassell)

In the autumn of 1852, two years after the crepe ring was discovered, its semi-transparency was detected independently by Jacob at Madras and by Lassell at Malta (*M.N.*, vol. 15, p. 220). By 24 August 1852 a new object-glass of 6·2 inches aperture had been fitted to the refractor at the Madras Observatory, and starting on that date Jacob and his assistants made a series of observations of Saturn with powers from 174 to 365. In the course of those observations they found that the faint ring where projected on the planet '. . . has a filmy appearance, and the planet is seen through it as through a film of smoke; it has an umber brown tint; off the planet it has a slaty hue. . . .' (*M.N.*, vol. 13, p. 240).

This excellent observation with a comparatively small telescope (but from a low latitude and no doubt in very good seeing conditions) was most probably the first in which the globe was seen through the crepe ring. A careful examination of the Harvard Annals indicates that the Harvard observers did not detect ring C's translucency until late November 1853, when both Tuttle and G. Bond noted that they could see the limbs of the globe through it. Lassell, however, observing with his 24-inch aperture reflector at Malta in late October 1852, and doubtless unaware of the Madras observation, which was not reported till the following summer, called attention to the phenomenon in a letter dated 1 November 1852 (*M.N.*, vol. 13, p. 11) as follows:

'. . . One of the most striking attendant phenomena which I now note for the first time, is the *evident transparency of the obscure ring.* . . . The ring C, crossing the ball, is of a much lighter texture or colour than the other parts, and both limbs of the planet can readily be traced through it. . . . The effect is precisely that of a band of crape stretched within the ring, which, projected on the white ball, would appear of a lighter hue than when projected on the dark sky. There is evidently, also, a sudden paleness of the main division of the rings where it crosses the ball . . . as if in some degree, at least, the ball were seen through the division. . . .'

Lassell had therefore been able to detect the translucency of both the crepe ring and Cassini's division, the latter a most remarkable observation in view of the obliqueness of the division where it crosses the globe, and only possible when the rings are fairly wide open and when at the same time the Earth and Sun happen to be at approximately equal elevations above the ring-plane. His greatness as an observer is shown by the fact that he noticed this paleness of the division in front of the globe when he had apparently not been expecting or looking for it, whereas half a century later, when it was expected, many observers in various countries tried in vain to repeat his observation (see Chapter 25).

CHAPTER 13

THE NON-SOLIDITY OF THE RINGS

The belief that Saturn's ring was thin, flat and solid (as supposed by Huygens) had survived the discovery of Cassini's division, had ignored the Cassini–Wright suggestion of a congeries of particles, and had persisted despite Laplace's mathematical theory of subdivision into many narrow rings. But the revelation of the semi-transparent crepe ring at last convinced astronomers that the whole question of ring structure needed a thorough re-examination. An additional puzzle was how the presence of the dusky ring in the ansae could have been completely overlooked half a century earlier by such eminent observers as Sir William Herschel and Schröter with their large telescopes, whereas from 1850 onwards it could usually be detected in the ansae with smallish instruments and be easily seen with large ones.

Roche's 'limit' (1848)

E. M. Antoniadi has pointed out (*L'Astronomie*, vol. 44, p. 54) that even before the discovery of the crepe ring, an able young French mathematician, Edouard Roche, professor of higher mathematics at Montpellier, had worked out a mathematical theory to account for the origin of the ring system through the disruption of a fluid satellite. He calculated that, within a circle of radius 2·44 times the radius of the planet, no fluid satellite with a density equal to that of Saturn could survive undamaged, because it would be disrupted by the tidal forces exerted by the planet. He therefore suggested that a fluid body, formerly a satellite, had approached to within the critical distance and been torn apart, the debris forming the ring system. To explain the wide spread of the rings Roche supposed that before its final disruption the fluid satellite had taken the form of an elongated ellipsoid; hence the remains extend over a distance of from 1·16 to 2·26 radii from the centre of Saturn. His paper, though published in 1848 (in *Mém. de l'Acad. des Sci ... de Montpellier*), seems not to have become generally known for a considerable time, and therefore not to have influenced the other investigations that will be related in this chapter. (For a re-examination of Roche's theory by Jeffreys see Chapter 38, and for an opposing theory of origin of the rings by Kuiper see Chapter 40.)

Supposed spread of rings (O. Struve; Main, Kaiser)

Otto Struve was impressed by the improved visibility of the dusky ring, and also by a discordance between his own results in 1851, and those of his father Wilhelm Struve in 1826, regarding the measurement of the breadth of the space between the inner edge of ring B and the limb of the globe. By comparing also the measures made by Huygens, Cassini, Bradley, Herschel, Encke and Galle he reached the remarkable conclusion that the inner edge of ring B was gradually approaching the planet, the inner diameter of that ring decreasing by about one second of arc per century, equivalent to about 60 miles a year, thus markedly altering in 300 years, while the total breadth of the two bright rings was constantly increasing. From a similar comparison he also deduced that, during the interval which elapsed between the observations of J. D. Cassini and those of Sir William Herschel, the width of the inner bright ring had increased in a more rapid ratio than that of the outer ring, while the subsequent measures seemed to indicate a reversal of this process. (Struve's original memoir in *Mem. of Acad. of Sci. of St. Petersburg*, vol. 5; abridgment in *M.N.* vol. 13, p. 22.)

Struve's theory gave rise to considerable discussion. Main (in *M.N.*, vol. 16, p. 30) made a detailed comparison between his own micrometrical measures of 1852–53 and early 1854 and those of Struve, and found that they disagreed as regards the interval between the inner edge of the bright rings and the globe; he felt compelled to express the opinion 'that no change whatever has taken place in the system since the time of Huygens.' Secchi, in a letter to the Astronomer Royal (*ibid.*, p. 50) while not challenging Struve's theory directly, did illustrate from discrepancies between his own various measures the difficulty of drawing any reliable conclusions from them. A more devastating criticism both of Struve's theory and of the sources and reasoning on which it was based came from the Dutch Professor Kaiser (*ibid.*, p. 66). After a detailed examination of the observations, and especially those of Huygens, Kaiser concluded by asserting that there exists no reason whatever for supposing that the compound ring of Saturn is gradually increasing in breadth.

As a result of a further series of measures of the ring dimensions which he carried out in 1881 and 1882, Otto Struve seems to have found himself obliged to abandon his own theory, so the idea that the rings were expanding inwards came to nothing. It was, however, a noteworthy attempt to prove that the bright rings, whether whole or subdivided into narrow components, were not the rigid, solid structures that had hitherto been assumed.

Fluid ring hypothesis (G. Bond, Peirce)

A different approach to the problem was needed, and it was found by following up and carrying further the attempt, initiated by Laplace more than half a century earlier, to determine from mathematical and physical considerations the true nature of the rings. This was brought, in the 1850s, to a triumphant conclusion. The first step in the re-investigation was taken by G. P. Bond very soon after his discovery of the crepe ring. His paper *On the Rings of Saturn* (*A.J.*, vol. 2, nos. 1 and 2) began by detailing the search for ring divisions (which has been related in previous chapters of this book, especially Chapter 11). Bond pointed out the disagreement between trustworthy observers as to the number and position of minor divisions in the rings, and the fact that with some of the best telescopes in the world and under perfect definition none at all had been seen; also that telescopes which had shown distinctly several markings on the bright rings had failed to reveal the inner dusky ring, 'while the latter is now seen, but not the former'. All this, he felt, suggested the possibility that the rings were in a fluid state, changing, within certain limits, their form and position in accordance with the laws of equilibrium of rotating bodies.

Bond then analysed mathematically the attraction of the ring, assuming its mass to be uniformly distributed and not exceeding Bessel's figure of 1/118 of Saturn's mass, its density to be similar to the planet's density, and its thickness, seen from the Earth, not to subtend an angle of more than 1/45 of a second of arc. Bond himself had estimated the thickness, during the disappearance of the rings in 1848–49, as not exceeding one hundredth of a second. From his analysis, he concluded, as Laplace had done, that the rings must be irregular in shape and density to avoid falling on to the planet, but he also found that if the irregularities were too great, mutual disturbances would result and the rings would fall upon one another, and that the smallness of the intervals between them and the near equality between the rotation periods of adjacent rings would make this danger 'imminent, if not wholly unavoidable'. The nearness of the rings would make impossible a position of permanent or nearly permanent equilibrium. Bond considered that the hypothesis that the whole ring is in a fluid state, or at least not cohering strongly, would present fewer difficulties. Particles on the inner edge of a ring would have a different revolution period from those on the outer edge and could continually flow past them, and even an accumulation of disturbances would not cause disaster. This hypothesis would also account for the fact that only one division is normally seen; occasional temporary subdivisions, which might arise from local disturbances, would naturally disappear with the removal of the sources of disturbance. 'Finally, a fluid ring, symmetrical in its dimensions, is not of necessity in a state of unstable equilibrium with reference either to Saturn or to the other rings.'

Professor Benjamin Peirce, after 1838 the foremost American mathematician, discussed the equilibrium of the rings with special reference to Laplace's conclusion as to the narrowness of the rings composing the system. Assuming that they are solid, Peirce found that the breadth of each must be much less than Laplace had stipulated, so that the rings would have to be very numerous and the insecurity of the system would be much increased. This led him, in 1855, to conclude, like Bond, that the rings must be of a fluid nature.

Nature of the rings explained, 1857 (Clerk Maxwell)

The University of Cambridge announced in 1855 that they had selected as the subject of the Adams Prize Essay the following problem: To determine the extent to which the stability and appearance of Saturn's rings would be consistent with alternative opinions about their nature—whether they are rigid, or fluid, 'or in part aeriform', or are made up of 'masses of matter not mutually coherent'. The prize was won in 1857 by James Clerk Maxwell, who, following up the work of Laplace, finally disposed of the possibility of the rings being solid. He showed that the irregularity of each ring would have to be such as to place the centre of gravity more than nine times as far from the lightest as from the heaviest side of the ring, and that the load at certain points on the rings would have to be so great as to be visible as a satellite. Also the eccentricity of position of each ring must be such that the system would have an appearance quite different from that actually seen. Moreover, he showed that even with such an arrangement the slightest cause would be enough 'to destroy the nice adjustment of the load, and with it the stability of the ring'. The idea of solid rings had therefore to be rejected altogether.

The hypothesis of fluid rings, which seemed to overcome most of the difficulties, and to explain the appearance of temporary markings supposed to be subdivisions, and also the translucency of the crepe ring, was next examined by Maxwell. He found that it also was untenable because the various disturbing attractions would be bound to lead to the formation of waves, which would cause the fluid rings to break up into fluid satellites.

Maxwell therefore concluded that although the rings had the appearance of consisting of continuous sheets of matter, solid or fluid, this could not really be anything more than an appearance caused by a vast multitude of small disconnected satellites, so minute and so closely packed that, at the immense distance of Saturn, they seem to form a continuous surface. His verdict was that 'the only system of rings which can exist is one composed of an indefinite number of unconnected particles, revolving round the planet with different velocities according to their respective distances'. The theory was to receive striking verification before the end of the century by Keeler's photography of the spectrum, and early in the present century by actual observations of the translucency of the bright rings. Thus the speculative suggestion that the rings might consist of small particles, made early in the eighteenth century by the Cassinis and later independently by Thomas Wright, was proved at last by mathematical physics to be the correct explanation of the nature of Saturn's ring system.

Kirkwood's 'gaps' (1867)

In 1866 the American astronomer, Daniel Kirkwood, propounded an interesting theory to explain why it was that at certain mean distances from the Sun there was a notable scarcity of minor planets, whereas at distances a little less or greater these small bodies were fairly plentiful. He pointed out that the gaps occurred at mean distances where, if planetoids had been revolving round the Sun, their revolution periods would be simple fractions of Jupiter's period. They would therefore pass fairly near to the great planet once in every few revolutions and be perturbed, and the cumulative effect of the repeated perturbations would force such minor planets into new orbits. The gaps among the planetoids are conspicuous at distances where their orbital periods would be 1/3, 2/5, 3/7, 1/2 and 3/5 of Jupiter's period, and there is a very large group of them with a period just under half that of Jupiter.

In the following year (1867) Kirkwood applied his theory to the particles of Saturn's rings. In this case the perturbing bodies would be the inner satellites Mimas, Enceladus, Tethys and Dione, which, revolving round Saturn comparatively near to, and almost in the plane of, the rings, and being of great mass in comparison with the tiny ring particles, could easily produce a very disturbing effect on them. For example, a ring particle having a revolution period equal to half that of Mimas, about a third that of Enceladus, a quarter that of Tethys, and a sixth that of Dione, would be perturbed by one or other of those satellites once or twice a day, and would very soon be forced into a different orbit. As a matter of fact a particle with such a period (11·3 hours) would be situated within Cassini's division, and so Kirkwood's theory gives a very attractive explanation of the existence of that prominent gap in the ring system, and also of the brightness of the outer part of ring B, where there may be a congestion of particles forced out of the division, with periods a little less than half that of Mimas, just as the very numerous Hecuba group of minor planets is found with periods just under half that of Jupiter.

According to Antoniadi (*L'Astronomie*, vol. 44, p. 163) Meyer extended Kirkwood's theory about the Cassini division to the next two satellites, suggesting that a particle in the division would have a period 1/9 that of Rhea and 1/33 that of Titan, but it seems evident that the perturbations by Mimas are the preponderating factor. In 1871 Kirkwood extended his theory to explain the existence of the marking a little outside the middle of ring A known as Encke's division, which though not a gap in the ring probably indicates a sparser distribution of particles (*Procgs. Amer. Phil Soc.*, vol. 12, p. 163).

It will be noticed that Kirkwood applied his theory only to Cassini's and Encke's divisions, but it has since been extended, especially by Lowell, to other places in the ring system where particles would have periods that are simple fractions of that of Mimas or that of Enceladus. Two other 'Kirkwood gaps' have been suggested (see Chapters 29, 37) in the middle and outer part of ring A; three somewhat outside and three inside the middle of ring B; and also one at the junction of rings B and C. The extension of Kirkwood's theory therefore supplied observers with a new motive for seeking faint markings that might indicate a certain sparseness of particles, on the bright rings; the former expectation of ring gaps based on Laplace's theory had been removed by the work of Clerk Maxwell.

CHAPTER 14

RING PHENOMENA IN THE 1860s AND 1870s, AND IDEAS ABOUT SATURN'S PHYSICAL CONDITIONS

Ring breaks and overlapping light, 1861–62 (Jacob; Wray)

The Earth passed through the plane of Saturn's ring on 22 November 1861 and until the Earth re-crossed the plane on 31 January 1862 to the same side as the Sun, the ring had theoretically disappeared. Capt. W. S. Jacob, in England at the time, made two drawings of Saturn, the first (Plate V, fig. 2) on November 12, shortly before the 'disappearance', showing the ansae of the almost edgewise ring broken into a number of short lengths (*M.N.*, vol. 22, p. 89). Jacob admitted that the seeing conditions were not very good, so that the inequalities in the ring might have been partly due to want of definition, but he found those inequalities most conspicuous at moments when the definition seemed at its best. His second drawing, made with a 9-inch aperture Cooke object glass on December 4, showed no sign of the ring's ansae, but he remarked: 'The excellence of its defining power may be judged of from the fact of its separating the dark streak across the planet into two parts, with an almost inconceivably fine line of light between, viz., the ring itself and its shadow. The breadth of the former could not have been much more than 0″04. . . .'

On 17 December 1861, when the dark side of the rings was turned towards the Earth and the Sun was elevated about 2° above the other side, William Wray directed his 7-inch aperture refractor at Saturn, expecting to be unable to detect any trace of the ring with so small a telescope, but he recorded: '. . . to my utmost surprise, with power 110, I instantly perceived the whole of the edge (*sic*) of the ring, not only where it crossed the dark shade on the body, but also extending on each side of the planet's margin. It was irregularly broken'. What he saw was the dark side obliquely, not the edge. Continuing his observations (*M.N.*, vol. 23, p. 85), he saw the nearly edgewise ring distinctly again on December 23, '. . . and suspected (it) to be thicker and somewhat nebulous about the region on either side, where it joined the planet's limb'.

On December 26 the image of Saturn was 'exquisitely steady and well defined'. At 16^h 30^m, the eastern ansa was scarcely visible, the western obvious and broken in two places. The brightness of the eastern arm gradually increased and after two hours the two ansae were equally bright; half an hour later still, the eastern was the brighter and the last to disappear in the morning twilight. This is rather reminiscent of Maraldi's observation in October 1714 (see Chapter 6). Wray also made the following interesting note on his observation of 26 December 1861: 'A prolongation of very faint light stretched on either side from the dark shade on the ball, overlapping the fine line of light formed by the edge of the ring, to the extent of about one-third its length, and so as to give the impression that it was the dusky ring, very much thicker than the bright rings, and seen edgewise, projected on the sky'. (See Plate V, fig. 3).

On four dates from 4th to 18th January 1862, he obtained further good views of the almost edgewise ring and the faint overlapping light. The western ansa was much broken but the brighter on January 4, whereas the next day it was at first much fainter and shorter than the eastern one, though in the course of three hours it became equally bright, and shortly after that looked decidedly the brighter of the two ansae.

Wray added: 'The fine thread of light seemed to me always of a bluish-white tinge, very different in colour to the ball of the planet'. His two drawings (of December 26 and January 5) both show the thin knife-edge of light of the ring not only in the ansae but also through the middle of the black trace across the globe, thus agreeing with the impression noted by Jacob on December 4. Commenting on Wray's observations of the faint overlapping light, Proctor expressed the opinion that it was 'the most singular phenomenon detected in the appearance of the rings, since the discovery of the dark ring'.

Disappearing ring and Titan's shadow transits (Dawes)

A transit of Titan's shadow across the disk of Saturn was due to take place on 15 April 1862, and the Rev. W. R. Dawes was anxious to observe it, because conditions had prevented his seeing similar occurrences in 1848 and 1849, and it did not appear that anyone else had been more successful since Sir William Herschel's observation of 2 November 1789 (see Chapter 8). Dawes, on 15 April 1862, was able to watch and time the progress of the transit from the ingress of the shadow onto the disk until after it had passed the central meridian, and to take a number of micrometric measures of the distance between Titan and the shadow. The latter seems to have moved along the southern edge of the north equatorial belt, so it was not close to the thin ring. Dawes was surprised at the shadow's large size which he estimated at not less than $0''8$, and from which he inferred that under good conditions a $3\frac{3}{4}$ or 4-inch aperture refractor should suffice to show it. He gave computed dates and times for seven further shadow transits of Titan due to take place that year (*M.N.*, vol. 22, p. 264). Confirming Dawes's opinion of the shadow's visibility, S. Gorton reported (*ibid.*, p. 294) that he had watched the transit quite easily with a telescope of $3\frac{1}{8}$ inches aperture.

Though clouds prevented a sight of the shadow transit of May 1, Dawes had a fine view of that of 17 May 1862, a date on which the Sun passed through the plane of the rings, causing a further disappearance. Though he had made no notes on the aspect of the ring on 15 April, on this occasion Dawes carefully examined the features of Saturn while waiting for the ingress of the shadow, and recorded (*ibid.*, p. 297) the following impressions: 'The arms of the ring were scarcely at all visible; a very faint gleam of coppery light, at moments of finest vision, being the only indication of its existence beyond the disk of the planet. It has since been seen, on several occasions, much more plainly. On the disk the projected ring appeared as a very dark line a little north of the equator, and of uniform breadth. But I was much surprised that, under the finest definition with this high power (620 on his $8\frac{1}{4}$-inch aperture refractor), I could discern no trace of the *shadow of the ring*. I expected to see it, if the atmospheric circumstances were sufficiently good, as an *exceedingly fine black line*, stretched across the disk about a quarter of a second to the south of the inner edge of the projected ring; and that the shadow of the satellite would travel almost centrally on the black line—a great part of it, however, falling on the southern portion of the ring. But no such thing was to be found. . . .'

He watched and timed the ingress of the shadow, finding that it projected rather more than half its diameter to the north of the ring, and a barely appreciable portion to the south of the trace. He was able to watch it until mid-transit of the disk. It seemed to him that the shadow looked much darker than the trace of the ring; the latter looked like a line drawn with a soft lead pencil, while the former by comparison resembled a spot of ink. He inferred from the position of Titan's shadow and from the positions of some of the inner satellites that their orbits did not lie exactly in the plane of the rings. He deduced from his observations of April 15 and May 17 that the synodic period of revolution of Titan is 15·96684 days, less than one minute shorter than the modern value. He considered that nothing could 'more fully prove the *almost inconceivable thinness of the ring* than the absence of all perceptible shadow'.

Under unfavourable conditions Dawes was able to watch and time the disappearance of Titan when eclipsed by Saturn's shadow on 25 May 1862; it was the first eclipse of Titan he had seen.

On June 2, he witnessed the earlier part of yet another transit of Titan's shadow; this time he found the shadow projecting about as much from the south side of the ring-trace as on May 17 it had projected from the north side. William Huggins, an eminent pioneer in astronomical spectroscopy, who had obtained on both occasions brief glimpses, between clouds, of the transit, confirmed Dawes's impressions as to the situation of the shadow with respect to the ring-trace (*M.N.*, vol. 22, p. 296).

Light beads and strange colours (Huggins, Birt, O. Struve, Carpenter)

It was computed that at 19^h on 17 May 1862 the Sun passed through the ring-plane from the southern to the northern side, and so there was another period of theoretical disappearance of the rings until August 12, when the Earth made its third passage of the plane, also to the north side, Earth and Sun thereafter remaining north of the plane until early in 1878. Huggins made a series of observations of Saturn with an 8-inch aperture refractor, using powers from 200 to 950, commencing on May 2, when he saw two bright dots of light on the eastern ansa, which he at first took to be satellites. These dots continued to be visible after the ansae had faded from view in the haze; they were not satellites as none were near those positions on that evening.

On May 12 the points of light were again seen, best with averted vision, their light being less dimmed by haze than that of the ansae. On May 13 Huggins distinctly traced them on the eastern ansa 'to correspond in position with the inner and outer edges of the extremity of the ring', one at what would be the inner edge of ring C, the other at 'the extremity of the visible portion of the ring'.

On May 16 he could still see the ansae, of a deep bluish purple colour, but no indication of the bright points. He thought he detected across the globe, close to the southern edge of the projected ring, 'a badly defined line of faint shading', in the position where the shadow of the ring would be expected, but in spite of careful search he saw no further indication of the shadow across the globe during the ensuing month. On May 17, 18 and 19 he could still see the ansae, though with increasing difficulty; on May 20 the ansae looked 'a beautiful dark blue colour, scarcely distinguishable from the dark blue of the sky'.

On several occasions, especially May 18, Huggins suspected 'little roughnesses, resembling miniature satellite-shadows' on the southern edge of the ring on the globe, but they disappeared when high powers were used, and with high powers at best moments the ring's edge appeared to be perfectly even and unbroken.

William R. Birt, best known as a lunar observer, on 13 May 1862 saw, with an 8-inch aperture refractor, '. . . very distinctly the fine ansae broken into minute points or beads of light' . . ., and he made the number of the most prominent bright points to be about three on the preceding ansa and four on the following one; another observer who was with him counted five on the following ansa (*M.N.*, vol. 22, p. 295). Birt considered that the following ansa was the longer, and added: '. . . On one occasion I obtained a very fine view, when they appeared as *irregularities or projections on the illuminated plane of the ring*, as if such projections were more directly illuminated, and consequently reflected more light than the obliquely illuminated plane of the ring'. He did not see this appearance again. He saw the trace across the globe as 'a somewhat narrow dusky line . . ., the northern portion being rather the darkest. Mr. Talmage, who was present, saw the fine line of light separating the two . . .', but Birt did not detect this owing to deterioration of the seeing conditions.

During the epoch of the Sun's passage through the ring-plane, Otto Struve and Winnecke, with the large refractor at Pulkovo Observatory, saw luminous appendages on the ansae of the rings similar to those seen by Wray in the previous December and January, although the elevation of the Earth above the ring-plane was then more than five times as great as it had been at the time of Wray's observations. Extracts from Struve's report published by the then Astronomer Royal (G. B. Airy) show that Struve on 15 May 1862 saw this luminous appearance only on the southern side of the ansae: 'On the north side it is more glittering, and limited by a sharp edge. On the south side, on the contrary, it appears as if clouds of a less intense light were lying on the ansae. . . .'

On May 17 and 18 Struve still saw the ansae distinctly, and on the 19th he wrote: 'I see distinctly traces of the ring, especially on the preceding side. The luminous appendages extend on the preceding side to 0·6, and on the following side to 0·3 of the diameter of the planet. Their colour differs much from the ordinary colour of the ring. They are not yellow, but more of a livid colour, brown and blue.' The aspect of the appendages was the same on May 20 except that the one on the following side seemed longer, about 0·5 of the diameter. On the 21st, the intensity of the light seemed much feebler on the following side, but Struve also noted: 'The size of these appendages increased in the neighbourhood of the planet, giving them the form of sharp wedges'. He saw them less well on May 22, and they were still distinctly visible on June 3 (*M.N.*, vol. 23, p. 87).

Alongside the extracts from Struve's report the Astronomer Royal had published notes of the observations on corresponding dates, made with the Greenwich equatorial by James Carpenter, then an assistant at the Royal Observatory, and later well-known for his book in collaboration with J. Nasmyth on 'The Moon', which expounded the volcanic theory of the origin of lunar craters. Carpenter and his colleagues observing at Greenwich during the same period as Struve saw no sign of 'luminous appendages'; Proctor has suggested that this may have been due to difference in atmospheric conditions between Greenwich and Pulkovo at the time. Both Struve and Carpenter agreed with Dawes as to the coppery colour of the ansae about mid May 1862. Carpenter noted on May 17: 'The ring beyond the planet at times just visible on the left (preceding) side of the disk, but on the right only a small faint spot could be seen in the plane of the ring, about 1/3 the diameter of the disk from the planet's limb. I should have suspected this was a satellite but for its elongated shape.. ..' On May 19 he found the preceding ansa more distinct than the other, and the under (north) edge of the trace across the globe much sharper and better defined than the other edge. The next night the ring appeared to be brighter; Carpenter could still see it on May 25 and June 1, while on June 3 it was very plainly visible, and he noted: 'During moments of very good definition I fancy I can see the flat surface of the ring very faintly illuminated.'

It is rather difficult to tell how much reliance can be placed on the colour notes quoted in this Chapter, since the observations were all made with refractors.

Crepe ring unusually bright, 1863 (Carpenter)

A brief note (*M.N.*, vol. 23, p. 195) refers to a very interesting observation made by Carpenter on 26 March 1863. He was watching the passage of Saturn across the field of view of the Royal Observatory's transit instrument, and wrote: '. . . it struck me that the dark space between the ring and the ball seemed much contracted. Upon looking at it with the Equatoreal I found that this arose from a great increase in the brightness of the dusky ring, which appeared nearly as bright as the illuminated ring, and might easily have been mistaken for a part of it'. At the time of the observation, the Earth was, according to Proctor, at an elevation of about $4\frac{1}{3}°$ above the ring-plane, but the Sun was at a lower elevation; hence the crepe ring was being very obliquely illuminated, and, although semi-transparent, would, under these conditions, reflect better than it usually does and hence not differ greatly in brightness from the bright rings.

Lull in observations

During the rest of the 1860s, and (apart from the work of Trouvelot) most of the 1870s, there seems to have been a temporary decline of activity in the observation of Saturn. Two of the main causes of this lull are not difficult to find. For about five years around 1869 the southern declination of the planet must have placed it too low in the sky for satisfactory observation from northern latitudes. Then again, the intense activity and interest in Saturn shown during the mid-nineteenth century had depended greatly on a few leading observers: of these, William Bond had died in 1859, his son in 1865, and Dawes in 1868, while Lassell seems, after 1864, to have given up regular observing. For a time it may have seemed that there were none to replace them, but Trouvelot started a renewed interest in Saturn which, during the last quarter of the century, brought valuable contributions from such great planetary observers as Asaph Hall, Holden, Hermann Struve and Barnard, using some of the world's largest telescopes, as well as Denning, Elger, Stanley Williams, Freeman and Antoniadi.

Bands in the spectrum, 1863 (Secchi, Huggins)

Another kind of observation at this period resulted in a discovery which eventually proved (see Chapters 25 and 35 and Plate XIV) to be of fundamental importance to knowledge of the atmospheres of Saturn and the other major planets. The great Italian astronomer, Father Angelo Secchi, a number of whose observations of Saturn have already been related, was also a pioneer in stellar spectroscopy and in the classification of stellar spectra. Applying the spectroscope to the major planets, Secchi found wide, dark absorption bands, especially towards the red end of the spectrum, and inferred that their atmospheres were 'not yet cleansed' and contained elements different from the Earth's atmosphere (Abetti: *History of Astronomy*, p. 189). According to Antoniadi (*L'Astronomie*, vol. 44, p. 10), Secchi made this discovery in the spectra of Saturn and Jupiter in 1863, anticipating by a year or two the announcement of the same discovery by another great pioneer of spectroscopy, Sir William Huggins. In 1869 Secchi found similar but even wider bands in the spectra of Uranus and Neptune.

Ring peculiarities, 1873–76 (Trouvelot)

Etienne Trouvelot, a French astronomer who spent thirty years in America, made a series of observations of Saturn devoting particular attention to the rings, during the years 1873–76, using the 15-inch aperture refractor of Harvard College, the 26-inch aperture Washington refractor, and his own refractor of 6-inch aperture; his results were summarised in *M.N.* (vol. 37, p. 191). As he used the letters A, B, and C for parts of the bright rings, the summary given here will be adjusted so as to use A for the outer bright ring, B for the inner bright ring and C for the crepe ring, in the usual manner. The results fall into four groups: (i) bright rings; (ii) shadow on rings; (iii) globe; (iv) crepe ring.

With regard to the bright rings, Trouvelot considered that during the four years of his observations the surface of ring A and that of the outer part of ring B had shown a mottled or cloudy appearance on the ansae. He also noticed on the inner border of ring A, adjoining the Cassini division, 'some singular dark angular forms' on the ansae which he thought might be due to 'an irregular and jagged conformation ... either permanent or temporary' of the inner border of A.

As to (ii) the shadow of the planet on the rings, he considered that the form of the shadow proved that the thickness of the ring system increases from the inner margin of ring C to the outer edge of B. Because of the rapid changes he observed in the indentation of the shadow, he was of opinion that the 'cloud-forms seen near the outer border' of ring B 'attain different heights and change their relative position, either by the rotation of the rings upon an axis, or by some local cause'.

Trouvelot considered that Saturn, like Jupiter, is less luminous near the limbs than in the more central parts, the brightness gradually diminishing towards the limbs.

Perhaps his most interesting conclusions are those concerning (iv) the crepe ring. He found that where ring C is projected on the disk the inner portion of the ring disappears in the light of the planet; that, contrary to all previous observations, the crepe ring is not transparent throughout, but 'grows more dense as it recedes from the planet, so that at about the middle of its width, the limb of the planet ceases entirely to be seen through it'; and that here and there the matter composing the crepe ring is condensed into small masses, which almost entirely cut off the planet's light in those places from the eye of the observer.

Some of the effects seen by Trouvelot are now considered to be illusory: a jagged edge of one of the rings suggests atmospheric unsteadiness at the time and place of observation; irregularities in the shape of the globe's shadow on the rings were proved (1914) to be an illusion, optical or atmospheric or both (see Chapter 29).

Nearly edgewise rings, 1877-78 (Trouvelot, Hall)

Trouvelot made 221 observations during 1877-78 prior to the disappearance and after the reappearance of the rings, using a 6·4-inch aperture refractor, with powers from 85 to 460. His main conclusions (summarised in *B.A.A.J.*, vol. 1, p. 343, from *L'Astronomie*, April 1891) were as follows. He found the rings less intrinsically brilliant than the globe for nine months before the ring-plane's passage through the Sun on 6 February 1878 and that they gradually acquired an orange tint. After the passage the brilliancy of the rings gradually increased until by the end of a further ten months they were again brighter than the globe. He considered that these changes were due to the varying position of the Sun relative to the ring-plane. Prior to the disappearance he also noted a gradual invasion of shadow over the illuminated part of the rings. As the Sun's rays became very oblique, from 18 December 1877, when the elevation of the Sun was not more than 45 minutes of arc, Cassini's division became decidedly more conspicuous in the eastern than the western ansa. He also noticed that for a period of about a week the nearer part of the ring by the east limb of the globe seemed only about a quarter of the width of the further part which adjoined the globe's shadow across the rings. On 5 February 1878 he detected the rings with great difficulty as a thin bright line; by 6^h on February 6 they were invisible; on the 7th traces of them had become visible again. He also noted apparent deformities of the planet's limb during the observations.

Professor Asaph Hall, observing with the Washington 26-inch aperture refractor, also completely lost sight of the rings on 6 February, 1878, and on the next day he noted: '... At times when nearly edgewise, the ring had the appearance of being broken up into small parts, but this was not steadily seen. . . .' His impressions on those two dates were therefore in complete agreement with those of Trouvelot (*B.A.A.J.*, vol. 1, p. 443).

Changing ideas about physical conditions, 1865–82 (Proctor)

It seems to have been during the period covered by this chapter that a fundamental change was taking place in the opinion of astronomers about the physical nature of the great planets. In his chapter on the 'Habitability of the Giant Planets' in his famous book *Saturn and its System*, R. A. Proctor has stressed the earlier revolution caused in the attitude of astronomers by the establishment of the Copernican theory and still more by the invention of the telescope: what had previously been regarded as mere lights in the sky were recognised as a group of worlds, of which the Earth was one, and it was generally assumed that the other planets also probably supported life, possibly intelligent life. This view was certainly held by the Herschels—witness the natural way in which Sir William Herschel referred to the 'Saturnian' in the vivid sentence quoted at the end of Chapter 9 of this book. When he published the first edition of *Saturn and its System* in 1865 Proctor was still subscribing to that opinion, and deeming it incredible that so splendid a world as Saturn should be devoid of inhabitants, but even then the study of the observations required for the writing of his book had caused him misgivings, and he added a note of warning that the physical conditions on Saturn did not seem to make the planet a suitable habitation for beings constituted like the inhabitants of the Earth.

Proctor went on to say that when he began his book *Other Worlds than Ours* (published 1870) he still had the idea that each planet is inhabited by some kind of living creatures, though he had recognised that it was impossible that all, and doubtful if any, of the other planets were populated by creatures similar to those on Earth. He mentioned this to show that his attitude was conservative; 'all my prejudices', he wrote, 'were likely to favour the old-fashioned views' rather than new ideas.

In an article on 'The Planet Saturn' written in 1870 (Proctor: *Essays on Astronomy*, 1872, p. 87) he carefully considered the many differences shown by observation between the atmospheres of Saturn and the Earth, including the lack of any evidence of diurnal and seasonal changes such as are displayed by the Earth and Mars, and he adopted the new view that under its veil of atmosphere

Saturn was a globe of incandescent heat. In the second edition (1882) of *Saturn and its System* he put forward still another powerful argument—the low mean densities of the major planets. He rejected out of hand the alternative explanation, revived with such powerful effect in the 1920s (see Chapter 33), that the major planets might be composed of different materials from the Earth, and decided in favour of the theory that Saturn and the other great planets were hot, distended, gaseous globes that had not yet cooled down and solidified sufficiently to support life. That this theory was so widely and firmly held during the ensuing half century must have been due in no small measure to the deservedly great popularity of the writings of R. A. Proctor.

CHAPTER 15

ROTATION OF GLOBE, AND NATURE AND ROTATION OF RINGS

(A) ROTATION PERIODS OF GLOBE SPOTS

It has been related how at the end of the eighteenth century Sir William Herschel had been able to deduce a rotation period for Saturn's globe from the observation of slight differences in the 'quintuple belt', and how Schröter had tried to do so from the observation of spots (see Chapters 9 and 10). It seems that during the nineteenth century no one had been able to observe any spots sufficiently to make a like attempt until 1876, when the American astronomer Asaph Hall succeeded—the year before he became renowned for the discovery of the two satellites of Mars.

Bright equatorial spot of 1876 (Hall)

It was on 7 December 1876, while observing Iapetus, that Hall noticed a bright, well-defined spot in the equatorial zone of Saturn, with an apparent diameter of 2 to 3 seconds of arc. It continued to be visible for some weeks, ultimately extending on one side into a streak. The day after he first saw it Professor Hall wrote to other astronomers asking for their co-operation in observing it, and enclosing an ephemeris which stated the approximate times at which it should be looked for on subsequent dates. Unfortunately (as stated in *M.N.*, vol. 38, p. 209) the ephemeris was computed by assuming a rotation period for Saturn which had been quoted in nearly all the nineteenth-century text-books on astronomy as that finally adopted by Sir William Herschel as the most accurate determination. In fact, however, the figure given in those text-books had never been suggested by Herschel and was more than 13 minutes longer than the period he genuinely deduced in 1794, which was 10^h 16^m 0^s4 and which Herschel considered to be correct to within much less than 2 minutes either way. At the time of the discovery of the white spot Herschel's genuine rotation period for the globe was unknown to Hall, who afterwards traced the origin of the incorrect figure to a value given by Laplace for the period of rotation, not of the globe, but of the rings.

The ephemeris was of course worse than useless because an error of over 13 minutes in the assumed rotation period would be quite sufficient to make all attempts to observe the spot in transit on subsequent dates miscarry. Seven rotations of Saturn take about 20 minutes less than three rotations of the Earth, so that a spot observed on Saturn should be looked for again at intervals of three days, but about 20 minutes earlier on each occasion, e.g. if it was central at 10 p.m. on the first day, it should also be central on the planet's disk at about 9.40 p.m. on the fourth day, 9.20 p.m. on the seventh day, 9 p.m. on the tenth day, and so on. An error in the assumed rotation period of about 13 minutes would amount to about 1½ hours in seven rotations of Saturn or three terrestrial days; hence the use of the incorrect value would produce an ephemeris suggesting to observers a time 1½ hours too late for the transit on the fourth day, 3 hours too late on the seventh day, and so on with a large and cumulative error. This is rather a good illustration of the trouble that can be caused to observers by the careless insertion of an erroneous figure in one text-book, since the error is likely to be perpetuated by being copied in other subsequent books.

Happily, as it turned out in this case, the spot was a long lasting one, its brilliant whiteness made recognition easy, and evidently some at least of the observers were on the watch for it in good time. Therefore, in spite of the faulty ephemeris, Hall was able to base his calculations, not only on his own observations with the Washington 26-inch aperture refractor, but also on those of five other American observers. The nineteen observations of the spot in transit over the central meridian of the disk extended over 61 rotations of the planet or nearly four weeks (*M.N.*, vol. 54, p. 297).

The value he deduced for the mean time of the rotation of Saturn, assuming that the spot had no proper motion on the surface of the planet, was 10^h 14^m 23^{s}8, with a probable error of only 2½ seconds. The spot was situated near the planet's equator. (For a fuller account of the observations see *A.N.*, No. 2146.)

Method of observing spots on Saturn (Williams)

Spots on Saturn are rather rare and the next outbreaks did not occur until the 1890s. Meanwhile Arthur Stanley Williams, an English solicitor and amateur astronomer, had gained years of experience (since 1877) in regular visual transit observations of the spots on Jupiter, in the course of which he had developed a simple but effective technique that has always been the chief basis of the work of the B.A.A. Jupiter Section. Through his skill, acuity of vision, accuracy and quick perception of delicate detail Williams succeeded by 1896 in identifying nine separate atmospheric currents on Jupiter, though nearly all his observations were made with as small a telescope as a 6½-inch aperture Calver reflector.* This is mentioned to show that, when spots began to appear on Saturn in 1891, there was an expert ready to deal with them.

In his paper *On the Rotation of Saturn*, dealing chiefly with the Saturn spots of 1893 (*M.N.*, vol. 54, p. 297), Stanley Williams gave an instructive account of his methods of determining the positions of spots:

(1) *Direct transits*: recording as exactly as possible the time at which a spot seemed to be bisected by the central meridian of the disk. This—his usual method with Jupiter—he found could not be relied on wholly with Saturn, because Saturnian spots were usually so delicate that a deterioration in seeing at the critical time could render them indistinct for ten minutes or more, so that many direct transits were lost. To compensate for this and increase the data and therefore the accuracy of the final results, he also used

(2) *Estimated transits*—An experienced observer, becoming familiar with the rate at which spots appear to move across the disk through the planet's rotation, can make quite an accurate estimate of the time of c.m. transit for a spot seen at a short distance (not much more than 20 minutes in time) preceding or following the centre; such estimate can be made at any instant when a spot, fairly near the c.m., is clearly seen.

(3) A third, less accurate method, only used in a few exceptional cases, is to record the time when it appears to transit some other meridian, e.g. one at 30° from the c.m.

Williams tried to get a direct transit and at least three or four estimated ones for a spot on each night when this was possible. He regarded the mean of the estimated transits as having the same value as one direct transit, and took the simple mean of the two as the transit time.

Williams stressed the importance of taking great pains to avoid any possibility of bias from recollection of previous observations: experience had shown him that errors of as much as a quarter of an hour in the time of a transit could arise if the observer had a preconceived idea from recollection of previous results of the time when a spot ought to be in transit. When observing therefore he deliberately tried to forget the records of previous nights. He also took precautions to avoid the influence of preconceived ideas in the final

* See B. M. Peek, *The Planet Jupiter* (Faber and Faber, 1958).

reductions of the observations. While observations were in progress he purposely avoided all reference to those previously recorded, and did not commence reductions till the observations of the period were completed. He reduced each observation separately, weighting it partly according to the satisfaction recorded at the time it was made and partly according to the number of observations. He did not attempt to compare the concluded times of transit till all had been separately worked out and weighted in a uniform manner, and in no case subsequently altered any of them.

The system of longitudes for Saturn used by Stanley Williams to chart his observations was provided by Albert Marth, who for over thirty years furnished ephemerides that were of great service to lunar and planetary observers; his ephemerides were at that period a regular feature of the *M.N.* Marth was German by birth and had been a pupil of Bessel; later he had assisted Lassell in Malta, and afterwards became director of an observatory in Ireland; to some extent therefore he was a link between the astronomical activities of the earlier, middle and late nineteenth century.

Bright equatorial spots of 1891 and 1892 (Williams)

Stanley Williams, in the introduction to his paper on the spots of 1893, stated that in the spring of 1891 he detected a number of bright spots in the bright equatorial zone, south of Saturn's equator, and that from the motion of four of these bright spots and one dark one, Marth had worked out an approximate rotation period of 10^h 14^m 21^s84. W. F. Denning made the period slightly different: 10^h 14^m 26^s6, deducing it from the observations of ten bright equatorial spots, using only the extreme observations of each, and an approximate method of reduction.

A great many observations of these spots were made by Williams during the 1892 opposition, and they were seen by two or three other observers. Marth's approximate rotation period for them was 10^h 13^m 38^s4, and he accordingly based his ephemeris for 1893 on that figure.

The 1893 spots were of two kinds: dark ones on a conspicuous double belt (presumably the N.E.B.) in the northern hemisphere; bright ones in the equatorial zone.

Dark northern spots, 1893

There were numerous small dark spots lying between saturnicentric latitudes 17° N. and 37° N., and Stanley Williams found them to be nearly all in pairs, one on each component of the double belt, one in each pair being almost due north of the other (see Plate VI for Williams's drawing). On good nights they were seen thus very distinctly, but with poor definition each pair appeared rather as one large diffuse spot extending over both components right across the double belt. Quénisset and Rudaux in France also observed some of these spots, and Henderson at Liverpool one of them. Williams considered that the darkest and best defined were comparatively easy to observe, and that the results were on the whole in very satisfactory accord. He gave in his paper a full table of the observations and rotation period of each spot.

Omitting from his summary those spots observed less than four times, he found that the others fell, as regards rotation period, into two main groups of five spots each, with one spot having a period (10^h 14^m 40^s3) intermediate between those of the two groups. The range of period for one group was from 10^h 14^m 27^s7 to 10^h 14^m 32^s1; that of the other group (on the opposite hemisphere* of Saturn) from 10^h 14^m 52^s2 to 10^h 15^m 3^s7. He obtained the following mean periods for the groups:

Class A	10^h	14^m 29^s1
Class B	10	15 0·7

showing that atmospheric material was rotating round Saturn in the same latitude half a minute faster on one hemisphere than on the other. He remarked that though similar differences were common on Jupiter, one was rarely found there so considerable and so well marked.

Bright equatorial spots, 1893

These spots were slightly brighter than the equatorial zone in general and much like those so characteristic of Jupiter's equatorial zone. Lying between saturnicentric latitudes 6° N. and about 2° S., they were roughly circular in shape but very indefinite at the edges and averaged perhaps a little less than 2 seconds of arc in apparent diameter. Occasionally they were recorded as very white but more usually slightly yellowish in tint and seemed variable in brightness like the spots on Jupiter. Stanley Williams made most of the observations, but received satisfactory data from four other observers: H. MacEwen (for many years Director of the B.A.A. Venus and Mercury Section) and G. L. Brown in Scotland, the Rev. P. H. Kempthorne in England, and F. Quénisset at Juvisy, France. The telescopes used ranged in aperture from 5 to $10\frac{1}{4}$ inches. As on the whole these bright spots were more indefinite than the dark ones, the observations did not accord quite so well.

* In the same northern latitudes.

Stanley Williams's paper (*M.N.*, vol. 54, pp. 297–314) gave full details for each spot, and the following is his summary (omitting margins of error—all less than 1 second) with the spots arranged in order of increasing longitudes:

Class	No.	Period of Rotation
	5	10^h 12^m 55^s0
	2	10 13 1·8
C	3	10 12 58·3
	1	10 12 59·4
	4	10 12 45·8

(The number of dates during 3 or 4 months on which transits were recorded for these spots ranged from six to ten.)

Williams regarded the first four of the spots as evidently forming a group, and gave the concluded rotation period for class C as 10^h 12^m 59^s4, over 13^s longer than that of the last spot, which appeared to be representative of a rather extensive region of longitude (about 220° in length). A comparison with the figures for 1892 and 1891 showed that there had been a continual acceleration in the motion of the bright equatorial spots:

Year	Period of Rotation
1891	10^h 14^m 21^s8
1892	10 13 38·4
1893	10 12 59·4

Williams pointed out that though similar changes are not uncommon on Jupiter, they rarely occur so rapidly: one such change in the corresponding region of Jupiter took eight years for an increase in the rotation period of half a minute. In his opinion such changes might occur on Saturn either by sudden alterations in the rate of motion of individual spots, or by old spots dying out and new ones successively appearing endowed with swifter or slower motion. What does seem clear is that the ascertained rotation periods are those of groups of spots or atmospheric currents on Saturn, and that they only give an approximate indication of the true rotation period of the equatorial region of the planet itself.

Spots of 1894 (Williams)

Stanley Williams gave a detailed account (*M.N.*, vol. 55, pp. 354–367) of the spot observations of 1894, stating that the other observers were MacEwen and Brown as before, and also L. Brenner (who sent measures made from drawings); agreement was even more satisfactory than in 1893.

The dark spots, like those of 1893, were situated on the north equatorial double belt, but in 1894 were noted independently by all observers more often to project south from the belt into the bright equatorial zone, and to a greater extent; also they were not so often seen double. Transits per spot ranged from recordings on three to eleven dates; for most about six dates each. They fell into three groups according to rotation period, the mean periods for the groups being:

Group A (4 spots)	10^h 14^m 57^s3
B (4 „)	10 14 44·2
C (3)	10 15 48·0

The spots within each group agreed closely in rotation period and the groups were quite distinct from one another: hence, as in 1893 widely differing rates were found in different regions of longitude in the same north latitudes (17° to 37°). The decreasing order of periods among group A spots might, Williams thought, be due to the influence of the faster moving group B. He compared the mean of classes A and B of 1893 with that of the three groups of 1894 as follows:

1893 (10 spots)	10^h 14^m 44^s7
1894 (11 „)	10 15 9·8

Though individual spots could not be certainly identified from 1893 to 1894, Williams noticed a great similarity in rotation period between group A (1894) and class B (1893), and by calculating back the position of group A he found it to be identical with that of the preceding part of class B (1893), the following spots of which had presumably disappeared. He traced a similar connection between group B (1894) and some of the spots of class A (1893), though with an increase of about 15 seconds in the rotation period.

The bright spots of 1894 were similar in aspect and position to those of 1893, but rather better seen owing to the rings being more open. A narrow dark belt (equatorial band) approximately at the equator was often seen cutting across the bright spots, dividing them into two. Their limits of latitude were about 6° N. and 6° S. Transits obtained for each spot ranged from three to ten. Spot 13, one of the nine in group (c), was of special interest, as in the course of a few days in the middle of the period of observations, it made a sudden shift to the east, causing a difference of more than a quarter of an hour in the transit time; ten transits of that spot were obtained. The mean rotation periods for the three groups were:

Group (a) (3 spots)	10^h 13^m 1^s7
(b) (3 „)	10 12 40·0
(c) (9 „)	10 12 25·8

The groups were quite distinct from each other: at about the middle of the period of observation the first two groups each covered about

206

80° of longitude and the third the remaining 200°. To find the average rotation period for the whole spot zone in 1894 he took the mean of the three groups, giving (c) treble weight because of its number of spots; for 1893 he took the simple mean of class C and spot 4 of that year. The resulting mean rotation periods of the equatorial spots for the four years were:

		Difference
1891	10^h 14^m 21^{s}8	
1892	10 13 38·2	43^{s}6
1893	10 12 52·4	45·8
1894	10 12 35·8	16·6

Between 1891 and 1894 the decrease in the period amounted to 1^m 46^s, meaning an increase in velocity of the spots of 66 miles an hour. As Williams described it: '... the great equatorial atmospheric current of Saturn was blowing 66 miles an hour more quickly in 1894 than it was in 1891'. He identified group (a) of 1894 with the preceding part of class C (1893), and also group (b) of 1894 with the region represented by spot 4 of 1893, which he thought was probably the same object as no. 6 of 1894.

Dark northern spots, 1896 and 1897 (Flammarion)

Camille Flammarion, the famous French astronomer and founder of the Société Astronomique de France, made a brief report (*M.N.*, vol. 61, p. 131) on the observation of Saturn's spots at Juvisy Observatory. Though systematic observations of the planet had been started at Juvisy in 1894, no spots were seen in that year and very few details on the globe. In 1895 the duplicity of the north equatorial belt was well marked, as was the narrow dusky band on the equator. A very small dark spot on that band was seen in transit once only.

Assisting Flammarion was a young Greek astronomer, Eugène M. Antoniadi, who had emigrated to France and was to become particularly eminent for his observations and studies of Mars, and Director of the B.A.A. Mars Section for two decades. The report stated that in 1896 Flammarion and Antoniadi detected several dark spots on the north equatorial belt and were able to follow them through several transits over the central meridian. In 1897 the dusky spots were easily seen, many being detected by Flammarion and others, but only four spots could be followed through more than one transit. The mean rotation periods found for each year were as follows:

1896 (3 spots)	10^h 14^m 14^s
1897 (4 ,,)	10 14 4

Flammarion stated that owing to the increasing southern declination of Saturn in 1898 and the next two years, it became more and more difficult to see the spots, and so they were seen too rarely to make satisfactory identification possible. He concluded that his results for 1896 and 1897 were satisfactory for 18° N. saturnicentric latitude, and in fair agreement with those of other observers. It will be noticed that the mean periods were a little shorter than those deduced by Williams for the corresponding dark spots of 1893 and 1894, possibly because the average latitude of Williams's spots was more northerly.

Objectivity of spots challenged (Barnard)

A formidable challenge to the reality of the dark and bright spots of 1891 to 1894 came from the great American astronomer, Professor E. E. Barnard, who declared: 'The black and white spots lately seen upon Saturn by various little telescopes were totally beyond the reach of the 36-inch (aperture Lick refractor)—as well as of the 12-inch—under either good or bad conditions of seeing. . . . Once or twice a very small and very dark spot was seen *at the North Pole*. . . .' (*M.N.*, vol. 55, p. 368). In 1895 he made a number of experiments, examining Saturn both with full and reduced apertures, to see whether he could find any of the spots or other markings reported by other observers.* His conclusion (*M.N.*, vol. 56, p. 165) was: '. . . I am convinced that everything that can be seen with this telescope diaphragmed down can be seen with the full aperture, and furthermore, such can be better seen with the full aperture when the air is steady'. He considered from his experience with the use of both large and small telescopes, that the smaller instrument is only preferable when the seeing is bad or very indifferent. He pointed out that with the 36-inch he had been able to see an extraordinary amount of small delicate detail on Mars and Jupiter.

There seems to be no possibility of reconciling the completely negative findings of Barnard with the long and detailed series of observations of these spots by Stanley Williams and other observers. Barnard's reference to 'black and white spots' does rather suggest that he was looking for something much more distinct than the diffuse darker areas on a dusky belt and the vaguely outlined brighter areas of a light zone identified by Williams. Moreover, the latter came to this work from much experience in the detection of faint spots and markings on Jupiter, and he could hardly have obtained results so mutually consistent, not only within each apparition but also from year to year, if the spots on which he based them had had no objective reality.

* But Williams does not seem to have reported spots in 1895.

FIG. 1.—G. P. Bond, 15 November 1850, Harvard College Observatory 15-inch refractor. Discovery drawing of crepe ring in ansae and across globe (from *Ast. J.* vols. 1 and II, plate).

FIG. 2.—W. S. Jacob, 12 November 1861: drawing of broken almost edgewise ring (from *M.N.* vol. 22, p. 89).

FIG. 3.—W. Wray: drawing of dark side of nearly edgewise ring with overlapping light and two satellites, 26 December 1861—from *M.N.* vol. 23, p. 86.

PLATE V

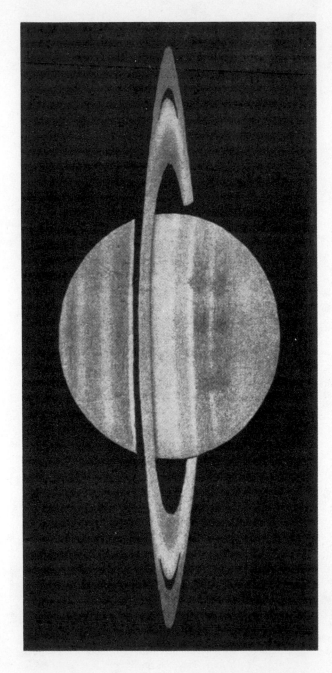

PLATE VI.—Saturn—general aspect in the spring of 1893, drawn by A. S. Williams (6-inch reflector), showing dusky spots in pairs on N.E.B. (from *M.N.* vol. 54, plate 6).

Antoniadi, who was an outstandingly able and very discriminating observer, is one whose opinion is of some value on this question, since, although the aperture of the refractor he used in the observations was not larger than 9¾ inches, and he had not detected any spots in 1894, yet he had observed the dark spots regularly in 1896 and 1897 and had deduced rotation periods in good agreement with those of Stanley Williams for the earlier years. In the course of a paper written in September 1909 (*Bulletin, S.A.F.*, vol. 23, p. 450), in which he expressed agreement with Barnard's negative opinion about the alleged outer crepe ring, Antoniadi remarked: '. . . on pourrait rappeler que M. Barnard n'est pas un observateur de détails planétaires délicats, étant donné que dans sa discussion au sujet des taches de Saturne, ses arguments ont été complètement refutés par M. Stanley Williams'. It is evident from this that Antoniadi was satisfied that Stanley Williams had met Barnard's challenge successfully.

(B) Spectroscope Proves Nature of Rings and Measures their Rotation, 1895

During the nineteenth century the camera and the spectroscope had been brought into use for astronomical work and improved and developed to such an extent that they had become, along with the telescope, essential tools of astronomy. This enabled Professor James E. Keeler at the Allegheny Observatory of Pittsburgh not only to tackle in an entirely different way the question of the rotation speeds of Saturn and its ring system, but also to settle once and for all, by photography of the spectrum, whether the rings rotate as solid bodies or as multitudes of separate particles. In April 1895 he achieved both these objectives by means of the camera, spectroscope and telescope combined, in one of the most brilliantly successful operations ever carried out in the observation of Saturn.

Keeler proves the meteoric constitution of the rings

In his paper (*Ap. J.*, vol. 1, p. 416) Keeler referred to the publication in 1859 of Clerk Maxwell's historic paper which proved mathematically that the rings must consist of a large agglomeration of separate particles (see Chapter 13). He stated that he had obtained the first direct proof of Clerk Maxwell's hypothesis, and that his spectroscopic proof also 'illustrates in a very beautiful manner . . . the fruitfulness of Doppler's principle and the value of the spectroscope . . . for the measurement of celestial motions'. At the same time he drew attention to the difficulty, namely, '. . . to find a method so delicate that the very small differences of velocity . . . may not be masked by instrumental errors'. His procedure was to arrange the spectrograph with the slit along the line of the major axis of a telescopic image of Saturn and the rings, and by an exposure of two hours to obtain a photograph of the spectrum. His first attempt in 1893 had failed partly owing to atmospheric conditions, and partly through technical troubles—a difficulty in guiding, and the fact that the particular colour correction of the objective of the 13-inch aperture refractor and the yellow colour of the planet rendered the negatives of the violet end of the spectrum too weak and granular. On 9 and 10 April 1895, however, he obtained satisfactory photographs of the spectrum with orthochromatic plates, and afterwards photographed a comparison spectrum of the Moon on each side of that of Saturn. An enlargement of one of these photographs was published (*M.N.*, vol. 55, p. 475), but the result can be more easily seen in the diagram reproduced as Figure 10. All the spectral lines of one half of the globe and the adjoining ansa of the rings are slightly shifted towards the blue end of the spectrum; all those of the other half of the globe and the other ansa are slightly shifted towards the red. This, in accordance with the well-known Doppler

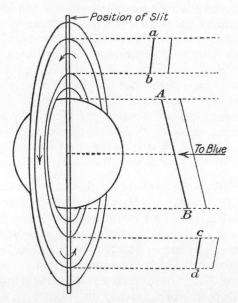

FIGURE 10. J. E. Keeler: diagram illustrating spectroscopic proof of rotation of rings. (From Sir Harold Spencer Jones: *General Astronomy*, 1951, Edward Arnold & Co., p. 253, fig. 87).

effect, shows that one half of the planet with that ansa is approaching the observer, the other half and the other ansa receding, so that the whole system—globe and rings—must be rotating.

Keeler commented as follows on the details: 'The planetary lines are strongly inclined, in consequence of the rotation of the ball, but the lines in the spectra of the ansae do not follow the direction of the lines in the central spectrum; they are nearly parallel to the lines of the comparison spectrum, and, in fact, as compared with the lines of the ball, have a slight tendency to incline in the opposite direction. Hence the outer ends of these lines are less displaced than the inner ends. Now it is evident that if the ring rotated as a whole the velocity of the outer edge would exceed that of the inner edge, and the lines of the ansae would be inclined in the same direction as those of the ball of the planet. If, on the other hand, the ring is an aggregation of satellites revolving around Saturn, the velocity would be greatest at the inner edge, and the inclination of lines in the spectra of the ansae would be reversed'. Therefore, he concluded, the photographs are 'a direct proof of the approximate correctness of the latter supposition'.

Rotation velocities from the spectra (Keeler, Campbell)

Assuming the limbs of the planet to be at 60,340 km. from its centre and the middle of the bright rings to be at 112,500 km., and using the rotation period of $10^h.23$ derived by Hall from the observations of a bright equatorial spot in 1876, Keeler computed mathematically a velocity of 10·29 km. per second for the limb and one of 18·78 km. per second for the middle of the bright rings. The following table summarises the velocities derived from measuring the shifts of the spectral lines, showing the mean result for five lines on each photograph, and also shows the excess of the computed over the observed value $(C-O)$:

	km/sec	$C-O$ km/sec
Photo 1. (9 Apr. 1895): velocity of limb	11·03	−0·74
mean velocity of ring	17·50	+1·28
Photo 2. (10 Apr. 1895): velocity of limb	9·58	+0·71
mean velocity of ring	18·52	+0·26
Mean of both photos: velocity of limb	10·3±0·4	−0·01
mean velocity of ring	18·0±0·3	+0·78

Keeler's result from the spectrum for the velocity of rotation of the equatorial region of the globe therefore confirmed the value derived by Asaph Hall from the observation of spots.

Keeler's investigation was speedily followed up (*Ap. J.*, vol. 2, p. 127) and confirmed by William W. Campbell at Lick Observatory, the directorship of which he held from 1900 to 1930. Campbell's speciality was the derivation of sight-line velocities of stars from measurements of their spectra, so that this investigation for Saturn came within his particular field. Using the new Mills spectrograph, he obtained four spectrograms on 10, 14, 15 and 16 May 1895 on a scale large enough to determine the excess of the velocity of the inner edge over that of the outer edge of the ring, which Keeler's spectra had not enabled him to do. He assumed the same computed values, 10·29 km/sec and 18·78 km/sec for the velocities of the limb and middle of the ring respectively, and derived a mean result from the measurement of ten spectral lines on each plate. These results may be summarised as follows:

		km/sec	$C-O$ km/sec
Velocity of limb:	plate 1 mean	9·6	+0·69
	„ 2 „	10·26	+0·03
	„ 4 „	9·4	+0·83
	mean of three plates	9·77	+0·52
Velocity of middle of ring:	plate 1 mean	17·38	+1·4
	„ 2 „	17·38	+1·4
	, 4 „	17·34	+1·44
	mean of three plates	17·37	+1·41
Excess velocity for inner edge over outer edge:	plate 1 mean	3·16	+0·71
	„ 2 „	3·06	+0·81
	„ 4 „	3·17	+0·70
	mean of three plates	3·13	+0·74

These results would give a velocity of rotation for the inner edge of ring B of 18·94 km/sec, for the middle of the bright ring (which would be near the outer edge of B) of 17·37 km/sec, and for the outer edge of ring A of 15·8 km/sec. Corresponding velocities in miles per second would be: globe 6·07; inner edge of ring 11·77; outer edge 9·82.

If the outer diameter of ring A is taken as 169,300 miles and the inner diameter of ring B as 112,600 miles, then particles at the outer edge of ring A would have a rotation period of about 15^h 3^m, and those at the inner edge of ring B a period of about 7^h 8^m.

Antoniadi, writing in 1930 (*L'Astronomie*, vol. 44, p. 56) stated that angular velocities had also been found from the spectrum by Deslandres in 1895, as follows: equatorial limb of globe 9·38 km/sec;

ring B 20·1 km/sec; ring A 15·4 km/sec (*Comptes Rendus de l'Acad. des Sciences*, vol. 120, p. 1157).* Antoniadi also quoted the following rotation periods calculated by André (*Les planètes et leur origine*, p. 142): ring A outer edge 13^{h}8, inner edge 11^{h}4; ring B outer edge 11^{h}0, inner edge 7^{h}5; ring C inner edge 5^{h}4. Hence almost the inner half of the ring system is rotating in a shorter period than the globe of Saturn.

Antoniadi also stated that in 1889 Keeler had shown from an examination of the spectrum that the rings have no atmosphere.

This chapter has traced the work done on *rotation* through the last quarter of the nineteenth century; other Saturnian studies during the same period are dealt with in Chapters 16–22.

* H. Deslandres was Director of Meudon Observatory.

CHAPTER 16

SATELLITE ORBITS AND SATURN'S MASS

Reciprocal mass and orbit (Le Verrier)

The masses of the planets are often expressed as fractions of that of the Sun (taken as unity), e.g. the Earth's mass is 1/329,400 of the Sun's, and for Mars, Mercury and the Moon the denominator runs to millions. Hence it is more convenient to use the reciprocal mass, which is the fraction turned upside down: the Earth's reciprocal mass is 329,400. Values for Saturn's reciprocal mass calculated by various mathematicians prior to 1875 were:

Date		R. Mass	Date		R. Mass
1726	Newton	3021	1802	Bouvard	3521·31
1782	Lagrange	3358·4	1811	Bessel	3379·12
1802	Laplace	3359·4	1821	Bouvard	3512

In 1875 Urbain J. J. Le Verrier, using more recent observations, suggested a small increase for the reciprocal mass (diminution of the mass) from Bouvard's value to 3529·6. (The value used at present by the *N.A.* and given in *B.A.A. Hdbk.* 1960 is 3501·6.)

The recalculation of Saturn's mass was only one item in the immense work done by Le Verrier over several years on the orbits of all the planets, his memoirs on Saturn being presented to the French Academy during 1872–75. Professor J. C. Adams, (the only mathematician other than Le Verrier who had been able mathematically to discover the planet Neptune in 1846) in 1876 as President of the R.A.S. presented Le Verrier with the Society's gold medal for his memoirs and tables of the planets, and in a long address (*M.N.*, vol. 36, p. 232) tried to convey an idea of the extraordinary mathematical intricacy and vast labour involved. Taking into account all the effects of mutual perturbations on all the elements of the orbits, Le Verrier had determined the values of the elements for Saturn and the other major planets for 2,000 years from 1850, and had constructed his formulae in an adaptable form, so that future astronomers with greater and more exact observational data would still be able to use his calculations.

Ring tilt and satellite conjunctions (Hall, Holden, Gledhill)

During 1877–79 Professors Asaph Hall and Edward S. Holden, using a filar micrometer with the Washington 26-inch aperture refractor, made a long series of observations of the major axis of the ring, and by using Bessel's formulae calculated for each observation the inclination of the ring-plane to Saturn's orbit (*M.N.*, vol. 42, p. 304). Hall also reported (*ibid.*, p. 308) a series of observations he made in 1881–82 of the dates and times of superior and inferior conjunctions of the satellites Enceladus, Dione, Tethys and Rhea.

Series of accurate observations of positions of satellites, though not spectacular, can contribute valuably to the correction of the elements of the orbits, and thus improve knowledge of satellites' motions. Much work of this kind was being done in the 1870s and 1880s, especially at leading observatories and with the use of large telescopes that could deal adequately with Mimas and Enceladus in spite of their apparent faintness through being swamped by the glare of the near-by planet. For example, there was one series made at the Washington Observatory in 1874, and another at the Royal Observatory, Greenwich, in the following year (*M.N.*, vol. 35, p. 327, and vol. 36, p. 266).

A note by Dr. A. A. Common (*M.N.*, vol. 40, p. 93) illustrates the difficulty of getting satisfactory observations of conjunctions of Mimas with the ring. In 1877 and 1878 he tried without success to observe Mimas with an 18-inch aperture reflector, though it several times showed Hyperion. With his 36-inch aperture reflector he could see Mimas fairly well when it was about 3″ or 4″ from the end of the ring, but even with that powerful instrument he failed to follow it up to conjunction in the autumn of 1879. On the other hand, he found that moonlight, which utterly obliterated Hyperion, had no effect on the visibility of Mimas.

In the same autumn E. S. Holden, using the 26-inch aperture refractor at Washington, also experienced much difficulty in following Mimas up to conjunction with the ring, and he seems to have achieved complete success only once in the series of observations (*M.N.*, vol. 40, p. 283). He obtained, however, a very satisfactory view of an occultation of Rhea by Saturn's globe on 30 September 1879, Rhea being 'nearly smooth with the limb' at 11^h 15^m, seen perhaps a third of the time at 11^h 22^m, seen for an instant at 11^h 23^m and 24^m, but not seen after 11^h 24^m (W.m.t.). He noted that Rhea seemed to grow smaller and of a more intense white as it came close to the limb; he thought this might have been due to contrast with the very dark dull olive colour of the south polar belt near which it impinged. Rhea certainly seemed to him smaller than usual, looking more diminutive than Dione, which was near and which usually looks the smaller of the two.

In November–December 1879, J. Gledhill, using an equatorial of $9\frac{1}{3}$ inches aperture, managed to time a conjunction of Rhea with the globe, and conjunctions of Tethys and Dione with the end of the ring (*M.N.*, vol. 40, p. 285). He also reported timing conjunctions of Dione and Rhea with the ring in November 1880, using the same telescope (*M.N.*, vol. 41, p. 295).

A very good series of conjunctions, 28 in number, including no less than 5 of Mimas, the others being of Enceladus, Tethys, Dione and Rhea, was reported by Hall (*M.N.*, vol. 44, p. 358). These observations were made in the winter of 1883–84, using a wire micrometer with the Washington refractor. Hall stated that on December 3, Mimas was observed in conjunction with the preceding end of the Cassini division of the ring, and again on the same date when in conjunction with the preceding limb of the globe, and expressed the following opinion about observing Mimas: 'On good nights, with the present opening of the ring, this satellite can be followed to conjunction with ease and certainty'. He acknowledged his indebtedness to Marth for the ephemerides of satellites which Marth was regularly publishing during that period. In 1883–84 the rings must have been open nearly to their fullest extent, and Saturn's glare very pronounced, so it was no mean feat even with a 26-inch aperture to have been able to follow Mimas so certainly.

Orbit of Hyperion (Hall, Newcomb)

A. A. Common, using his 36-inch aperture reflector, recorded a good series of times and positions of Hyperion during the autumn of 1879 (*M.N.*, vol. 40, p. 94), but it was Asaph Hall who made a series of observations of this satellite extending over several oppositions and investigated its motion. Hall's short note (*M.N.*, vol. 39, p. 517) pointed out that Hyperion had the most eccentric orbit† of any satellite known at that time, and that owing to its distance from Saturn combined with the large eccentricity, it can make very near approaches to Titan, the largest satellite in the Saturnian system. This peculiar orbit not only makes its motion interesting but, in Hall's opinion, should enable an accurate knowledge to be obtained of the mass of the ring and of Titan. He added that his own observations in 1878 'seem to put it beyond doubt that the line of apsides* of the orbit of Hyperion has revolved at least a semi-revolution, or 180°, since the time of the discovery of this satellite in 1848, by the Bonds

---

* The line joining the perisaturnium (nearest point to Saturn) with the aposaturnium (furthest point from Saturn) of the satellite's orbit, i.e. the orbit's major axis. † See Fig. 14, p. 272.

and Lassell'. This means that in the course of thirty years the position of Hyperion's nearest point to Saturn had swung round to the opposite direction from Saturn, with a corresponding change in the direction of its furthest point from Saturn.

He went on to state that, apart from the observations made at Washington since 1874, the only other observations of Hyperion that he knew of were those made by the Bonds in 1848 and 1849 and some made subsequently by Lassell and Marth.

A few years later in a paper entitled 'The Motion of Hyperion' (*M.N.*, vol. 44, p. 361), Hall investigated the problem mathematically, using the only complete series of observations made during the quarter of a century following the satellite's discovery—those made by Lassell at the opposition of 1852 (*M.N.*, vol. 13, p. 181)—as well as the Washington ones of 1875, 1876 and 1879–83.

By comparing Lassell's observations with his own of 1880 and 1882, he obtained for the sidereal and synodic periods of revolution of Hyperion respectively the values: 21·276742 days and 21·318901 days; the present-day figures agree with the former to four places of decimals and with the latter to the nearest minute of time.

Hall pointed out that Hyperion can approach very near to Titan, and since three times the period of the one is nearly equal to four times the period of the other, Titan probably greatly perturbs the motion of Hyperion, the perturbation having a period of 19·185 years. He worked out mathematically the probable effect of the perturbation, but felt unable to make an exact determination in the absence of exact knowledge of Hyperion's mean motion. He did, however, derive an estimate that the mass of Titan would be 387 times that of Hyperion, assuming their densities to be the same, and a basis for 'a tolerably accurate ephemeris of Hyperion for the next opposition'.

Asaph Hall's inference that the line of apsides of Hyperion's orbit had made rather more than a half revolution in thirty years was an under-estimate. Attention has been drawn in Chapter 14 to the comparative dearth of Saturn observations during the epoch of the eighteen-sixties: Hyperion is a case in point. Had several good series of observations been made at regular intervals between 1852 and 1874, Hall would have found that the line of apsides revolves at thrice the rate he supposed, making actually over half a revolution in every *ten* years, so that, in the thirty-year period he was considering, it must have gone through one complete revolution and rather more than half a second one. The resulting position after thirty years was of course similar to that which intervening observations, had they been made, would have revealed after ten years.

In 1884 Professor Simon Newcomb, an eminent mathematical expert on the solar system who became head of the American N.A. Office, cleared up all the doubts and omissions regarding Hyperion's motion, showing that in the case of Hyperion the apsides move in a *retrograde* direction at the rate of $18°56$ every year. This is an exception to the usual rule that in a disturbed orbit of a body having direct motion, the line of apsides advances. Newcomb investigated the apparent exception and showed that the cause is the close commensurability of the motions of Titan and Hyperion, closer than that of any other pair of satellites in the solar system (*A.P.* of *A.E.*,* III, part 3, 1884). The mean motions of Titan and Hyperion are almost in the ratio $4 : 3$, as $3T - 4H = +0°051$.

Dr. J. G. Porter (*B.A.A.J.*, vol. 70, p. 56) has shown the significance of the small angle. If the two satellites are in conjunction at a certain time, after three revolutions of Hyperion they will again be in conjunction at almost the same point. In the interval of $63·6377$ days between conjunctions, Titan moves through $1436°75$ and Hyperion through $1076°75$. These angles are $3°25$ short of 4 revolutions of Titan and 3 of Hyperion, so the point of conjunction retrogrades through $3°25$ in $63·6377$ days, which is $0°051$ per day or $18°56$ per year. The line of apsides has to follow the same law, and so the conjunctions of these two satellites always occur at the same place in the orbit, and this coincides with the aposaturnium of Hyperion, where it is more than a million miles from Saturn.

Saturn's mass from a satellite orbit

From Sir Harold Spencer Jones's *General Astronomy* (1951), p. 215, it is clear that the mass of Saturn (or of any other planet with a satellite) can be determined as a fraction of the Sun's mass by means of the law of gravitation, provided that the radii of the orbits and the periods of revolution are known. If S is the Sun's mass; M, m the masses of Saturn and the satellite; T, t their periods of revolution; and A, a the radii of their orbits respectively, then the formula will be:

$$\frac{M}{S} = \frac{M+m}{S+M} = \left(\frac{a}{A}\right)^3 \left(\frac{T}{t}\right)^2$$

Since m is negligible compared with M, and M is negligible compared with S, $(M+m)/(S+M)$ is practically equivalent to M/S. If the formula is turned upside down the reciprocal mass of the planet S/M (a whole number instead of a small fraction) is obtained.

* Astronomical Papers of American Ephemeris.

Orbit of Iapetus and Saturn's mass (Hall)

About 1885 Hall fully investigated the orbit of Iapetus, using his own observations of 1875–84 and similar ones of Newcomb made in 1874 (*Washington Obsns.* for 1882, App. I, summarised in *M.N.*, vol. 46, p. 235). The observations were measures of differences of R.A. and Dec. of Iapetus and Saturn, and measures of position angles and distances. To correct the periodic time of Iapetus, Hall compared his observations with one by Sir W. Herschel in 1789 and those of Sir J. Herschel made at the Cape in 1837. He calculated the chief perturbations of Iapetus due to the Sun, and the long period effect in the motion of Iapetus due to Titan, arising from the fact that five periods of Titan are nearly equal to one of Iapetus. He deduced two sets of elements for the orbit, in very good mutual agreement, from the two different kinds of observations. Hall's results for the periodic time and the eccentricity agree to three places of decimals with the values used at present: 79·331 mean solar days and 0·028 respectively. His result for the reciprocal mass of Saturn, 3481·3, differed rather considerably, however, from Bessel's value, as corrected by Hermann Struve.

Orbit of Titan and Saturn's mass (Hall Jnr.)

This difference caused Asaph Hall, Junior, to make a new determination of Saturn's mass, using the heliometer at Yale for observations of Titan during the oppositions of 1885–86 and 1886–87 (*Trans. Astron. Observatory of Yale University*, vol. i, pt. ii, summarised in *M.N.*, vol. 51, p. 249). As the ring was then open to nearly its widest extent, Hall (like Bessel) was able to bring Titan quite accurately to the extremities of the major axis of the ring-ellipse, and read off the position angle and distance at the same time. He not only took perturbations into account but made elaborate checks on the accuracy of the instruments he was using.

Hermann Struve, a son of Otto Struve, had shown that the corrected value of the reciprocal mass of Saturn found by Bessel from Titan should be 3502·5. Struve himself had found the values 3500·2 and 3495·7 from Iapetus and Titan respectively. Asaph Hall, Jnr., obtained a value 3500·5, in very good agreement with those of Bessel and Struve, and with the figure now used by the *N.A.*: 3501·6.

Orbits of inner satellites (H. Struve)

Hermann Struve made some important investigations of the orbits of Mimas, Enceladus, Tethys and Dione (*M.N.*, vol. 51, p. 251, from *A.N.*, No. 2983–84). He had the advantage of being able to use the

30-inch aperture Pulkovo refractor which, with its great optical power, steady mounting and excellent measuring apparatus, enabled the positions even of Mimas to be obtained with very great accuracy. His observations showed that the deviations of Enceladus and Dione from the plane of Saturn's equator can only be small, but he found an inclination of the orbit of Mimas to that plane of 1°26', and of that of Tethys of 1° 5'. These are very close to G. Struve's (1930) values used at present by the *N.A.*: Mimas 1° 31', Tethys 1° 5'.6, and a nominal tilt of 0° 1'.4 each for Dione and Enceladus.

There is a close relationship between the periods of revolution round Saturn of Mimas and Tethys, the period of the former satellite being a little less than half that of Tethys. Struve found from his observations that the line of nodes* of Mimas's orbit had an annual retrograde motion of about 365°, and that there was a similar motion for the nodes of Tethys of about 72°. He noticed that the sum of these motions of the nodes is approximately equal to double the excess of the mean motion of Mimas over twice that of Tethys, and this gave him a probable clue to puzzling variations in the motion of Mimas which had been indicated by previous observations and further verified by his own.

Struve found the eccentricity of Enceladus's orbit to be about 0·0047 (*N.A.* value, *B.A.A. Hdbk.*: 0·0044) and that there is an annual progression of about 120° of the line of apsides. Now there is a similar relationship between the revolution periods of Enceladus and Dione as there is between those of Mimas and Tethys, but in this case the period of Enceladus is a little greater than half that of Dione. Struve noticed that the annual progression of the line of apsides of Enceladus is approximately equal to the excess of twice the mean motion of Dione over the motion of Enceladus.

From the brief indications given in this chapter it will be realised that by the end of the nineteenth century such work as the investigation of satellite orbits had attained a high degree of accuracy, and that further refinements to the results could only be achieved by highly skilled astronomers using the very powerful and accurate instruments available only at great observatories.

* The nodes of a satellite's orbit are the two points at which the orbit intersects the plane of reference (in this case the plane of the planet's equator). Apsides could not be considered in the case of Tethys whose orbital eccentricity is nil; on the other hand, the positions of the nodes would be indeterminate for Enceladus and Dione since their orbits show no appreciable inclination.

CHAPTER 17

GLOBE AND RINGS IN THE 1880s

Saturn in 1880 (Denning)

From late September to early November 1880 a series of observations of Saturn with a 10-inch aperture reflector yielded to William F. Denning a number of interesting features (*M.N.*, vol. 41, p. 82). Denning, a Bristol accountant, was a prominent amateur planetary observer but is best known for pioneer work in meteor observation and investigation of radiants.

In Saturn's southern hemisphere, then on view, besides an indication of white patches on the bright equatorial zone, Denning seems to have seen something resembling the 'quintuple belt' noted by Sir William Herschel, and also a series of very dark belts in the south polar region, the one adjacent to the pole being the darkest on the disk. The polar area, though light by comparison, he found to be much enveloped in shade. The belts, unlike those on Jupiter, could be traced nearly out to the limbs, and a bright zone in the temperate region even appeared to project beyond the limbs.

Ring A seemed rather faint, especially at the outer edge, and Encke's division, though at best moments dark and distinct, was sometimes seen only as a pencil line for a very short distance in the ansae. Faint lines of shading on ring A were momentarily glimpsed now and then. The parts of rings A and B bordering Cassini's division were much brighter than other parts, and the interior shading of B, which had appeared gradual to Dawes and others, seemed always quite abrupt to Denning and to show a faint pencil line where it began. The depth of shading was greater further in the interior of B but the extreme inner border lighter, perhaps by contrast with the darkness of ring C. The crepe ring seemed extremely dark at its outer border, possibly through contrast with the luminous vein of B, but brighter on its inner margin bounded by the dark sky. He occasionally had the impression of a difference in width of the dark spaces on either side of the globe.

The most striking feature, Denning said, and one repeatedly and clearly seen, was that 'square-shouldered' aspect of the globe which had astonished and then convinced Sir William Herschel and had bothered so many astronomers of the early nineteenth century. But one might almost say that Saturn had tried that old trick once too often; Denning, armed with his knowledge of the micrometric measures which had proved beyond doubt the truly spheroidal shape of the planet, was not deceived. Indeed he offered a simple explanation: 'The singular figure is due to the contrasting effects of the belts. While the bright belt near the pole causes a very evident "shouldering" out of the two limbs at its extremities, the dark belts at the pole and towards the equator act in a precisely contrary way. They apparently compress the planet where they come up to the limb. . . .' Denning tried to get rid of the illusion by very careful focusing but even in moments of superb definition the true spheroidal form of the planet would not appear.

Suspected transparency of ring A, 1880–81 (Trouvelot)

During the apparition of 1880–81 Trouvelot noted that on several occasions the shadow of ring A was not as usual 'black as ink' but grey, sometimes the inner zone of the shadow being black and the outer grey. Similar observations had been made by Bond, Webb, Dawes, Huggins and Lassell; in some of these no shadow at all was visible when theoretically it should have been seen. According to a paper by P. H. Hepburn in 1914 (*B.A.A.J.*, vol. 24, p. 480) Trouvelot did not favour the idea that this and other phenomena of rings A and C might be caused by refraction owing to a supposed atmosphere of the rings. In regard to an alternative theory that the ring was transparent, Trouvelot said: 'For my part I hardly see how transparency of rings A and C can account for it. But if ring A is transparent, it is evident that it is not always so, it is even supposable that it is rarely so, since the shadow it casts on Saturn is generally black as ink, and its border is sharply defined'. Trouvelot then argued that even if ring A is transparent, one could not hope to see the limb of the planet through it, and concluded: 'Rather would the occultation of bright stars by the ring be what is required to indicate its transparency.'

These opinions of Trouvelot are of extraordinary interest in view of the discussions and observations that took place more than thirty years later (see Chapter 30).

The globe glimpsed through Cassini's division, 1883 (Young and Hall)

According to a letter from Dr. C. A. Young (*Observatory*, vol. 7, p. 115), at Princeton on 19 or 26 November 1883, using the new 23-inch aperture equatorial of Halsted Observatory, he with Professor Hall, who happened to be visiting the observatory at the time, managed to detect a narrow strip of the globe through Cassini's division, thus repeating Lassell's observation of October 1852 (see Chapter 12). Young added that the seeing was very fine, Mimas was conspicuous not far from conjunction, and so were the reddish south equatorial belt and the olive-green polar cap.

Saturn in 1884–85 (Green)

A very well known amateur astronomer, Nathaniel E. Green, who a few years later became the first Director of the B.A.A. Saturn Section, made many observations of Saturn during the winter of 1884–85, using an 18-inch aperture reflector (*M.N.*, vol. 45, p. 401). He found the south polar cap dark grey, and the edge of ring C across the globe reinforced by a dark belt north of the equatorial zone, but otherwise he does not seem to have noted anything unusual about the globe. Ring C appeared to extend nearly half-way to the globe and had a definite inner boundary. He did not apparently see any shading or markings on ring B, but noted that the broadest middle part was dull yellow in contrast to the brightness of the outer part. Ring A he found difficult, light at both edges, while in the central part there was an appearance of shade but no division to be seen.

Saturn in 1887 and 1888 (Elger)

Thomas Gwyn Elger, an amateur astronomer who was originally an engineer and is best known for his lunar observations and lunar map and book on the Moon, made a series of observations of Saturn with a Calver reflector of 8½ inches aperture in the early months of 1887 and another series from January to April 1888 (*M.N.*, vol. 47, p. 511 and vol. 48, p. 362). As Elger had the lunar observer's keen eye for minute detail, the reports are lengthy and difficult to abridge, and much repetition will be avoided if a single summary is made of his findings during the two oppositions.

Ring A. The colour of this ring appeared on the best nights to be lavender-grey, with much variation in uniformity and depth of tone. Sometimes it was wholly or partly divided about midway between the two edges into an outer dark and an inner light zone, the latter sometimes evidently the narrower. In 1887 Encke's division was frequently observed as a faint, delicate dark line, occasionally being glimpsed all round the visible part of the ring, but in 1888 it was only glimpsed twice and then only in one ansa. In both years a narrow, light streak concentric with the ring was frequently observed about midway between the outer and inner edge but generally a little nearer the outer edge.

Cassini's division never impressed Elger as being perfectly black, and though the inner edge was well defined, the outer one seemed 'fuzzy' on both ansae and without a definite limit. On many nights in 1888 the division was much darker round the south side and barely visible on the north side.

Ring B. A characteristic of almost every fine night in early 1887 was the 'gradual gradation of shade' from the dark inner edge of this ring for about two-thirds of the width of the ring to the outer bright zone, but this appearance was not prominent in 1888. The brilliant outer zone seemed brighter and more conspicuous in 1888 and to show distinct differences in brightness in different parts, e.g. it often seemed much more brilliant on the following than on the preceding ansa. Its width seemed to vary on different nights from 1/3 down to 1/5 of the width of B, but Elger felt that no great weight should be given to these estimates because the inner edge of the bright zone was often barely distinguishable from the adjoining part of the ring. In 1887 (but much less so in 1888) 'the delicate rounding off of the shading from within to the outer bright zone' sometimes gave B the appearance of a solid ring. In 1888 Elger often saw very distinctly the same sort of crescent-shaped shading on the inner edge of ring B at the ansae as had been noticed by Tuttle in 1851 (see Chapter 11).

Ring C. Though appearing generally dark in both ansae in 1887, the crepe ring in 1888 several times seemed light grey and distinct in the p. ansa, but dark and very indistinct in the f. ansa. In March 1888 Elger often saw the inner edge of the crepe ring in the p. ansa scalloped or ragged, as he had noted on some occasions in February 1887.

Globe Belts and Zones. He noticed a very marked change in the number, position and visibility of the belts, especially in the south temperate region and towards the south polar cap, in 1888 as compared with 1887, similar to the variation of the fainter belts on Jupiter. A very wide dusky zone covering the south temperate region in the former year was in 1888 of a pale cinnamon hue, and a wide rather dark belt on its southern edge had become extremely faint. The broad bright equatorial zone was occasionally seen to be mottled with delicate white fleecy spots. On several good nights the northern half of the equatorial zone seemed distinctly duller than the southern half. The greyish blue south polar cap seemed to vary considerably in darkness on different nights. He noted that the south equatorial belt, the darkest on the disk, often seemed harder and sharper on the north edge than on the south.

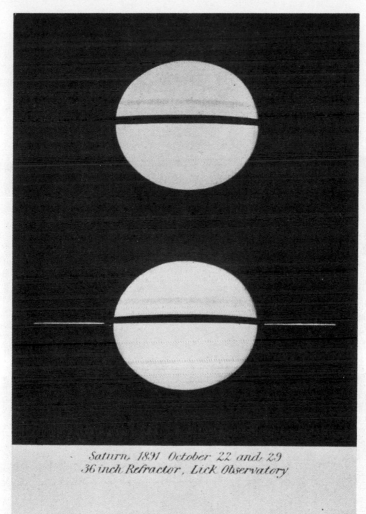

Saturn, 1891 October 22 and 29
36 inch Refractor, Lick Observatory

Diagram Oct. 29 1891 Saturn
Showing the positions of the two irregularities on the Ring
17ʰ.30ᵐ

PLATE VII.—E. E. Barnard's drawings (36-inch Lick Observatory refractor) showing (1) 'round phase' of Saturn, 1891 October 22; (2) the reappearing ring, October 29; also diagram of two small ring projections, October 29—(from *M.N.* vol. 52, plate 1).

PLATE VIII.—Drawing of Saturn by Barnard, 2 July 1894 (36-inch Lick refractor) showing no spots (from *M.N.* vol. 55, plate 8). The crepe ring is prominent (see text p. 254).

Shadow of globe on rings. Elger included in his report a series of sketches showing the irregular shape of this shadow and the decided notch in it often seen at the point where it crossed Cassini's division.

In the concluding part of his 1888 report he referred to the scepticism concerning his Saturn observations that had been expressed by certain critics, who had suggested that the abnormal appearance of ring C, the peculiarities of the shadow and other unusual details noted by Elger might be due either to bad definition, distorting eyepieces, or defective vision! Elger refuted the first two suggestions, pointing out that the strange appearances had been best seen (or in some cases only seen) on the best nights, and that his excellent Cooke eyepieces had given very good results on other objects, such as the Moon and double stars, and as regards his vision, he was able to quote similar impressions obtained by many other observers who were highly skilled and some of whom observed with large telescopes. He named Trouvelot, Terby and several others, quoting Trouvelot's summing up, as follows: 'Les anneaux de cette planète, loin d'être stables, sont, au contraire, essentiellement variables et subissent des changements continuels.'

It may seem curious that the critics did not suggest the most likely explanation of these anomalous appearances, namely, that they are optical illusions due, e.g. to irradiation and contrast.

Terby's white spot, 1889 (Common)

On 6 March 1889 a well-known amateur astronomer, Dr. François Terby at Louvain, using his 8-inch aperture telescope, observed a curious phenomenon on Saturn's rings which was telegraphically announced as follows: 'région blanche sur anneau Saturne, contre ombre globe'. This famous white spot was situated on the rings adjoining the shadow of the globe, and it was also seen by other observers in different parts of the world. A. A. Common, in a paper (in *M.N.*, vol. 49, p. 388) gave an account of his telescopic search for the object which he had not seen on March 6, although he had been carefully observing Saturn at the time with his 5-foot reflector, and had been able to detect many details on Saturn and see several of the satellites, including Mimas. Common, with another observer, searched especially for the spot with his powerful telescope on eleven evenings after receiving the news, but without success. He pointed out that the ring near the shadow always appears slightly brighter by contrast than other parts of the ring, an effect that is much more marked with a smaller telescope. He found that when a dark wedge was used the ring disappeared at the same time all round, whereas if the white spot were really brighter than the other parts of the ring it would have remained visible after the rest of the ring had been cut out by the dark wedge. He said that an apparent increase of brightness could be produced on any part of the planet with his large or small telescope by letting it pass partly out of the field, when all parts of the globe and ring in contact with the edge would appear brighter than the rest of the planet. He photographed Saturn several times in the following month with his large telescope, but the photographs did not show the slightest indication of any bright spot on the rings near the shadow. He concluded that Terby's white spot is an optical effect due to contrast, and that it is more prominent sometimes than at others owing to atmospheric conditions and to the positions of the planet and rings.

Terby's white spot on the rings alongside the globe's shadow has often been seen since by many observers, but it is now considered to be a mere contrast effect.

Saturn 1876 to 1889 (Hall, Holden)

Professor Asaph Hall made many observations and measurements of Saturn during this period, using the 26-inch aperture refractor of the U.S. Naval Observatory (the same telescope with which he had discovered the two satellites of Mars) (*M.N.*, vol. 51, p. 250; more fully in *Washington Obsns.*, 1885, App. II).

Apart from the white equatorial spot of 1876 (recounted in Chapter 15), Hall saw very little that was remarkable on the globe. He found the curvature of the globe's shadow on the rings generally regular and observed no notch. He could detect no division between rings B and C. He was never able to see Encke's division as an actual division 'notwithstanding especial pains were taken to do so', but he sometimes noticed what he described as 'a kind of marking' on ring A. He made an elaborate series of accurate measurements of the globe and ring system, and in regard to Otto Struve's idea that the inner edge of the ring would reach contact with the globe in 1976, Hall concluded, from a comparison of his own measures with those of Bradley, Herschel, Encke, and F. G. W. and O. Struve, that no change had occurred in the dimensions of the ring system, at any rate since 1719, so Struve's expected catastrophe was a wholly visionary one.

Professor Edward S. Holden, who after service at the U.S. Naval and Washburn Observatories, became the first Director of the Lick Observatory, also made a series of observations and drawings of Saturn during the period 1879–89, using refractors with apertures of

26, 15½ and 36 inches, the last-named being at Lick (*P.A.S.P.*, vol. 3, p. 11; summarised in *B.A.A.J.*, vol. 1, p. 339). Holden remarked that his drawings, made with powerful telescopes, did not show certain peculiarities reported by others with smaller instruments; he considered that some at least of these differences are due to Saturn phenomena often being too difficult and at or near the limit of vision for small telescopes whereas they can often be seen with perfect precision in a larger one. In November 1879 he found the southern hemisphere nearly all olive-green and dark with a darker patch near the south pole, the rest of the globe being light. In August 1880 he considered the shadow on the rings less black than the spaces within the ansae, and he could detect no notches at the ends of Cassini's division such as were drawn by Trouvelot in 1874. He saw a trace of Encke's division about 1/3 the distance from the outer to the inner edge of ring A; ring B seemed to be composed of three rings, distinguished by difference in shading, the inner one being about 1/3 and the outer one about 2/5 the width of the ring. He noticed that the crepe ring just inside the limbs seemed wider than, and not in alignment with, its continuations outside the limbs. He could see the outline of the globe through the crepe ring on the east side but not on the west. In October 1880 he witnessed the illusory appearance of the shadow on the rings on both sides of the globe, and saw Encke's division well at 2/5 the width of ring A from the outer edge. On 15 October 1880 large light areas on each limb in the south temperate region, with darker limb shading to north and south, produced an impression rather like the 'square-shouldered' figure reported by Sir William Herschel in 1805; the equatorial zone was full of rose tints.

In 1881 he again found much shading on the inner 2/3 of ring B. In September 1882 Holden observed, drew and explained a remarkable optical illusion produced by Saturn when the rings were getting towards their widest opening: on the east side of the planet the outer ellipse of ring A prolonged towards the west seemed almost exactly tangential to the south limb of the globe, but on the west side the intervening shadow across the rings produced the illusion that the rings were lifted up and that the Cassini division was tangential to the south limb of the globe, while the outer ellipse of ring A seemed to pass south of it.

In November 1882 he detected a narrow dark band across the middle of the equatorial zone, and later in that month when the rings were wide open with the far part of ring A beyond the planet's south limb, the Cassini division touching that limb gave it a flattened appearance. In 1889 on one occasion a minor division was seen on ring B at about one-quarter the distance from the inner to the outer edge of that ring, but the appearance was not subsequently confirmed.

Holden's observations are important as he was a very experienced observer using large telescopes.

CHAPTER 18

RING AND SATELLITE PHENOMENA
1889–92

Rings' reappearance, 1891 (Oudemans, Barnard, Freeman)

Professor J. A. C. Oudemans gave a brief account (in *M.N.*, vol. 52, p. 157) of his observations, with the 9½-inch aperture refractor at Utrecht Observatory, of the reappearance of the rings. At 18^h on 28 October 1891, Saturn was decidedly without a ring, and on the next day at 17^h 30^m the ring was also invisible, but there was a thin dark line crossing the disk along the equator. October 30 was overcast, but on October 31 at 17^h 30^m Saturn was visible for a moment with the ring as a thin bright line on both sides of it.

Edward E. Barnard, in his youth an ardent comet seeker, later became a professor and a most celebrated planetary observer at the Lick and Yerkes Observatories. He witnessed the reappearance of the rings with the 12-inch and the 36-inch aperture Lick refractors and gave an account of his observations (in *M.N.*, vol. 52, p. 419) accompanied by drawings (see Plate VII). He wrote that, while neither the reappearance of the rings in 1878 nor their disappearance in 1891 had been observable owing to the proximity of the planet to the Sun, the conditions for observing the reappearance in 1891 were somewhat better than those at the disappearance of 1878. With the most careful scrutiny and good seeing he found it impossible to detect the slightest trace of the rings projected on the sky on any occasion up to 29 October 1891, though they appeared as 'a heavy black band crossing the planet and apparently cutting it nearly in halves'. On October 22 he had carefully examined Saturn with the 36-inch aperture refractor: the rings were invisible; their projection on the globe was perfectly black with sharp edges and free from any irregularities but slightly convex at the southern edge. No evidence of the crepe ring was seen and the only belt visible was some 3 seconds of arc south of the ring. He obtained the following micrometric measures (corrected for the thickness of the wires):

Width of the middle of the projection	0″51
distance of centre of projection from polar limbs of Saturn: from N. limb	7″40
from S. limb	6″56

At the date and time of these observations the Earth was 1° 36′ north of the ring-plane, and the Sun 0° 8′ south.

On October 26 Barnard carefully examined Saturn with the 12-inch aperture refractor in nearly perfect seeing. There was no trace of the ring against the sky with powers from 150 to 500. He was impressed at this observation with a phenomenon that needs to be taken into account in observations of the ring at or near disappearance: 'Looking at the black trace on the ball, and then glancing at the sky near the sides of the planet, I could, apparently, see the rings for a moment as a faint line of light on the dark sky. I satisfied myself beyond question that this was an optical phenomenon'. He explained this as an 'after image' on the retina of the eye; he said the same sort of thing could be produced by gazing for some moments from a darkened room at the sash of a window projected on a bright sky; if the eye is then turned to a dark part of the room one sees a perfect image of the window sash but with white lines instead of black.

October 27 and 28 were cloudy, but on October 29 he observed the planet with both telescopes. With the 36-inch at 17^h the rings were 'easily and distinctly visible as slender threads of light' against the sky on either side of the planet, but it was not possible to trace them right to the limbs: owing to the nearness of Saturn's bright disk they were invisible for a space of about 2 seconds of arc from each limb. The lines of light were in an alignment slightly north of the north edge of the trace of the rings on the globe. He again measured the width of the projection of the rings on the globe and found the middle width 0″65, which again closely agreed with the calculated value. At 17^h 30^m with the 12-inch aperture he found the rings faint but distinctly visible, though much less conspicuous than in the large telescope. According to the ephemeris, at the observation of October 29 the Earth was 1° 58′ north of the ring-plane, while the Sun was exactly in that plane. He also noted that the following ansa appeared the brighter, and that there were two little lumps, possibly two satellites, on its south edge.

Barnard found on November 3 that the preceding ansa was the brighter and appeared uneven; the rings were very pale and their light weak and they could not be discerned up to the limbs; the trace across the disk was very black. On November 5 and 6 the following ansa seemed possibly the brighter. No spots, dark or light, were seen on the globe at any time during the observations. He considered that, although under more favourable circumstances of observation it was possible that the rings might have continued to be visible in the 36-inch, they were certainly beyond its reach at this disappearance, and therefore their extreme thinness was confirmed; he estimated their thickness at less than 50 miles. From Oudemans's observations and his own, correcting for differences in local time, Barnard concluded that the time of reappearance of the ring must have been between 9^h 3^m and 17^h on October 29 1891, Mt. Hamilton mean time.

The Rev. Alexander Freeman, who two years later became for a time Director of the B.A.A. Saturn Section, made observations of Saturn at the period of the reappearance of the ring, using a 6-inch aperture refractor. His account (*M.N.*, vol. 52, p. 19) states that on October 29 he saw a dark band, which he presumed to be the unilluminated north side of the rings, crossing the planet with its north edge along the equator. His eye estimate of its width was about one-tenth of the planet's polar diameter (Barnard's micrometric measurement made it not much more than 1/28). He also saw the south equatorial belt, and a dusky appearance at the poles.

His small telescope did not enable him to detect the reappearing ring on that date, nor was he able to do so on October 30 or 31. He, like Barnard, noticed the slight convexity southwards of the south edge of the dark band and Freeman also saw two dusky belts south and one north of it. The equatorial zone appeared pale yellow, but most of Saturn pale white. On October 31 the equatorial zone seemed as wide as the dark band, which had appeared a good deal wider than the zone on the two preceding days. Freeman was still unable to see any trace of the ansae of the rings, though he detected a bright round spot on the equatorial zone similar to those which Stanley Williams had been observing in the previous spring (see Chapter 15).

As the sky was overcast on November 1, the first view Freeman had of the fine bluish line of the reappearing ring was at 5^h 30^m in the morning of November 2. Each ansa seemed to him to be about 5/8 of the diameter of Saturn in length, but the following ansa seemed the narrower and less distinct. At the limb the ansae appeared to be nearly as broad as the dark band across the disk. The ansae seemed streaked with white, more or less in line with the equator, and the dark band to be divided near each limb by a fine bright line, which he thought might have been an indication of the bright outer rim of ring B.

When satellite phenomena are observable

The usual satellite phenomena comprise transits of satellites and their shadows across the planet's disk, and the disappearances of satellites into, and their reappearances from, occultation behind the planet and eclipse by its shadow. There are also occasional occultations and eclipses of one satellite by another, requiring the use of large telescopes for satisfactory observation. As the seven moons nearer to Saturn all revolve, like the four bright satellites of Jupiter, in orbits which lie nearly in the plane of the planet's equator, it might be thought that for Saturn, as for Jupiter, many of the usual phenomena would occur at each apparition, but this is not so in Saturn's case. A moment's consideration will show that if, from the point of view of the Earth, the apparent tilt of the planet's equator to the sightline is more than a very few degrees, satellites will appear to pass clear of the planet to the north or south of it and will not transit the disk or be occulted or eclipsed. The apparent tilt of Saturn's equator and ring-plane as seen from the Earth is usually very considerable, and it is only for about two years before and two years after the edgewise position of the rings that their tilt is small enough for satellite phenomena in connection with Saturn to take place. This means that they occur only during five successive apparitions of that planet out of every fifteen. Moreover, the moons of Saturn are far more distant from the Earth than those of Jupiter, and therefore their apparent brightness is much less than that of Jupiter's four bright ones. Hence observers with small telescopes find the phenomena even of Titan and Rhea rather difficult to observe and those of most of the others very difficult if not impossible.

Transits across the planet's disk of a satellite's shadow are easier to observe than those of the satellite itself, because the shadow looks larger, blacker and more clearly defined. A satellite, bright against the sky background, may show up as a little, round, light spot when crossing in front of a dark belt or shadow on the disk, but is apt to look dusky by contrast when silhouetted against a light zone. Satellites and their shadows pass across the disk from the following (east) limb to the preceding (west) limb. The direction of the Sun causes the shadow to precede the satellite in a passage before opposition, the satellite partly to cover its shadow at opposition, and the shadow to follow the satellite after opposition.

Phenomena of Iapetus

Iapetus when at western elongation is brighter than Rhea, but phenomena of Iapetus are rare for two reasons: its great distance from Saturn, and, more important, the fact that the plane of its orbit is inclined at 18° to the ecliptic and at 14° 43′2 to Saturn's equator and ring-plane. Only when Saturn reaches heliocentric longitude about 145° (or 325°) and Iapetus's orbit plane is at ecliptic longitude about 141° (the ascending node of the orbit of Iapetus) or about 321° (its descending node) will the satellite's orbit become edgewise to the Sun, and for a few hours at some date within a period of a few weeks Iapetus may be in conjunction with Saturn, and a phenomenon such as a transit, occultation or eclipse of Iapetus may occur. But no phenomena will occur if Iapetus happens to be at greatest elongation at the crucial time. The chance comes once in about every fifteen years, and on the most recent occasion in 1948, Saturn and Iapetus were too near the direction of the Sun for any phenomenon to be observable.

Figure 11 gives recent illustrations of the apparent path of Iapetus from diagrams in *B.A.A. Hdbks.*: (*a*) in 1947 when the orbit of Iapetus was nearly edgewise to the line of sight; (*b*) in 1958 when the orbit was fairly wide open from the viewpoint of the Earth.

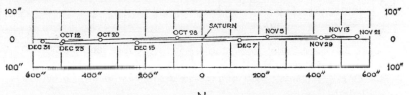

FIGURE 11a. Apparent path of Iapetus 1947 October–December. (Redrawn after lower diagram *B.A.A. Hdbk*, 1947, p. 28).

Saturn reaches a longitude at which the orbit of Iapetus becomes edgewise to the Earth about two years before reaching one (172° or 352°) at which the ring-plane becomes edgewise or nearly so to Sun and Earth. Hence when phenomena of Iapetus may occur the rings are partly open and it is possible for the satellite to pass into the shadow of the rings and be eclipsed by them. This actually happened on the night of 1–2 November 1889, and fortunately Barnard succeeded in observing this very rare phenomenon.

Eclipse of Iapetus by crepe ring, 1 November 1889 (Marth, Barnard)

Great credit is due to the able planetary computer, A. Marth, for predicting this event and its date some five months before it happened (*M.N.*, vol. 49, p. 427). It afforded a unique opportunity for testing the translucency of the crepe ring, one that no other satellite could provide, since the orbits of the seven nearer to Saturn are all too

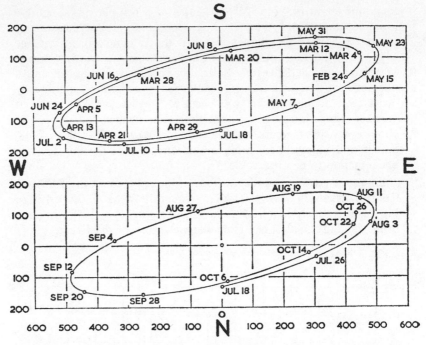

FIGURE 11b. Apparent path of Iapetus 1958 (*from B.A.A. Hdbk. 1958*, p. 38).

nearly in the ring-plane for the rings ever to occult or eclipse them. Barnard's extremely important observation (described in D. G. McIntyre's *The Translucency of Saturn's Rings*, p. 6 and *M.N.*, vol. 50, p. 107) may be summarised as follows:

Soon after Saturn rose above his horizon Barnard picked up Iapetus at about the time of its emergence from behind the globe into one of the spaces between the globe and rings. At first it was very faint, about as faint as Enceladus, but brightened very rapidly, and he decided to make a record of its light-change by comparing its brightness at frequent intervals with the brightness of Enceladus and Tethys, so as to test the effect of the crepe ring's shadow on its visibility. He used the 12-inch aperture refractor as Saturn was unsuitably placed for using the 36-inch. Afterwards he was able to draw a light curve which showed that the brightness hardly varied at all while Iapetus was passing across the space, showing that the space between globe and rings is free or nearly free of obscuring matter. But the satellite's light steadily lessened from the time of its

234

entry into the crepe ring's shadow and continued to diminish for about 1¼ hours. Then, as Iapetus entered the shadow of ring B, the satellite suddenly disappeared. Soon after that the onset of daylight put an end to the observation, so that later accounts suggesting that the observation proved the invisibility of Iapetus in the shadows of the bright rings but its visibility behind Cassini's division and the crepe ring, are inaccurate.

Although the disappearance on entering the shadow of ring B suggested that that ring is opaque, it was not really a good test of ring B's opacity, because the observation ended just as that stage of the eclipse began, and also because, Iapetus not being self-luminous, the sunlight had to pass through the ring and be reflected back through it for the satellite to be seen. Hence ring B could be translucent to a limited extent and yet the satellite disappear as surely as if the ring were completely opaque.

In his description of the physical appearance of Saturn at the 1894 opposition (*M.N.*, vol. 55, p. 369) Barnard pointed out that his observation of the eclipse of Iapetus had proved that the supposed division between rings B and C sometimes shown on drawings, has no real existence, and added: 'There is no distinct junction between the two rings, though the bright ring is not well defined where it meets the Crape ring. It looks as if there was a very quick diffusion of the inner bright ring into the dusky ring, as if, indeed, the latter were formed at the expense of the bright ring'. He also considered that the light curve showed that ring C is very thin at its inner edge, and that it grows much denser where it joins the bright ring. In this respect, therefore, Barnard's observation confirmed the impression gained by Trouvelot fifteen years earlier, as related in the latter part of Chapter 14.

Transits of Titan and Rhea and their shadows (Freeman)

Other satellite phenomena were observed by the Rev. A. Freeman in 1891 and 1892 with a 6½-inch aperture Grubb refractor. Though these were of more ordinary type the accounts he gave of them (in *B.A.A.J.*, vol. 2, pp. 83, 131 and 274) contain certain interesting features. In the early morning of 21 November 1891, he saw the shadow of Titan situated nearly centrally upon the dark rings across the globe, and clearly past the central meridian. 'The shadow was round and black, the colour of the rings, on which it was, a very dark indigo blue, the colour of the ansae beyond the planet a bright light blue.' He continued to watch the shadow and time its progress until, 55 minutes later, only its own breadth separated the shadow from

Saturn's preceding limb. Titan itself was off the disk but seemed likely to pass just in contact with the south pole of the planet when it reached it. Freeman noted among other things the dark spaces between ansae and globe, the shadow of Saturn on the preceding ansa, and that the northern part of the dark rings seemed much lighter than the southern part.

He explained that his observations of the transits of the shadows of Titan and Rhea in December 1891 were made difficult by the small elevation of the Sun, Titan and Rhea above the ring-plane as seen from the centre of the planet. According to H. Struve's then recent determinations the orbits of Titan and Rhea are inclined only 36' 24" and 20' 30" respectively to the plane of the rings, while according to Marth's ephemeris the Sun (as seen from Saturn) was only 45' 2", though the Earth was as much as 3° 31' 50" north of the ring-plane on 16 December 1891. This meant that the greater part of the shadows of both satellites were thrown upon the relatively dark rings crossing the planet; the difficulty was increased by the unclear state of the air. In the early morning of 7 December 1891, Freeman saw the shadow of Titan on the central meridian at approximately $3^h 51^m$ and had it in view for some minutes before and after. About 2/5 of the shadow seemed to be on the globe south of the rings.

At approximately $7^h 29^m$ on December 11 he saw Rhea's shadow on or very close to the C.M., and had seen it at times for nearly an hour before; about 1/3 of it seemed to fall on the disk south of the rings. He saw a bright yellow spot enter the limb on the equatorial zone just south of the rings just after $6^h 32^m$, and observed this for 27 minutes; presumably the yellow spot was Rhea itself in transit.

At approximately $7^h 28^m$ on December 20 he saw Rhea's shadow all but central as a darker spot on the dark rings. It had been seen at intervals for 47 minutes previously and, at first, part of it was slightly south of the rings, but it seemed to pass gradually on to them as it advanced.

On 11 March 1892, when the Earth was at an elevation of 2° 5'.9 and the Sun about the same, north of the ring-plane, Freeman, although much hampered by moonlight and cloud, saw the shadows of Tethys and Dione once or twice when they were in transit across the disk; Rhea and Iapetus were the only satellites seen off the disk. He did, however, get a most favourable view of the passage of Titan's shadow and of the satellite itself as a dark object following the shadow. He carefully noted the times at all stages of the passage, starting from $8^h 51\frac{1}{2}^m$ in the evening, when Titan was a bright object on the limb and the shadow partly within the limb a little further south, until 11^h 26^m when both were well past the C.M. and clouds prevented further observation. His estimates of the times of C.M. passage were: 10^h 33^m for the shadow, and 11^h 13^m for the dark Titan. He estimated the diameter of the shadow as 1" of arc and that of Titan rather less (about 2/3). The satellite appeared during most of the passage to be following its shadow on about the same parallel of latitude across a light part of the disk not far from the south pole. Titan, though not so black as its shadow, appeared dark in contrast with the light part of the disk it was crossing. In addition to recording times Freeman made careful estimates of the positions of the two objects from time to time and also notes of the appearance of the rings and belts.

Other shadow transits (Freeman, Williams)

Long lists of other shadow transits said to have been observed in 1891 and 1892, mostly by Freeman and Stanley Williams, are given in the first Memoir of the B.A.A. Saturn Section (*B.A.A. Mem.*, vol. 2, pt. 1, pp. 9, 17). Both used 6½-inch aperture telescopes, Freeman a refractor, Williams a reflector. Freeman claimed to have observed 26 shadow passages: 6 of Mimas, 5 each of Tethys and Titan, 4 each of Enceladus and Rhea, 2 of Dione; in several cases C.M. passage was within 2 or 3 minutes of the predicted time. His most emphatic statement was that on 25 April 1892 the shadow of Enceladus was seen on the C.M. 'black and certain', touching the S. edge of the ring-shadow.

During a discussion in 1951 (*B.A.A.J.*, vol. 61, p. 90) Dr. W. H. Steavenson referred to Freeman's claim to have seen the shadows of all the inner satellites with a 6½-inch aperture telescope as follows: '. . . this has always seemed to me an almost incredible performance. The shadow of Titan can be seen with an aperture of three inches, or even less, but I have found that of Rhea a delicate object with 20-inches. . . . ' He added that the smaller satellites are much more easily seen if well removed from the planet, and that steadiness of air is an important factor affecting their visibility.

Stanley Williams claimed to have seen about a dozen shadow passages in 1892. He glimpsed Rhea's shadow once or twice under very unfavourable weather and seeing conditions, he obtained a few transits of that of Tethys but found it difficult to observe, and Dione's shadow even more difficult. On one very fine night he obtained a satisfactory transit of the shadow of Mimas, and observed it a little uncertainly on two other nights; on one night he saw the shadow of Enceladus a little doubtfully. He concluded that, with the exception of those of Titan and Rhea, 'these shadows require a larger aperture than $6\frac{1}{2}$ inches (on a reflector) for their satisfactory observation ... or an exceptionally fine night'. One of his best observations was a dark transit of Titan and its shadow on 12 April 1892: 'When in mid-transit the shadow seemed to be about half its own diameter from the south limb of Saturn ... Titan itself was seen beautifully on dark transit, in the chief S. dark belt ... It was a round, dark, but not black spot, about three-quarters of the diameter of its shadow; perhaps a little more, but distinctly smaller than its shadow. It was darkest in the middle part, fading off somewhat on the edges, where it was nebulous looking, not sharp and clean-cut like the shadow. ...'

On April 4 he noted that the shadow of Mimas 'was very difficult to locate'. On April 10 he described the shadow of Enceladus as 'a minute black dot in the south equatorial zone, and separated from the shadow of the ring'. On April 14 he glimpsed the shadow of Tethys 'just on the S. edge of the ring-shadow, and in apparent contact with it'. In good definition on April 22 Williams believed he caught Mimas's shadow as a dark spot on the south edge of the ring-shadow but could only glimpse it now and then.

In the first B.A.A. Saturn Memoir (p. 12) Freeman claimed to have observed several conjunctions of pairs of satellites in the spring of 1892, exactly at or within 3 minutes of the predicted times. On May 16 (he stated) Dione and Enceladus 'formed a mass as large as Rhea, and as bright as Iapetus. ...' He made similar comments on conjunctions of Titan with Dione, and Tethys with Dione.

At the beginning of the next chapter a number of very accurately timed observations of eclipses and transits of inner satellites, by Professors H. Struve and Young, using large telescopes, will be referred to in connection with their use in determining the dimensions of Saturn.

The almost edgewise rings, 1892 (Bigourdan)

It was reported (in *B.A.A.J.*, vol. 2, p. 401) that in the spring of 1892 G. Bigourdan obtained an unusual view of the almost edgewise rings at a time when the Earth's elevation above their illuminated plane was only 23' of arc. His impression was that both ansae tapered towards the globe instead of showing increased thickness near the limbs of the planet such as Otto Struve recorded in 1862 (see Chapter 14). Bigourdan considered that the following ansa showed the tapering in a much greater degree than the other ansa, beginning at about 2/3 of its length from the outer edge, continuing very gradually, and terminating in an acute angle whose apex abutted on the limb. He observed this on several dates in a period of a fortnight, at the end of which the following ansa showed a protuberance near Cassini's division nearly doubling the apparent thickness of the ansa at that point. The protuberance, he considered, was not a projection from the ring's surface so much as a gradual increase of thickness and brilliancy.

Chapter 19

DIMENSIONS OF GLOBE AND RINGS

H. Struve's investigations, 1889–94

In Hermann Struve's long and important paper of 1894 (published in *M.N.*, vol. 54, p. 452) he first of all called attention to the great difficulties, especially optical, of making an exact determination of a planet's diameter: (1) The imperfect union of the rays of light passing the object-glass causes the planet's disk to be surrounded by a secondary spectrum, so its boundaries are not sharply defined; this indistinctness is further increased by (2) diffraction, (3) irradiation,† and (4) unsteadiness of the atmosphere; (5) the dispersion of light in the atmosphere produces different coloration on the planet's northern and southern boundaries; (6) the boundaries will also be made less distinct by the decrease of brightness from the centre to the borders of the disk and the planet's atmosphere; (7) it has also been proved by experience that determinations of diameter depend in some degree on the type of micrometer used. To show that there had been differences between the results of almost all previous determinations of the equatorial and polar diameters of Saturn and the length of the major axis of the ring he gave the following list, in which those determinations based on observations made on ten or more nights are starred:

Observer	Epoch	Equat. diam.	Polar diam.	Axis of Ring	Aperture (ins.) and Micrometer
W. Struve	1826	17″99		40″10	fil. 9-inch refr.
*Bessel	1829–33	17·06	15″39	39·33	heliometer
Encke	1837	17·68	16·49	40·94	fil. 9-inch refr.
Galle	1838	17·91		40·91	fil. 9-inch refr.
*Main	1848	17·50	15·60		Airy's mic. 6¾-inch refr.
O. Struve	1851	17·59		39·73	fil. 15-inch refr.
Lassell	1852	17·45		40·88	jaw mic. 24-inch refl.
Jacob	1853	17·87	16·52	39·93	fil. 6-inch refr.
*Main	1852–5	16·22		39·73	Airy's mic. 6¾-inch refr.
*Secchi	1854–6	17·66		40·89	fil. 9-inch refr.
Jacob	1856	17·94		40·00	fil. 6-inch refr.
Maedler	1860–3	17·18	15·85		fil. 9-inch refr.
*Main	1862	16·88	15·11		heliometer

† 'Irradiation': a bright object appearing to the eye somewhat larger than it actually is.

Observer	Epoch	Equat. diam.	Polar diam.	Axis of Ring	Aperture (ins.) and Micrometer
*Kaiser	1856,				
	1862–3	17″27	15″39	39″72	Airy's mic. 7-inch refr.
*Kaiser	1865–6	16·83	15·47	39·47	Airy's mic. 7-inch refr.
*W. Meyer	1880	17·40	16·18	40·43	fil. 10¾-inch refr.
*W. Meyer	1881	17·77	16·12	40·35	fil. 10¾-inch refr.
O. Struve	1882	17·76		40·16	fil. 15-inch refr.
*A. Hall	1884–7	17·72		40·45	fil. 26-inch refr.

(In the foregoing list the diameters are all reduced to the mean distance of Saturn, 9·5389 A.U.)

Struve pointed out with regard to this list that for all three diameters the differences amount to 1 to 2 seconds of arc, and even if some results are rejected as inferior or as based on too small a number of measures, there still remains an uncertainty of at least half a second; this at the mean distance of Saturn is equivalent to about 2,100 miles. He then drew attention to the fact that there was a substantial difference between (a) the mean of the results of Bessel and Kaiser deduced from many careful observations made on the heliometric method:

equatorial diameter 17″1, axis of ring 39″5;

and (b) the mean of the best filar micrometer observations:

equatorial diameter 17″7, axis of ring 40″3.

In his opinion it could not be decided à priori which of these results is less affected by systematic errors.

As he had only made a small number of direct measures of Saturn's diameter in the usual way with the filar micrometer applied to the 30-inch aperture Pulkovo refractor, but had made long series of observations of the satellites Rhea and Titan to correct the elements of their orbits, and could deduce values for the equatorial and polar diameters of Saturn from these observations, he used this indirect method and derived the following results:

		Equatl. Diam.	Polar Diam.	No. of Obsns.
Rhea series	1889	17″230	15″333	19
	1890	17·352	15·650	23
Titan	1891	17·579	15·798	26
	1892	17·722	15·758	25
	Mean	17·471	15·635	Total 93

He wrote that he did not expect such large differences in the perception of the boundaries as are indicated by the results of the different series, since all the observations were made by the same observer, with the same instrument and in the same manner. He discussed possible causes of the discrepancies.

As he had determined, by observations over several years, the elements of the satellites' orbits within very narrow limits, he felt he would be justified in using the observation of eclipses of satellites as another method of determining the planet's dimensions. He commented on the small number of such observations in former periods: W. Herschel, a reappearance of Titan in 1789; Jacob, a disappearance of Tethys in 1853; Lassell and Marth, a reappearance of Tethys in 1863. The small number recorded was, he thought, due to the fact that astronomers formerly did not possess predictions of these phenomena, like the excellent ephemerides of Marth, and therefore could only observe such occurrences by accident. He pointed out the infrequency of these phenomena, and the fact that accurate observations could only be obtained if the phenomenon happened at a sufficient distance from the limb of the planet and the ring. Hence even in 1890–92 the total number was limited: six observations by himself, four by Professor Young at Princeton, and two by the Rev. A. Freeman at Sittingbourne. These phenomena (for each of which he gave the outline particulars) comprised four each of Dione and Rhea, three of Tethys and one of Titan. From the observed times of disappearance into, or reappearance from, eclipse, and from the orbital elements of the satellites he found very close agreement between the results for three different satellites obtained at Pulkovo and Princeton. From them and the more accordant of Freeman's two observations, Struve obtained the following mean values:

equatorial diameter of Saturn 17″500
polar　　　　　,,　　,,　　,,　　　　15″775
ellipticity　　　　　0·0986

This value of the ellipticity of the globe agreed perfectly with that obtained by Bessel, while Bessel's result for each of the two diameters was too small by 0″4.

The following list of values for the equatorial semi-diameter shows how close to each other were the results from this method of three different observers: it was from this set of figures that the mean of 17″5 for the equatorial diameter was derived:

Date	Phenomenon	Equat. semi-diam.	Observer
1891 Apr. 11	Reapp. of Tethys	8″725	Struve
13	,,	8·729	,,
30	,,	8·735	,,
30	Dione	8·795	,,
8	Rhea	8·767	Young
17	,	8·758	,,
26	,,	8·782	,,
27	Dione	8·772	,,
1892 Jan. 4	Disapp. of Rhea	8·691	Freeman

Struve considered that the values obtained for the diameters of Saturn from eclipse observations of satellites were confirmed by those derived by another indirect method which, however, had larger limits of uncertainty. This was to derive them from his observations of times of egress and ingress of Titan and its shadow in transits that took place on 12 and 28 April 1892. The values he thus obtained were:

Equatorial diameter of Saturn   17″28 and 17″54
Polar diameter of Saturn   15″68 and 15″62

He considered the larger value for the equatorial diameter the more exact, but felt that both these results for the polar diameter and even those found by the eclipses were too small.

Finally, from his exact knowledge of the satellite orbits and his own 1892 observations of conjunctions of the ansae of the ring with satellites (4 with Enceladus, 3 with Tethys, 2 with Dione and 2 with Rhea) he deduced for the major axis of the ring the mean value: 39″2, which agreed well with the mean of Bessel's measures and was considerably smaller than all the other determinations listed by Struve. He found that Rhea and Dione gave somewhat smaller values for the axis of the ring than did Tethys and Enceladus, but could not find any appreciable difference in the extent of the ansae east and west.

Thus by deriving his values through indirect methods Struve tried to get rid of the margin of error that seemed inherent in those obtained by the usual direct measures.

Measures made at Royal Observatory compared with Hall's, Barnard's and others

Professor Barnard carried out long series of careful measures of Saturn and the ring system during the oppositions of 1894 and 1895, using a filar micrometer with the 36-inch aperture Lick refractor (full details in *M.N.*, vol. 55, p. 367 and vol. 56, p. 163). In 1895 a similar series was obtained* at the Royal Observatory, Greenwich, with the filar micrometer on the 28-inch refractor used at full aperture. The report of these results (in *M.N.*, vol. 56, p. 14) contained the following excellent summary, including for comparison the results in 1895 of Barnard at Lick Observatory, Mount Hamilton, as well as those obtained by Professor Hall at Washington 1885–87:†

* By T. Lewis and F. Dyson (afterwards Sir Frank Dyson, Astronomer Royal).

† *Note.* In the first table (the report stated) the measures have all been reduced to a mean distance of 9·53885, and the polar diameters have been corrected for the elevation of the Earth above the plane of the ring. For the second table the mean distance of the Earth from the Sun was assumed to be 92,797,000 miles.

DIMENSIONS OF GLOBE AND RINGS

	Greenwich	Mt. Hamilton (Barnard)		Washington (Hall)
Saturn's Equatl. diameter	17″754	17″744	(17″800)	17″72
Saturn's Polar diameter	16·793	16·307	(16·241)	—
Ring A—outer diameter	40·590	40·249	(40·108)	40·45
Ring A—inner diameter	34·870	34·864	(35·046)	34·95
Centre Cassini division	34·349	34·306	(34·517)	34·53
Ring B—outer diameter	33·828	33·748	(33·988)	34·11
Ring B—inner diameter	25·647	25·522	(25·647)	25·75
Ring C—inner diameter	20·765	20·737	(20·528)	20·52
Width of Cassini division	0·521	0·558	(0·529)	0·42
Width of Ring A	2·860	2·693		2·75
Width of Ring B	4·091	4·113		4·18
Width of Ring C	2·382	2·393		2·625
	(miles)	(miles)	(miles)	(miles)
Saturn's Equatl. diameter	76,190	76,150	(76,470)	76,050
Saturn's Polar diameter	72,066	69,980	(69,770)	—
Ring A—outer diameter	174,190	172,730	(172,310)	173,590
Ring A—inner diameter	149,640	149,620	(150,560)	149,990
Ring B—outer diameter	145,170	144,830	(146,020)	146,380
Ring B—inner diameter	110,060	109,530	(110,200)	110,500
Ring C—inner diameter	89,110	88,990	(88,190)	88,060
Width of Cassini division	2,240	2,400	(2,270)	1,800
Width of Ring A	12,270	11,560		11,800
Width of Ring B	17,560	17,650		17,940
Width of Ring C	10,220	10,270		11,260

The alternative sets of Mt. Hamilton figures, shown above between brackets, were not included in the Greenwich report, but are derived from Barnard's paper of January 1896 (*M.N.*, vol. 56, on pp. 171 and 172); these are included because Barnard regarded them as his final results for the two years 1894 and 1895; he based them on assumed values for Saturn's mean distance and the astronomical unit of respectively 9·538861 and 92,879,000 miles.

The Greenwich report justifiably considered the agreement between the results of the three observatories to be very satisfactory except for the polar diameter, of which only a comparatively small number of measures were made at Greenwich.

The foregoing results were obtained by filar micrometers applied to what were at that time three of the world's largest telescopes. Such large telescopes could, by comparison with small ones, provide a considerably larger image to measure without loss of distinctness in its boundaries. Their determinations should therefore be more accurate than most of the earlier ones among which Struve found such substantial differences, since most of those earlier results had been obtained with small or fairly small apertures.

At Greenwich the edges of the rings were measured from the nearer and further limbs of the planet, both on the preceding and following sides; then the mean of each pair of measures gave the radius of a ring, and the difference the equatorial diameter of Saturn. The polar diameter and the width of Cassini's division were measured directly.

Having compared his measures with previous ones, Barnard considered them to show that no change had taken place in the Saturnian System since the first systematic measurements were made. He added: 'They also thoroughly prove (as did the measures of Hall) the fallacy of the supposition that the rings are sensibly closing in on the planet'.

Barnard also wanted to settle the question of a possible eccentricity of the rings with reference to the globe. To decide this he made a series of measures from the ends of the ring to the preceding and following limbs of the planet. He selected the outer edges because they seemed more definite than the inner edge of ring B, though he also devoted two nights to measuring the latter, and found the result agreed with the others.

He gave details of all these determinations in his paper and concluded that they 'do not show any sensible deviation of the ball from the exact centre of the rings during the opposition of 1894. Hence, if any eccentricity exists it must lie nearly in the line of sight. . . .'

He also made a set of measures of the distance from the centre of the Cassini division in both ansae to the nearest limbs of the planet. He found that these results agreed in placing the globe 0″25 towards the preceding portion of the division, but though they showed this consistency, he was not prepared to accept the difference as real without further verification. He therefore made a further series of such measures in 1895 to settle the question, and concluded that the difference found in the former year must have been due to peculiarities in the measures themselves, as the 1895 ones showed that the distances preceding and following were identical. Hence his final verdict was: '. . . measures therefore go to show that the ball is symmetrically placed in the rings.'

᾽ It may be of interest to compare the dimensions of the system derived from the measures of Barnard and the Royal Observatory with two sets based on some made half a century or so earlier, and with a set based on values in current use in the *N.A.* One of these earlier sets was given on p. 110 of J. R. Hind's *Solar System* (1851); he stated that the figures were based on the most accurate micrometrical measures, but did not specify when, by whom or where they were made. The other set comes from Sir J. Herschel's *Outlines of Astronomy* (1859 edition), p. 342, a book originally published in

1849, and he stated in a footnote that these dimensions were calculated from the micrometrical measures of Professor Struve, presumably Otto Struve. These books were published too early to include dimensions of the crepe ring.

Mid-nineteenth Century Dimensions

	Hind miles	J. Herschel miles
Saturn's Equatl. diameter	(77,224)	79,160
Ring A—outer diameter	172,130	176,418
Ring A—inner diameter	151,500	155,272
Ring B—outer diameter	148,000	151,690
Ring B—inner diameter	114,480	117,339
Width of Cassini division	1,752	1,791
Width of ring A	10,316	(10,573)
Width of ring B	16,755	(17,175)
Interval between Saturn and inner edge of ring B	18,628	19,090

(The figures in brackets are deduced from the published figures)

N.A. Dimensions from B.A.A. Hdbk. 1960

	miles		miles
Saturn's Equatl. diameter	75,100		
Saturn's Polar diameter	67,200		
Ring A—outer diameter	169,300	Width of ring A	(10,150)
Ring A—inner diameter	149,000	Width Cassini div.	(1,750)
Ring B—outer diameter	145,500	Width of ring B	(16,450)
Ring B—inner diameter	112,600		
Ring C—inner diameter	92,900	Width of Ring C	(9,850)

[The figures in brackets are deduced from those published, the width of a ring or division being half the difference between the diameters of its boundaries.

The *N.A.* figures are based on an assumed mean distance of Saturn of 9·538843 A.U. and a value for the astronomical unit of 92,900,000 miles.]

The statistics quoted in this chapter show the colossal proportions of the ring system: a total span of nearly 170,000 miles, and individual rings whose width is about 10,000 miles each in the cases of rings A and C, and about 16,000 miles in the case of ring B. But the most extraordinary feature of the system is the extreme thinness of the rings in proportion to their diameters and surface areas.

Estimated thinness of rings

This can be illustrated by quoting some estimates that have been made at various dates but have not always been correctly stated in astronomical books and papers:

W. Herschel	(1789)	(less than 280 miles) ·	(see Chapter 8)
J. Herschel	(1833)	not more than 250 miles	
G. P. Bond	(1848–9)	not more than 42 miles	(see Chapter 13)
E. E. Barnard	(1891)	less than 50 miles	(see Chapter 18)
H. N. Russell	(1908)	less than 13 miles	(see Chapter 28)
L. Bell	(1919)	less than 10 miles	(see Chapter 28)

Sir W. Herschel's figure is inferred from his observation of the ring's thickness being certainly less than 1/3 the diameter of Rhea, which is now believed to be about 850 miles. Sir John Herschel's value is stated in his *Outlines of Astronomy*, p. 342, a footnote explaining that it was deduced from the total disappearance of the ring in 1833, in a telescope that would certainly have shown a line of light of breadth 0″05; he could therefore have put it as low as less than 210 miles. Bond's estimate during the disappearance of the rings in 1848–49 was a thickness not more than 0″01. The last two estimates are from *Ap. J.* (vol. 27, p. 233 and vol. 50. p. 1).

CHAPTER 20

THE BELTS IN THE 1890s

The B.A.A. and its Saturn Section

The R.A.S., founded in 1820 with the aged Sir William Herschel as its first president, had, as is evident from preceding chapters, made an enormous contribution to the study of Saturn (as well as to that of other members of the solar system), and was continuing to do so, though under increasing pressure from other rapidly growing branches of the science, e.g. stellar astronomy and astrophysics. By 1890 there seems to have been a feeling that it had become mainly a body for professional astronomers and mathematicians, and that the increasing number of amateur astronomers, many of them beginners, had a need which could best be met by an 'association of observers, especially the possessors of small telescopes, for mutual help, and their organisation in the work of astronomical observation'. This was one of the prime objects of the B.A.A., which, with the support of several leading amateur astronomers, was founded in 1890 by E. Walter Maunder, himself a professional and head of the solar department of the Royal Observatory. Within a few years the new society had a world-wide membership of over 1,000, now increased to more than 2,500. From the first it was organised in observing sections, each in charge of a director, the Saturn Section being one of those launched in 1891. The remainder of this book will show how usefully, albeit intermittently, the B.A.A. and its Saturn Section have contributed to the study of the planet during the past seventy years. Directors of the Section since 1891 are named in the following list:

1891–1893	N. E. Green	1929–1931	no director appointed
1893–1895	A. Freeman	1931–1934	M. A. Ainslie
1895–1899	N. E. Green	1934–1935	B. M. Peek
1899–1912	G. M. Seabroke	1935–1939	T. E. R. Phillips
1912–1917	P. H. Hepburn	1939–1946	M. A. Ainslie
1917–1919	W. H. Steavenson	1946–1951	A. F. O'D. Alexander
1919–1929	P. H. Hepburn	1951–	M. B. B. Heath

Equatorial mottled belt and temperate belts, 1892 (Freeman, Williams)

A paper by the Rev. A. Freeman in January 1892 (*B.A.A.J.*, vol. 2, p. 172) stated that the *N.A.* had predicted that the elevations

248

of the Earth and Sun with respect to Saturn's ring-plane would be almost equal on March 12 of that year, so that no ring shadow would be visible on the disk of the planet, which would be within four days of opposition. This, he suggested, might be a good opportunity of testing a theory put forward twenty years before by Professor G. A. Hirn, that the crepe ring particles were much more sparse than those of the other rings, each particle retaining its own atmosphere, and that if the narrow strip of the disk behind the crepe ring were viewed through that ring when the particles were covering their own shadows, the particles' atmospheres might modify the light reflected, so that that small part of the disk might look mottled.

On 12 March 1892 Freeman, observing under excellent conditions, saw a mottling effect as expected, but apparently not just where he had thought to see it. The following is abridged from his description (*B.A.A.J.*, vol. 2, p. 272): '... The crape ring in crossing Saturn was sharply defined on S. edge, and also steadily darkest on that edge and graded down equably to its N. edge... From the N. edge of the crape ring to the true equator of Saturn a narrow bright portion of the planet's disk was visible. Then from the ... equator northwards to points on the limb, a little within the very short portions of the limbs conterminous with the ... boundaries of the bright inner ring beyond Saturn, there was a ...band mottled with both bright and dark patches of circular and oval form ... I am convinced that this ... band, rather less dark than the crape ring, but quite as dark as the chief south belt, *was a ... north equatorial mottled ... belt having its S. edge on the ... equator of Saturn. ...,*'

Freeman therefore could see no mottling on the strip of disk behind the crepe ring; in any case, the elevation of the Earth was too low and the ring too narrow for any sign of the disk to be seen through the ring. The narrow mottled belt he did see was clearly separated from the crepe ring by a bright narrow strip of disk. He still maintained the probability of an atmosphere over the crepe ring, citing the observations of Wray, Struve and Carpenter in 1861–62 (see Chapter 14), but he admitted that this hypothetical ring atmosphere '... did not dim the narrow bright portion of Saturn's disk ... between the crape ring and N. equatorial dull ... belt, ...' and also that Clerk Maxwell had considered that the crepe ring had no atmosphere.

As regards the rest of the disk, Freeman saw the south equatorial belt dark, wide, steely-grey and not divided into two components; the north equatorial belt less dark; both poles dusky; a narrow, dark south temperate belt with traces of two fine streaks between it and the S. pole; a sort of 'quintuple belt' in the north temperate region, consisting of three dark north temperate belts (the southern and middle ones very conspicuous) separated from one another by narrow light zones.

Freeman's observations were made with a 6½-inch aperture refractor; another member of the Saturn Section, Stanley Williams, using a reflector of the same aperture, found Saturn during that apparition similar in general to, but differing somewhat in detail from Freeman's impression (*B.A.A. Mem.*, vol. 2, pt. 1, p. 13). Williams saw two delicate south temperate belts, the southern one being intermittently visible; on any good night the duplicity of the south equatorial belt was very clear; the southern part of the equatorial zone was very bright with brighter condensations here and there; the crepe ring across the disk uniform as to intensity with no spots or mottlings; north of the ring-streak a narrow bright zone of the disk (as seen by Freeman). Williams found the narrow dark band on the equator '. . . perhaps the most curious and interesting one on the planet. Viewed on a fine night with high powers, this dark belt appeared composed entirely of light and dark spots and patches. The northern edge was particularly broken and irregular, and was frequently invaded or indented by bright spots. . . .' from the northern part of the equatorial zone—this irregularity and overlapping had also been observed by Freeman. Williams, however, considered this mottled belt to be right on the equator rather than just north of it.

On very fine nights Williams found the whole surface of the northern part of the equatorial zone to be covered all over with very faint and delicate light and dark mottlings. He also observed a sort of 'quintuple belt' in the northern hemisphere, but formed of the north equatorial belt and two rather wide and more conspicuous north temperate belts, the northern one being the darkest, the belts being separated by lightish zones. Including the ring-streak and ring shadow, Williams distinguished no less than 21 distinct degrees of shading on Saturn's globe, a very unusual variegation for Saturn.

The other members of the Section saw some of the chief features observed by Freeman and Williams, but only one or two of them glimpsed any of the temperate belts, and only one (A. Mee) noted any sign of mottling.

Belt latitudes, 1893 (Williams)

From measurements on two of Stanley Williams's drawings Freeman calculated the saturnicentric latitudes of Saturn's belts

(*B.A.A.J.*, vol. 4, p. 68). In the particulars given here the names of the belts are altered to the modern system of nomenclature of Jupiter and Saturn, and the figures are given to the nearest tenth of a degree:

Belt	North edge	South edge
N. Temperate	N. 36°5	N. 28°6
N. Equatl. (N. Component)	N. 24·4	N. 17·2
N. Equatl. (S. Component)	N. 12·2	N. 6·3
S. Equatorial	S. 14·5	S. 18·0
S. Temperate	S. 34·5	S. 43·6
S.S. Temperate	S. 45·1	S. 49·6

The paper also stated that Williams had made micrometric measures of the middle of the northern component of the S.E.B. on three nights in 1892 the mean result being S. lat. 16°7, which is in good agreement with the figures for 1893. As the belts of Saturn, like those of Jupiter, are liable to vary in position, records of this kind are useful for comparison with those at other oppositions. The formulae for calculating saturnicentric latitudes of belts are given in Appendix II.

Belts in 1894 and 1895 (Cammell, Roberts; Barnard)

In March 1894 B. E. Cammell obtained a fine detailed view of Saturn, using his 12½-inch aperture Calver reflector (*B.A.A.J.*, vol. 4, p. 362). Owing to the tilt of the rings what he saw was mostly in the equatorial region and the northern hemisphere. He found the southern part of the equatorial zone much the brighter, and dividing it from the northern part a faint spotted line, representing the equatorial mottled belt. North of that he saw three belts and two zones forming a quintuple belt, the faintest being the north equatorial belt and the darkest the one on the north polar side (the N.N.T.B.), very much as Williams had seen in 1892. North of these belts was a much lighter zone and then a very distinct dusky north polar cap.

In January 1895, C. Roberts, with a 6½-inch aperture reflector, had a similar view (*B.A.A.J.*, vol. 5, p. 219) to Cammell's of 1894 but found the belt on the equator a very narrow dark one without mottling, and saw four belts in the northern hemisphere: the two nearest the equator presumably the north equatorial belt in two well separated components; the others two north temperate belts to the north of a fairly wide bright zone.

It was at the oppositions of 1894 and 1895 that Barnard made his micrometrical measures and tried in vain to detect the spots seen by Stanley Williams and others (see Chapters 19 and 15). He was using the 36-inch aperture Lick refractor, made a drawing (reproduced in Plate VIII), and described Saturn as he saw it at both oppositions (*M.N.*, vol. 55, pp. 368, 381). After commenting on the total absence of spots, Barnard said of the globe in 1894: '. . . One narrow dusky belt alone was visible upon the ball. This was placed near the middle of a broad light equatorial zone, and just north of the trace of the rings.

'The light zone was bordered to the north by a rather heavy dusky diffusion extending to the north limb. The southern edge of this—where it met the light zone—was of a warm colour, and gradually diffused into a colder or ashy-hued region that formed the true polar cap, and which was almost definitely bounded at its southern edge. Once or twice a very small and very dark spot was seen *at the North Pole. . . .*'

Hence the only belt Barnard distinguished from the general shading was the one on the equator, the 'mottled' belt of 1892. Barnard stated that the finest view he had ever had of Saturn was on the morning of 1 April 1895, but no abnormal features were visible. He found the narrow dark belt near the equator, drawn by him in 1894, very distinct, but the space between it and the 'north dusky region' seemed narrower than before. This region he described as 'a very broad warmish dusky belt distinctly terminated at its north edge'. North of that he saw a wide light yellow-green region, but 'at the pole there was a small, very heavily marked dark cap, which was nearly as dark as the crepe ring against the sky' . . . and 'strikingly and sharply defined in outline. . . .' Though he had seen this dark cap before, and in the previous year, it had never seemed so distinct.

Belts 1896–1900 (Antoniadi, Flammarion, Barnard)

Owing to the increasing southern declination of Saturn, the planet was difficult to observe during this period. The report of the Saturn Section for 1896 (*B.A.A.J.*, vol. 7, p. 236) showed little change in the globe features from 1895; numerous hazy spots were drawn on the double north equatorial belt. Antoniadi drew attention to the distinct duplicity of that belt the width of which he estimated at some 20°, and to the very marked darkness of the north polar region, though he was unable to detect any distinct 'black cap' at the pole such as Barnard had seen in the previous year. He found the dusky spots on the north equatorial belt comparatively easy to see in 1897 (*B.A.A.J.*, vol. 8, p. 203), but very few in 1898 and the dark polar cap not very obvious (*B.A.A.J.*, vol. 9, p. 292).

Flammarion reported (*M.N.*, vol. 60, p. 441) observations in 1899 by Antoniadi and himself with the Juvisy 10¼-inch equatorial: the polar cap was not particularly dark, certainly lighter than in 1895; the double north equatorial belt was very obvious, but only vague traces of dark spots were glimpsed on it on two occasions. The equatorial zone 'did not show its ordinary "wool pack" structure'; the narrow equatorial band was invisible. 'An interesting "black drop" appearance, due to irradiation, was repeatedly detected in the shadow cast by the planet on the ring, where it met the Cassini division'. Antoniadi's drawing at that opposition is reproduced on Plate IX.

Flammarion's report on the Juvisy observations in 1900 (*M.N.*, vol. 61, p. 129) with the 9¾-inch aperture telescope stated that the north polar cap was not very dark at that apparition, and no certain traces could be seen of a north temperate belt; on the other hand, the great double north equatorial belt was a very striking feature, its duplicity being recognised even when definition was poor. The dark spots on that belt were much better seen than in 1899. The equatorial zone, though easily the brightest and yellowest part of the disk, was very uneven in tint and mottled in appearance. The faint dusky belt almost on the equator was twice seen with certainty. 'The decreasing luminosity of the globe towards the limb was very marked.'

Barnard reported a few observations of the belts 1897–99 (*M.N.*, vol. 68, p. 366). In May 1897 he found the north polar cap dark bluish, bordered by a light zone. On July 1 with the 12-inch aperture refractor the shadow of the globe on the rings seemed to him 'squarey' where it struck the Cassini division; he attributed this to a kind of 'black drop' effect by the junction of the black shadow with the black space. The polar cap, dark grey and not large, was bordered by a light space, then a delicate dark narrow belt; there was no light equatorial zone.

In April 1898 there was no longer a definite dark polar cap but the polar regions were dark; the shadow on the rings still looked 'squarey'. On 7 July 1898 with the Yerkes 40-inch aperture refractor he obtained one of the finest views he had ever had of Saturn, and saw more detail than ever before; he made a drawing showing no less than five diffuse belts in the northern hemisphere, interspersed with light zones. He saw all these belts with certainty, and a north polar cap darker than anything else on the globe. In April 1899 he found the north pole not very dark, with no well-defined cap, while a few years later, in 1904, the north polar region had become light yellow in colour.

Chapter 21

THE OPENING RINGS IN THE LATER 1890s

Rings in 1894 and 1895 (Barnard)

Barnard's remarks on the aspect of the rings in 1894 and 1895, as seen with the 36-inch aperture Lick refractor, were of a somewhat negative nature (*M.N.*, vol. 55, pp. 369, 382): the crepe ring in 1894 was steely blue, uniformly even in shade and not strongly contrasted with the sky. 'No markings whatever were seen upon it. The inner edge was a uniform curve; the serrated or saw-toothed appearance of its inner edge, . . . previously seen with some small telescopes, was . . . beyond the reach of the 36-inch. No division was seen between the Crape ring and the inner bright ring, as . . . sometimes . . . shown on drawings . . . The bright ring and Crape ring so gradually merge into each other that no one can tell where one begins or the other leaves off. . . .' (This verdict is worth noting in view of the alleged discovery of a division between them in 1897.)

Barnard also stated that his Iapetus eclipse observation (Chapter 18) had proved the increasing density of ring C towards ring B, and this could readily be seen in examining the trace of the crepe ring across the disk: '. . . In the present observations the ball could faintly be seen through a small portion of the projection of the Crape ring, but not up to the bright ring. This trace was, on account of the bright ball behind it, sensibly lighter than where the ring was projected against the dark sky.

'The inner portion of the inner bright ring is exactly the same depth of shade as the outer ring. About $\frac{1}{4}$ of its outer edge, next the Cassini division, however, is very bright . . . brighter than any other part of the ball or rings.

'From the southern declination of the planet one or two features previously seen were not readily made out. The division in the outer ring seen by Professor Keeler in 1888 was not visible at any time, though I had seen it in 1889.

'The Encke "division" was not seen at the time of my drawing, though I have seen it before . . . I do not think this marking is a real division like that of Cassini. It has always appeared to me to be a rather feeble dark line *on* the ring. But, as I understand, it has been seen on both the south and north faces of the ring. It must therefore penetrate through the ring. It would seem probable that at this point in the rings all the particles for some reason have not been removed, as appears to have occurred in the case of the Cassini division; the "division" would therefore be only partly free of them, or in a similar condition to the Crape ring.'

Barnard's remarks on the rings in 1895 were briefer: 'The Encke "division" was seen faintly at both ansae, but more distinctly following. It was nearly as broad as the Cassini division. . . . The trace of the crape ring was pale and the ball was easily seen through it up to the bright ring. On the sky it was clearly defined, very distinct, and of a steely blue colour.'

The shadow of the globe on the ring on the preceding side he found 'narrow, slightly concave, and concentric with the ball. At the following side there was a darkening of the ring and the ball suggestive of the faintest trace of the shadow on that side.' This optical effect has even appeared on photographs.

Ring markings, 1896 (Antoniadi, Roberts, Brenner)

On 25 April 1896, C. Roberts, observing with a $6\frac{1}{2}$-inch aperture Herschelian reflector and power of 160, found the definition quite perfect and was easily able to see Cassini's division all round the visible part of the rings. He saw Encke's division well on both ansae, finding it decidedly outside the centre of ring A, dividing that ring in the ratio of about 2 : 3 (*B.A.A.J.*, vol. 6, p. 337). His report and drawing make it clear that he saw no sign of unusual markings on either bright ring.

It appears, however, from an announcement accompanied by a sketch (see Figure 12), that E. M. Antoniadi at Juvisy had detected very unusual markings on both bright rings on 18 April 1896 (*ibid.*, p. 339) as follows: '. . . a new and well-defined division and two fainter ones on the inner bright ring . . . The first and darker of these objects lies nearly midway between the Cassini division and the crape ring, somewhat nearer the crape ring very likely. It is an obvious feature, and, I believe, quite accessible to a $6\frac{1}{2}$-inch reflector. I cannot say as much of the other two divisions, which I could do no more than glimpse with our $9\frac{3}{4}$-inch equatoreal, magnifying some 300 diameters. Instead of the Encke division, ring A shows (just now) some enormous white spots separated by dusky intervals. This ring appears broken (as it were) into fragments. . . .'

N. E. Green, Director of the Saturn Section, drew attention at a meeting of the B.A.A. to the sharp disagreement between two observations so close together in date, and asked for further observations to be made. The President (E. W. Maunder) suggested that owing to the rapid rotation and the fact that the rings consisted of clouds of particles, changes in the disposition of the particles and the appearance of the rings might take place in an interval of a week (*B.A.A.J.*, vol. 6, p. 326).

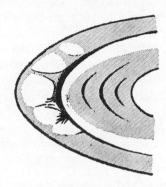

FIGURE 12. Antoniadi: sketch of ring markings (northern face) 18 April 1896. (Redrawn after Antoniadi's sketch in *B.A.A.J.*, vol. 6, p. 339.) The radial streaks, at least, were probably illusory.

Antoniadi had suspected the three divisions on ring B in July 1895, and after observing them on 18 April 1896 had written to Leo Brenner at Lussin asking if he could confirm them. Brenner and Philip Fauth (a well-known lunar observer) made a joint observation on April 26 and were able to see a marking rather like Encke's division at about the middle of ring B, but not the two fainter markings Antoniadi had detected on that ring; on ring A they saw Encke's division clearly but no unusual markings (*ibid.*, p. 384). They noted that Cassini's division was bordered on both sides by brilliant zones, that on ring B being a little broader, that on ring A a little narrower, than Cassini's division.

The day before he received a copy of the *Journal* containing Antoniadi's drawing C. Roberts had noted that the outer edge of Cassini's division was 'curiously indefinite and indented'; on May 8 he was able to confirm this indented appearance as seen by Antoniadi and also a 'pretty dark' division on ring B about 3/5 of the distance from the outer to the inner edge; he also noted two 'curious triangular bright patches' on the preceding ansa of the crepe ring (*ibid.*, p. 389). On May 25 he also glimpsed a faint marking on ring B corresponding to the outermost of Antoniadi's three markings (*ibid.*, p. 435).

A substantial part of the Saturn Section's report for 1896 (*B.A.A.J.*, vol. 7, p. 236) consisted of a report from Antoniadi mainly concerned with the rings, and accompanied by a drawing (Figure 12), showing his three subdivisions on ring B, the outer two appearing as edges of duskier shading, and also showing half a dozen dark radial marks on the inner part of ring A at each ansa, which 'seemed to shoot forth in the direction of radii emanating from Saturn's centre'. Antoniadi stated that though he had seen Encke's division the year before and had made a 'rigorous and systematic examination', no trace of it was to be seen in 1896. He considered that the central division he had seen on ring B divided that ring in the ratio of 3 : 4 from the inner edge outwards. The other two very faint lines were about equidistant from it on either side.

Antoniadi went on to report that while the central division was obvious on April 18, it had already become much less distinct by April 24, and visible only on the following ansa from May 9 to June 12. The other two lines, glimpsed with certainty by him on April 18 and 24, had not been satisfactorily seen since.

He expressed the opinion that as the three divisions corresponded to the edges of gradations of shading, the ring might be regarded as composed of four concentric rings, though the separations were very probably not quite free of matter, being filled up at times through the mutual attraction of the particles as well as the attraction of the satellites of Saturn. This is reminiscent of Dawes's and Secchi's 'step-like concentric bands of shading on ring B' mentioned in Chapter 11.

Antoniadi may have encountered some scepticism and have felt that the announcement of three subdivisions in ring B required some justification, or it may have been purely for historical interest that he stated that divisions on ring B had been seen 'on rare occasions by some of the best observers of the planet', and then proceeded to give a fairly long list of past instances, which may be summarised as follows:

1780   Herschel's 'black list', near inner edge of B.

1838   De Vico—two divisions in ring B.

1851   Tuttle—'a new division not far from the exterior edge' of B; 'a great many narrow rings occupying more than half the interior surface' of ring B.

1855 ⎱ Coolidge—three or four dark divisions on interior part
1857 ⎰   of B.

1875   Asaph Hall suspected divisions in B.

1876   Asaph Hall: '... At times there seemed to be a division of the inner bright ring, but it was not certain.'

New outbreak of division-finding, 1897 (Brenner, Schaeberle, Fauth)

The summary of the report of Flammarion and Antoniadi on Saturn in 1897 (in *B.A.A.J.*, vol. 8, p. 203) said that the radial markings on ring A were obvious, and Encke's division, 'a difficult feature', was seen about three-eighths of the distance from the outer to the inner edge of ring A. 'No trace of divisions on ring B was to be found by Antoniadi in 1897. Nor did any separation exist between rings B and C.'

But Antoniadi's discoveries had started a new pursuit of ring subdivisions that was not to be called off even by the original discoverer finding that the objects of the chase had disappeared.

The discovery of a new division, separating rings B and C, had been claimed by a Mrs. Manora, although of course Dawes and others had drawn it many years before, and in 1889 it had been shown to be non-existent by Barnard. Brenner sent to the B.A.A. a short report with a drawing of Saturn dated 19 July 1897 (*B.A.A. J.* vol. 7, p. 513). The drawing shows not only Encke's division but also Antoniadi's chief one on ring B (which Antoniadi himself could no longer detect) and the Manora one; also no less than 25 bright and dark spots on the globe, as if Saturn had developed a dreadful rash all over the northern hemisphere from pole to equator. The report admitted that the Antoniadi division was not always visible on both ansae, and Encke's only when the image was steady. Brenner also stated that Krieger at Trieste had twice seen the Manora division with a 10-inch aperture refractor. Brenner measured this division in June 1897, finding it $0''56$ wide and stated that it looked very similar to Cassini's division (*B.A.A.J.*, vol. 8, p. 155).

The following statement from J. M. Schaeberle (*B.A.A.J.*, vol. 7, p. 519) presumably relates to 1897: 'Observations at the Lick Observatory during the opposition of Saturn show a division in the B ring, $0''7$ from its inner edge. It is not complete, as it reflects some light. No division was seen between rings B and C'.

Fauth's observation on 15 June 1897, with a refractor of about $7\frac{1}{2}$-inches aperture showed Encke's division easily; ring B divided into three parts, with Antoniadi's division seen as a faint line; ring C clear but very weak and divided from ring B by a distinct dark shading (*B.A.A.J.*, vol. 8, p. 156).

Brenner announced the discovery, on 27 August 1897, of a new division 'inside of Encke's division in ring A. The Manora, Struve, and Antoniadi's divisions were also visible, but more distinct' on one ansa (*B.A.A.J.*, vol. 8, p. 259).

As had happened half a century earlier (Chapter 11), there was much contradiction among the various observations of ring markings, and sceptics might be inclined to dismiss them all as due to optical and atmospheric causes, differences between depth of shading in adjoining ring areas, or even the bias of preconceived ideas: the difficulty is to decide which of these temporary appearances were truly objective and which were not.

Eccentricity of rings (Antoniadi, Flammarion)

Antoniadi stated (*B.A.A.J.*, vol. 7, p. 242) that in 1895 Flammarion and others saw distinctly that the rings were eccentrically placed, invariably finding the eastern space between globe and rings wider than the western. In 1896 this was checked by micrometer: two sets of accordant measures gave Antoniadi 4".3 for the distance from the planet's limb to the interior edge of ring B on the eastern side, and 4".0 for the corresponding distance on the west. Flammarion later stated that this effect was very marked indeed in 1900. But Barnard's opinion after investigation in 1894 had been negative (see Chapter 19), and so was the result of a thorough investigation by Stroobant in 1934 (see Chapter 35).

Rings in 1898 (Barnard)

In February 1898 Barnard was able to see the globe clearly through the crepe ring. On 7 July 1898 he had an exceptionally fine view of the widely open rings: the trace of the crepe ring across the globe was hazy at both edges, so that the inner edge of ring B was ill-defined where it crossed the planet, but the rings were well-defined against the sky. With the 40-inch refractor he could not see with certainty any division in ring A; there seemed to be merely a dusky shading where the Encke division is usually shown (*M.N.*, vol. 68, p. 367).

Rings in 1899 and 1900 (Flammarion)

Flammarion's report on the rings in 1899 (*M.N.*, vol. 60, p. 442) said that Encke's division was seen only on July 30, the best night of the season, when it was perfectly visible on both ansae (see Antoniadi's drawing, Plate IX). An easier feature of ring A was 'a series of dusky indentations emerging from the Cassini division. The outer edge of A was in no wise sharply defined, but seemed to shade off rather gently into space'. Cassini's division could be traced easily all round the ring, even under very poor seeing; it was dark grey, not black, and seemed tangential to Saturn's north limb. Rings B and C seemed to shade into each other 'without any intermediate separation', but on July 30 ring B was 'split into two rings by the certain visibility on both ansae, of a narrow and faint dusky line' (as shown in Antoniadi's drawing).

At Juvisy in 1900 (*M.N.*, vol. 61, p. 129) Encke's division, though carefully looked for, could not be seen, while the indentations along the ansae of A by Cassini's division were easily seen on many occasions, appearing triangular in shape, one side of each triangle resting on the inner edge of A. Cassini's division seemed as usual grey, not black, so there is probably some matter in it. Ring B was destitute of detail throughout 1900, the gradations of shading which Trouvelot had shown being invisible, though they had been so marked a year or two before. B's inner edge was quite indefinite, and at times it was impossible to say where B ended and C began. C was fainter than usual in the ansae, probably through Saturn's low altitude; on October 2, ring C across the globe seemed certainly wider on the limbs than the part projected on the sky and therefore at its inner edge out of alignment with its inner edge in the ansae.

Constitution of crepe ring (Antoniadi)

Around the turn of the century Antoniadi gave much careful thought to the nature of the crepe ring and its particles. In his 1896 report (*B.A.A.J.*, vol. 7, p. 241) he stated that whenever definition was good, ring C looked purplish on the ansae, but across the planet grey and quite transparent for Saturn's limb. He then made the following interesting suggestion: 'What we really see of the crape ring on the planet, a dusky belt, is not, I believe, the ring itself, whose particles, ... though brighter than the sky, might be less luminous than the globe, but the (penumbral) shadow on the planet of these same particles, which are quite as luminous as those of rings A or B, but more rarefied in the case of ring C, a shadow which we see through the interstices separating these flights of meteors. The fact that the crape ring is darker when the rings are most widely open, and lighter some two or three years after disappearance, is a confirmation of this view, as the gaps being greater in the first case, allow us to see the shadow of the meteors to its best advantage, while in a very oblique position we can scarcely see more than the little bodies themselves, hiding a great part of their shadow from our view.'

Antoniadi's idea therefore was that the albedo of the crepe ring particles is comparable if not equal to that of the particles of the bright rings, but that in the crepe ring the particles are more sparsely distributed; also that the 'dark' ring across the planet is invisible as far as the particles themselves are concerned, the trace on the planet consisting of the shadows of the particles, which probably being individually very small would cast a penumbral shadow on the globe.

He examined this theory in more detail three years later (1899, in *M.N.*, vol. 59, p. 498). He first of all referred to Barnard's observation of the passage of Iapetus behind the crepe ring (see Chapter 18) which showed that ring to be very thin at its inner edge and much denser where it joins ring B, which is confirmed by the fact that ring C is more transparent to the planet's limb towards its inner edge than close to ring B. The fact that with smaller telescopes the reverse has often been noted in the ansae, is, Antoniadi considered, merely due to contrast, the brightness of B making the part of the crepe ring close to it seem much fainter than the inner edge of C seen against the sky background.

(1) He then showed that it was in accordance with his theory that the outline of the dusky shadow projected on the globe 'would not usually be a rigorous continuation of the "nebulous" ansae'. Should the Sun be higher than the Earth above the ring-plane, the breadth of the shadow across the planet would shrink. On the other hand, should the Sun be lower above that plane than the Earth, the breadth of the shadow would be increased by the addition of the shadow of the inner edge of the bright ring, but its darkness would be much attenuated as it would be viewed through the outer part of the swarm of the ring C particles where they are thickest.

(2) His next deduction was that as the real intensity of the crepe ring's shadow is an inverse function of the Sun's altitude above the ring-plane, the transparency of the so-called dark ring ought to diminish with the closing of the rings, because (*a*) the 'perspective grouping of the particles would, in this case mask more effectively the planet's limb', and (*b*) the transparency would be further reduced by the strengthening of the shadow of the particles owing to their closer apparent grouping.

Antoniadi then showed that both these deductions had been confirmed by observation. Regarding the second deduction he quoted Proctor (*Old and New Astronomy*, p. 632): 'As the ring system closes up, the distinction between the dark ring and the neighbouring bright ring becomes less marked, the dark ring appears greyish or slate-coloured, . . . while the outline of the planet is either not seen at all through the dark ring, or only seen with difficulty and indistinctly'.

With regard to the first deduction, he showed that Dawes (*M.N.*, vol. 11, p. 52) had found the projection of ring C at its minor axis considerably narrower than would be expected from its breadth at the major axis,* at an epoch when the Sun's elevation above the southern face of the ring-plane was nearly $2\frac{1}{2}°$ more than the Earth's.

Antoniadi considered that were the crepe ring really a 'dark' ring, it ought, whenever the Sun's elevation above the ring-plane is considerable, to be possible to distinguish the dusky projection from the shadow it would cast on the globe. Dawes looked for this in 1862, but without success,† and his failure cannot be explained by assuming a dark ring, but follows logically from Antoniadi's theory. Trouvelot had tried to explain the phenomena observed by Dawes and himself (narrowing of the dark ring at the minor axis) as an effect of irradiation from Saturn's globe, but did not feel satisfied with his explanation, and thought there must be some other cause. The idea that the albedo of the crepe ring particles might be similar to that of those of the other rings did not occur to Trouvelot. Antoniadi admitted that irradiation would affect the breadth of the shadow cast on the globe by the crepe ring, but only to a slight extent, as the intensity of the shadow is also slight. As, however, the planet's luminosity is greatest towards its centre, decreasing very rapidly towards the limbs, the effect due to irradiation would be greatest at the minor axis and least by the limbs; this would tend to exaggerate the concavity of the shadow's outline with respect to the centre of Saturn.

By this theory of the crepe ring consisting of brightly reflecting particles loosely scattered Antoniadi made a notable contribution to the understanding of the phenomena of Saturn.

* See Chapter 12.   † See Chapter 14.

CHAPTER 22

PHOEBE

As the last seven chapters have been devoted to the observations and studies of Saturn made during the last quarter of the nineteenth century it might be supposed that the tale of the remarkable achievements of that period was complete, but there is still one more important item to add—the discovery, by photography, in 1898 of Saturn's ninth satellite. In the very early days of the observation of Saturn with telescopes Kepler is said to have suggested that the planet would probably be found to have six or eight moons, a prediction that had been completely fulfilled by 1848 and was now, half a century later, to be exceeded.

Photographic discovery, 1898 (Pickering)

The discoverer was Professor William H. Pickering of Harvard Observatory who became one of the foremost lunar and planetary observers of the early twentieth century, and whose mathematical calculations on the motions of Uranus and Neptune were eventually to lead to the discovery of the planet Pluto. Pickering gave the name Phoebe to Saturn's ninth satellite, and the most remarkable thing about his achievement was that it was the first instance of a satellite being found by photography and from an observatory in the southern hemisphere. The photographic discovery of minor planets had begun in 1891 but until 1898 attempts to find satellites by this method had failed owing to the lack of a suitable lens. According to a summarised account of the discovery (in *B.A.A.J.*, Vol. 9, p. 291) a special lens of 24 inches aperture and 160 inches focal length, which had been presented to Harvard Observatory, was used at its southern station at Arequipa, Peru, on August 16, 17 and 18, 1898, and four plates were exposed for two hours each, 100,000 stars appearing on each plate. The plates were examined in pairs, placed together, so that each star was represented by two contiguous dots. The suspected object, which seemed to share Saturn's motion relative to the stars, showed as a single dot. It was found that the object's motion (in relation to the background stars) was direct, less than Saturn's and nearly in the same direction. It therefore could not be a minor planet, but must be either a satellite of Saturn or a more distant planet, and its proximity to Saturn—distance on August 17: 24' 40"—

made it more probably a satellite. Its apparent orbit was found to be a very elongated ellipse, and from comparisons with Hyperion, the new satellite's magnitude was estimated at 15·5, though it was thought likely to have a magnitude of about 6 if observed from Saturn. Assuming it to have a reflecting power equal to that of Titan, its diameter was estimated at about 200 miles. Nowadays the stellar magnitude of Phoebe is estimated at 14 and its diameter at only about 150 miles.

Retrograde motion (Crommelin)

In 1904 a very interesting paper on the orbit of Phoebe (*B.A.A.J.*, vol. 15, p. 32) was presented to the B.A.A. by A. C. D. Crommelin, who had just been elected its President. Crommelin, a very able computer on the staff of the Royal Observatory, Greenwich, who later became very well known for his mathematical work on the orbits of Halley's and other comets, stated in his paper that widespread interest had been aroused by the announcement of the discovery of this tiny satellite far outside the others, but that as months and years passed on with no further news, many astronomers had begun to feel sceptical about the existence of Phoebe. In July 1904, however, Pickering had stated that further photographs had been obtained, and had shortly afterwards given three more positions of the satellite. Then on September 12, Barnard had succeeded in detecting Phoebe visually with the great refractor at Yerkes Observatory, determining its position and noting its motion and apparent magnitude, so that its existence could be regarded as beyond dispute.

Crommelin then proceeded to discuss the five Harvard and Yerkes positions of the satellite—one of 1898 and four of 1904—and from this meagre information calculated that Phoebe must have a revolution period of more than 400 days, be at a distance of some 7 million miles from Saturn, and have an orbit inclined at only a few degrees to Saturn's orbit-plane and the ecliptic, but at a much greater inclination to Saturn's equator. He made his calculations on the two alternative assumptions of direct and retrograde motion by the satellite in its orbit. He found that on the former assumption the position angle of the minor axis of the apparent ellipse would have worked out at 6°6 for August 1898, whereas on the latter assumption it worked out at 15°4. Since the value of that angle deduced from the photographs of August 1898 was about $14\frac{1}{2}°$, he concluded that the retrograde assumption alone seemed tenable, 'unless we suppose most extravagant errors in the 1898 position. This implies the startling conclusion that Phoebe is going round Saturn in the opposite

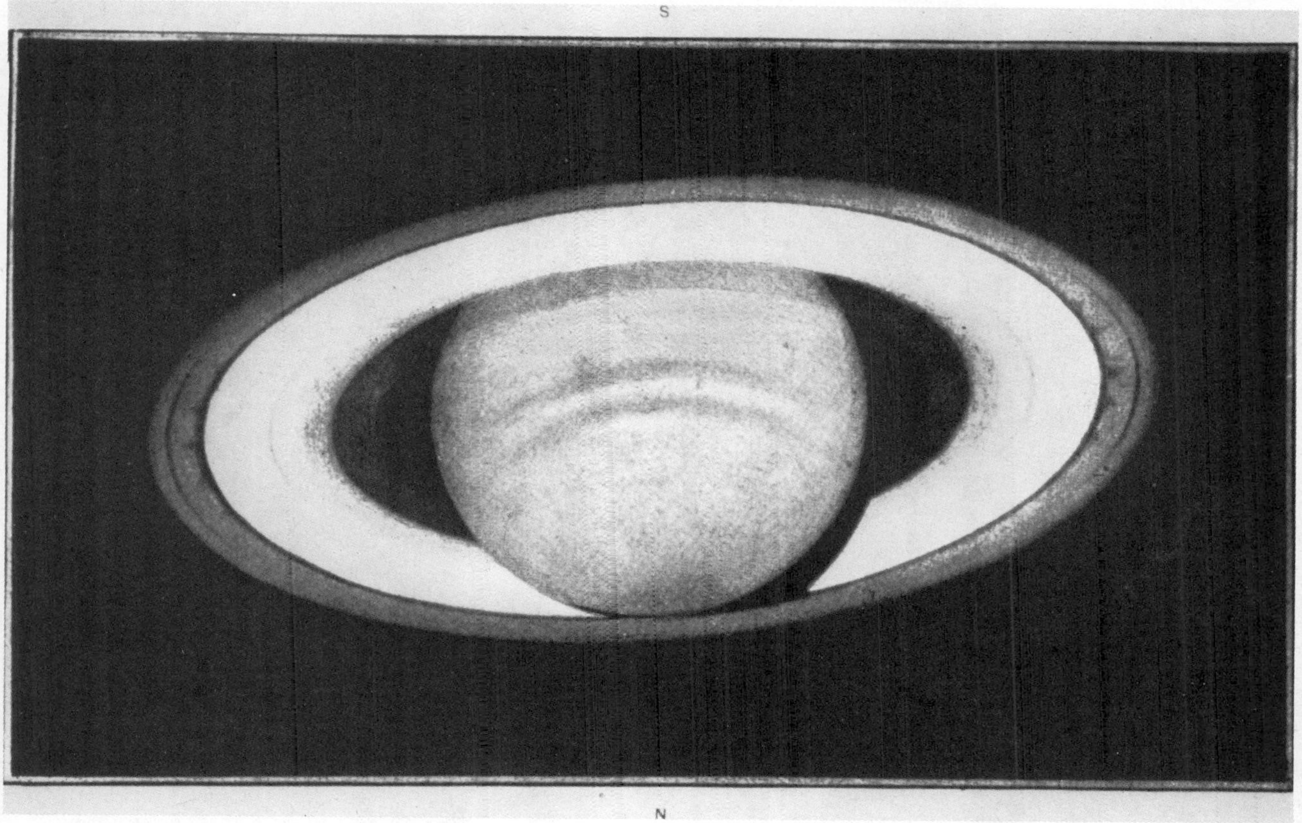

PLATE IX.—E. M. Antoniadi's drawing of Saturn ($10\frac{1}{4}$-inch refractor at Juvisy Observatory) on 1899 July 30, showing Encke's division (ring A) and a faint linear marking on ring B (from *M.N.* vol. 60, plate 12).

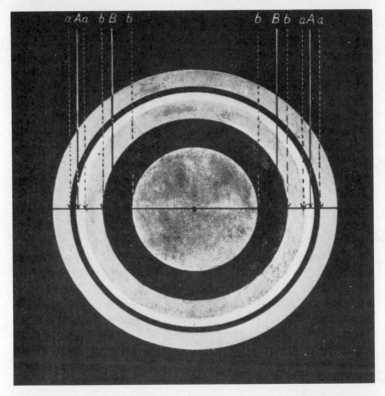

PLATE X.—Barnard's drawing of Saturn, 1907 November 25 (40-inch
Yerkes Observatory refractor), showing condensations on the ring's
unillumined face, with his diagram of their positions in the ring system
(from *B.A.A.J.* vol. 30, plate 1). The white round object below the f.
ansa may represent Enceladus, with which he often compared the
condensations in brightness.

direction to all the other satellites of that planet. We are indeed familiar with retrograde satellites in the families of Uranus and Neptune, but this is absolutely the first case on record of a family being divided against itself'.

Crommelin went on to point out that this result was unfavourable to the nebular hypothesis, and that therefore upholders of that hypothesis would probably argue that Phoebe was not an original member of Saturn's family but a later capture 'which is, of course, possible, though the chances of such a capture are slender indeed'.

He estimated the diameter of Phoebe at about 150 miles. In view of its immense distance from the planet, he considered that the perturbations by the other satellites would be very small, but those by the Sun probably quite appreciable. He had had to assume the orbit to be circular for the purpose of his calculations, as he had insufficient material for deducing the eccentricity.

After writing his paper Crommelin received Pickering's discussion of the orbit, based no doubt on more ample data. This fully bore out his brilliant deduction that the direction of Phoebe's motion in its orbit is retrograde. Pickering, however, made the distance about 8 million miles and the period 547 days, and had deduced a considerable eccentricity, but gave the position of the orbit-plane very nearly in agreement with that computed by Crommelin.

It may be mentioned here that Jupiter's four outer satellites, discovered in 1914, 1938 and 1951, were all found to have retrograde motion in their orbits, and that Neptune's two moons revolve in opposite sense to each other, so that Saturn is now no longer to be considered unique in having a family 'divided against itself'.

Phoebe's peculiarities

The most remarkable points about Phoebe are its vast distance from Saturn and the excessive tilt of its orbit with respect to the plane of the planet's equator. The seven inner satellites are all at mean distances of less than a million miles from Saturn, and revolve round the planet in periods ranging from about $22\frac{1}{2}$ hours to 21 days, and their orbits lie nearly in the plane of Saturn's equator. Iapetus, with a mean distance exceeding 2 million miles and a revolution period of $79\frac{1}{3}$ days, has an orbital inclination of nearly $14\frac{3}{4}°$ to the ring-plane of Saturn and $17°86$ to the ecliptic (1950·0). Phoebe's mean distance from the planet exceeds 8 million miles, and one revolution of the satellite takes almost $550\frac{1}{2}$ days; moreover, its orbit is even more eccentric than that of Hyperion, and is inclined at $150°$ to the plane of Saturn's equator and $173°95$ to the ecliptic

(1950·0). Where the inclination of a satellite's orbit is more than 90° its motion is considered retrograde with respect to the planet's rotation. Phoebe is therefore circling round in a retrograde or clockwise sense, while the motions of the planet and all its other satellites are direct or anticlockwise. Another interesting thing is that, since the planet's equator is inclined nearly 27° to its orbit-plane, and Phoebe's orbit 150° to the plane of Saturn's equator, the inclination of Phoebe's orbit-plane to that of Saturn is about 177° which might be regarded as equivalent to only about 3°. Hence, while the seven inner satellites revolve nearly in the plane of the planet's equator, Phoebe is revolving nearly in the plane of Saturn's orbit.

Royal Observatory photographs, 1907 (Melotte)

At the request of Dr. F. E. Ross, who was investigating the orbit, P. J. Melotte of the Royal Observatory, Greenwich, took a series of 16 photographs of Phoebe during the opposition of 1907, using the Greenwich 30-inch aperture reflector, and giving exposures of from one to two hours. Melotte, who in the following year was to discover the eighth satellite of Jupiter, managed to drive the telescope to follow Phoebe exactly during each exposure, so that the satellite's image appeared on the plates as a circular dot. Phoebe's position was measured in relation to some faint stars surrounding it, and their positions in respect of certain standard reference stars on plates taken with the astrographic 13-inch aperture refractor. Similar measures were made of Saturn's position. As a result an extremely fine series of positions of the satellite was obtained, and one of these photographs of Phoebe taken exactly at elongation was expected to be of value for determining the mass of Saturn, as well as the exact dimensions of the satellite's orbit (see *B.A.A.J.*, vol. 18, p. 192; *M.N.*, vol. 68, pp. 127, 211 and 671). A further series of photographs was obtained at the 1908 opposition. Figure 13, based on the Royal Observatory diagram of positions in 1907 (*M.N.*, vol. 68, plate 16) shows the extraordinary size and tilt of Phoebe's orbit in relation to the rest of the Saturnian system.

Since Saturn's rings were more or less in the edgewise position during the 1907 opposition, the orbits of the seven inner satellites were also almost edgewise to the sightline and are represented in the Figure by the straight line HH', the points H and H' being the positions (to scale) of the western and eastern elongations of Hyperion. The western elongations of Titan and Rhea (between H and Saturn) are shown; those of the four innermost satellites are between that of Rhea and Saturn. The orbit of Iapetus, in a different

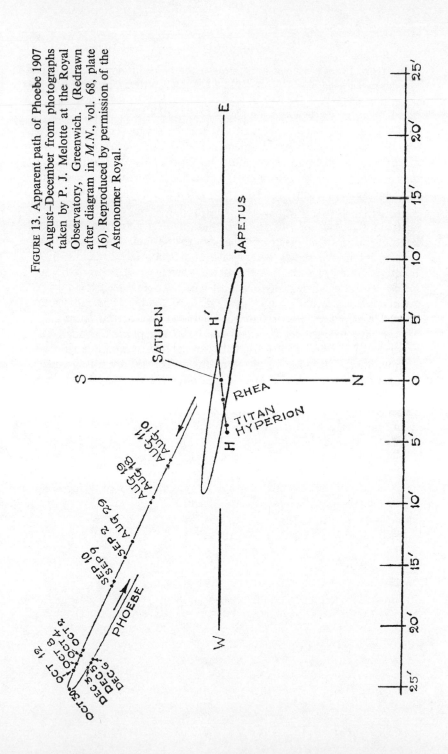

FIGURE 13. Apparent path of Phoebe 1907 August–December from photographs taken by P. J. Melotte at the Royal Observatory, Greenwich. (Redrawn after diagram in *M.N.*, vol. 68, plate 16). Reproduced by permission of the Astronomer Royal.

plane from Phoebe's orbit and from those of the inner satellites, is also shown to scale.

Themis, a supposed tenth satellite

Not long after the discovery of Phoebe, Pickering claimed to have found a tenth satellite and named it Themis, but it seems to have disappeared before the claim could be thoroughly substantiated; it may possibly have been a minor planet that happened to be in the field and seemed at first to share Saturn's motion.

Dr. J. G. Porter (*B.A.A.J.*, vol. 70, p. 57) has explained that the delay in the issue of further information about Phoebe after the announcement of its discovery was due to the fact that it was hard to identify in the plates taken with the Bruce telescope in 1899, because Saturn was then in the dense star field of Scorpio–Sagittarius, and a great deal of time and work was required to identify positively the satellite's faint image among the crowded masses of stars. He also says that it was during the course of this work that Themis was found and identified on 13 of the plates. Though the existence of Themis has never since been confirmed, and it is ignored in most modern text-books, the elements of its orbit are still included in the *Connaissance des Temps*, the period being given as 20·85 days and the mean distance from Saturn as 908,000 miles. The orbit therefore is of similar size to that of Hyperion, but has an eccentricity of 0·23, and an inclination to the ecliptic of 39°.

CHAPTER 23

SATURN AND SATELLITES: GENERAL REMARKS AND STATISTICS

By the year 1900 most of those fundamental facts about Saturn that could be ascertained by normal observation with the telescope had been found out, and all the satellites at present known had been discovered. This therefore seems an appropriate point at which to supply some information not covered in previous chapters, including statistics of the planet additional to those in Chapters 1 and 5, and full statistics of the satellites. The orbit, ring cycle, ascertainment of mass and dimensions have been dealt with in Chapters 1, 5, 16 and 19 respectively, while the physical appearance of Saturn has been sufficiently described in the narratives of the many observations. Much has already been said of the investigation (observational and theoretical) of the ring system, and subsequent chapters will show how much further work has been done on this subject in the present century; the brilliant progress achieved in recent times in regard to the atmosphere, temperature and internal constitution of Saturn will be related in Chapters 33, 35, 38 and 40. By comparing Saturn with the Earth and Jupiter what needs to be said about mean density, escape velocity, surface gravity and albedo can now be given.

Comparisons with the Earth

The colossal size of Saturn compared with the Earth may not be fully realised, so it is perhaps worth noting that Saturn's equatorial diameter is $9\frac{1}{2}$ times the Earth's diameter, and that the span of the ring system from tip to tip is more than 21 times the width of the Earth, which in fact is small enough to fit into either of the spaces between the globe and crepe ring. Saturn has 763 times the Earth's volume but only 95·2 times the Earth's mass. Therefore the great planet's mean density is extremely low, being only 0·69 grams per cubic centimetre or just under 7/10 the mean density of water, and of course substances with a mean density of that order will float in water quite easily. By contrast the mean density of the Earth, 5·52 times that of water, is greater than that of any other major planet.

The attraction of Saturn (or surface gravity leaving out of account centrifugal force) differs little from the Earth's, being 1·15 times the latter at the poles and 1·13 times at the equator. On the other hand, the escape velocity for Saturn is 22 miles per second or more than three times that of the Earth, so that Saturn must have retained all the gases originally in its atmosphere, including hydrogen, which the Earth has lost.

From the observations described in Chapter 15 it seems evident that in spite of its huge bulk Saturn's globe has a rotation period in its equatorial region of about 10^h 14^m, which implies a rotation rate of some 23,000 miles an hour, about 21 times as fast as the rotation of the Earth at its equator. Since the length of the day at Saturn's equator is only about $10\frac{1}{4}$ terrestrial hours and the planet takes about $29\frac{1}{2}$ terrestrial years to travel once round the Sun, the Saturnian year in the equatorial region is made up of some 25,000 Saturnian days.

Subsequent chapters will show that in regard to temperature and atmosphere also Saturn is entirely different from the Earth; in fact one would be hard put to it to find any resemblance between the two planets except that of a rather considerable tilt of the equator of each with respect to the plane of the orbit. There is however a quite fortuitous numerical connection that is helpful for remembering the statistics: if the values for the Earth are taken as unity, approximate values for Saturn are mean equatorial diameter $9\frac{1}{2}$, mean distance from Sun $9\frac{1}{2}$, mass 95, revolution period $29\frac{1}{2}$.

Comparisons with Jupiter

Apart from its ring, the tilt of its axis, and its much lower mass and gravitational attraction, Saturn could be regarded as a slightly smaller replica of Jupiter. They are alike in shape, both having considerable polar compression, in colours and general appearance, and in albedo—Jupiter reflects about 44 per cent of the light received from the Sun, Saturn 42 per cent. They both have considerable atmospheres, now known to be of like composition, with dusky belts variable in position but more or less parallel to the equator, outbreaks of bright and dark spots, rapid and differential rotation, very low temperature, and a large entourage of satellites. Among the major planets, Jupiter has the next lowest mean density after Saturn, Jupiter's being 1·34 times that of water, and in fact the mean density in proportion to mass is so much lower for these two than for any of the other major planets that they are now considered to be of a different chemical composition from any of the others, even Uranus and Neptune.

The Satellites

The following is a list of the satellites, their discovery, and some *N.A.* statistics (from *B.A.A.J.*, vol. 70, p. 33, and *Hdbk.* 1948); other statistics are given at the end of the chapter:

	Satellite	Discoverer and date	Distance from Saturn (*miles*)	Mean daily motion	Stellar magnitude
I	Mimas	W. Herschel, 1789	115,000	381·°994	12·1
II	Enceladus	W. Herschel, 1789	148,000	262·732	11·7
III	Tethys	J. D. Cassini, 1684	183,000	190·698	10·6
IV	Dione	J. D. Cassini, 1684	235,000	131·535	10·7
V	Rhea	J. D. Cassini, 1672	328,000	79·690	10·0
VI	Titan	C. Huygens, 1655	759,000	22·577	8·3
VII	Hyperion	{ W. and G. Bond, 1848 { W. Lassell	922,000	16·920	15
VIII	Iapetus	J. D. Cassini, 1671	2,213,000	4·538	10·8
IX	Phoebe	W. H. Pickering, 1898	8,050,000	?	14

[*Note.* The distances are to the nearest thousand miles from the centre of Saturn, and the stellar magnitude means the apparent stellar magnitude at mean opposition distance. The brightness of Tethys, Dione and Rhea is slightly variable, while that of Iapetus varies within a range of 2 magnitudes.]

From the above table it will be seen at once that the satellites may be classified in three distinct groups according to distance from Saturn: the first five form an inner group, Titan and Hyperion a more distant one, and Iapetus and Phoebe an outer group.

The statistics of orbital inclination at the end of this chapter show that the orbits of the satellites of the first two groups—from Mimas to Hyperion—all lie nearly in the plane of Saturn's equator, none being inclined to it more than about $1\frac{1}{2}°$ and most much less than one degree; Figure 14 gives an *N.A.* diagram of their apparent paths in 1959. The orbits of Iapetus and Phoebe, besides being much larger, are inclined about $14\frac{3}{4}°$ and $150°$ respectively to Saturn's equator (see Figure 13).

Difficulties in connection with the satellites (Porter)

Dr. J. G. Porter of H.M. Nautical Almanac Office, Royal Greenwich Observatory, Director (until 1959) of the B.A.A. Computing Section and editor of the B.A.A. Handbook, has written a very valuable paper on 'The Satellites of the Planets' (*B.A.A.J.*, vol. 70, p. 33) which has been heavily drawn upon for this chapter, and has contributed useful information to others, e.g. Chapters 5, 16, 18 and 22.

He has pointed out that most of the difficulties in connection with the satellites of the other planets arise from their distance from the Earth, so that their diameters and even their positions with respect

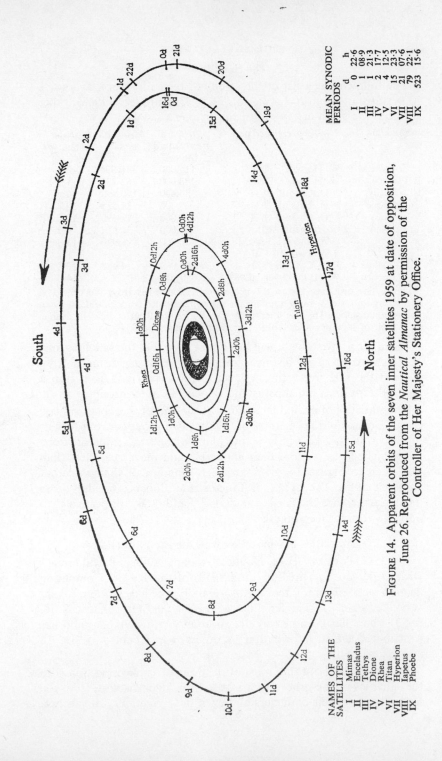

FIGURE 14. Apparent orbits of the seven inner satellites 1959 at date of opposition, June 26. Reproduced from the *Nautical Almanac* by permission of the Controller of Her Majesty's Stationery Office.

South

North

MEAN SYNODIC
PERIODS

	d	h
I	0	22·6
II	1	08·9
III	1	21·3
IV	2	17·7
V	4	12·5
VI	15	23·3
VII	21	07·6
VIII	79	22·1
IX	523	15·6

NAMES OF THE
SATELLITES

I	Mimas
II	Enceladus
III	Tethys
IV	Dione
V	Rhea
VI	Titan
VII	Hyperion
VIII	Iapetus
IX	Phoebe

to the planet can only be determined approximately. Also, it is almost impossible to detect the eccentricity of an orbit that is almost circular, and generally speaking the only elements that can be measured with any real accuracy are the longitude of the node and the inclination of the orbit.

He classifies satellites according to their orbits into two well-marked kinds: (a) regular satellites, moving in almost circular orbits in the plane of the planet's equator: in the case of Saturn the six inner satellites, Mimas to Titan, are of this type; (b) irregular satellites, whose orbits may be quite eccentric and inclined at any angle, even in some cases retrograde: to this class belong Saturn's three outer satellites, Hyperion (the inclination of whose orbit, though small, is variable), Iapetus (with a highly inclined though nearly circular orbit), and Phoebe (whose orbit is very highly inclined and retrograde). He points out that the usual kind of tables cannot be used to express the motion of irregular satellites, which, like comets, have to be dealt with by numerical integration, and that the tables in use for the regular satellites are half a century old, and are at present under revision, especially in U.S.A. and U.S.S.R., utilising modern observations for this purpose.

As most satellites are so close to their primary, and so distant from the Sun and the other planets, perturbations by those bodies can be ignored, though solar perturbations are included in the ephemeris of Titan. In the case of Saturn's entourage, he points out that perturbations of Titan on the neighbouring satellites (particularly Rhea) are important, and also that the motion of close satellites is modified by disturbances due to the oblateness of the planet, the effect being to cause the line of nodes to regress, while the line of apses (joining pericentre and apocentre) advances; the advance and regression are at the same rate only for satellites which revolve in the plane of the planet's equator.

Although the position of the node and the inclination will completely define the orientation of the plane of a satellite orbit, Porter states that there is no standard system to which these angles are referred, and therefore it is very necessary in quoting any of these angles to specify exactly what system is to be used.

Size and visibility of the satellites

Some amateur astronomers incline to exaggerate the difficulty of observing Saturn's moons. One reason, as explained in Chapter 18, is that owing to Saturn's tilt, the satellites appear as a rule scattered all round the planet and not neatly aligned with the equator like Jupiter's four bright moons. But the main reason is that, owing to Saturn's distance from the Earth being so much greater than Jupiter's, Saturn's satellites, though by no means insignificant in size, all appear much fainter than Jupiter's four bright ones which have apparent magnitudes of about 5 or 6. For example, Titan, which is much larger than the Moon and similar in size to the planet Mercury and to Ganymede and Callisto, the largest moons of Jupiter, has a stellar magnitude of only 8·3; the five inner satellites and Iapetus, ranging in diameter from 300 up to 800 or 900 miles, have magnitudes of about 12 up to 10. Yet all Saturn's satellites, even Phoebe, are larger than Amalthea (Jupiter V), the innermost and fifth largest of Jupiter's family; while Jupiter VI to XII are not only extremely faint but very diminutive, even compared with Phoebe.

No hard-and-fast pronouncement can be made as to the visibility of Saturn's satellites with various apertures, because observations are so greatly affected by such circumstances as the elongation of the satellite from the planet, the tilt of the rings at the time, the seeing conditions, and the experience of the observer. Subject to these factors, the *B.A.A. Hdbk.* (1948) states that Titan should be easily seen with any telescope whose aperture exceeds 1-inch; Rhea can be seen (if well clear of the planet's limb) in a 2½-inch instrument; Iapetus when brightest, near western elongation, in a 3-inch; Dione and Tethys in a 4-inch; Hyperion has been well seen with 6-inch aperture; Enceladus, more difficult because of its nearness to the brilliant planet, can sometimes be detected with a 6-inch telescope. Mimas is not fainter than magnitude 12·1 and has been easily seen in very good conditions with a 10-inch o.g., but it is deeply involved at all times in Saturn's glare, since the little satellite, even at its greatest elongation, appears to be less than half the length of one ansa of the ring from the ring's extremity. Lassell considered Mimas incomparably fainter than Enceladus and an extremely difficult object even with his large telescopes, and this opinion was fully borne out by the experiences of Common and Holden with powerful instruments in 1877–79, as related in Chapter 16, although Asaph Hall in 1883–84 was more fortunate. Phoebe, not only very faint but also as a rule very distant from the planet and the other satellites, and therefore doubly difficult to find, is also quite beyond the scope of small apertures.

The inner group

From the foregoing table of mean distances it will be noticed that the three innermost moons are nearer to the centre of Saturn than the Moon is to the Earth, Mimas being less than half as far; that Dione is at about the Moon's distance; and that Rhea, the brightest and largest of these five, is about 33 per cent farther from its primary. Owing, however, to the very great mass of Saturn compared with that of the Earth, the linear orbital velocities of these satellites have to be much greater and therefore their revolution periods much less than is the case with the Moon in its motion round the Earth. From the tables at the end of this chapter it will be found that the revolution of Mimas round Saturn takes a little less than one day, that of Enceladus about $1\frac{1}{3}$ days, Tethys less than 2 days, Dione less than 3 days, and Rhea about $4\frac{1}{2}$ days.

These five inner satellites, all lying within 300,000 miles of the apparent surface of Saturn's globe, have orbits that are almost circular and very nearly in the plane of the rings, and they form a remarkably rigid system. The foregoing table of mean daily motion will make obvious a fact referred to in Chapters 10 and 16, namely, that the motions of Mimas and Tethys, and also those of Enceladus and Dione, are very nearly in the ratio 2 : 1. Moreover, the motions of Mimas and Enceladus, and also those of Tethys and Dione, are approximately as 3 : 2, and so the motions of the four satellites could be approximately expressed by the single ratio 6 : 4 : 3 : 2. If n_1, n_2, n_3, n_4, represent respectively the mean daily motion of Mimas, Enceladus, Tethys, Dione, then the relations between their motions can be very neatly and exactly stated as:

$$n_1 - 2n_3 = +0°599$$
$$5n_1 - 10n_2 + n_3 + 4n_4 = -0°509$$
$$n_2 - 2n_4 = -0°338$$

The mass of Rhea is indeterminate because its mean motion is not even approximately commensurate with any of the others and the perturbations it produces are very small. Moreover, the eccentricity and inclination of Rhea's orbit are variable, because the orbit is affected by librations due to the action of Titan. The longitude of Rhea's perisaturnium is close to that of Titan, oscillating about it in a period of 38 years, and the eccentricity of Rhea's orbit is also produced by Titan, having variations of the same period with a range from 0·00068 to 0·00128.

A great deal of work has been done during the present century on the orbits of the various Saturnian satellites and to try to determine their diameters, masses and densities. Numerous references to this work and the difficulties encountered will be found in later chapters, e.g. Chapters 29, 32, 33, 35, 37, 38, 40.

Titan and Hyperion
Rhea is separated by a gap of over 400,000 miles from the next in order of distance, Titan, the largest and brightest of Saturn's satellites, greater in size and mass than all the others combined, and large enough to show a distinct disk in big telescopes. Two remarkable discoveries about Titan were made in the 1940s: in Chapter 37 is related, with illustration in Plate XVII, fig. 1, the detection of markings on the disk by the Pic du Midi observers, and in Chapter 38 how proof was obtained by Dr. G. P. Kuiper of the presence of an atmosphere, the first case of one being discovered on any satellite in the solar system.

Titan, 759,000 miles from Saturn, is more than thrice as far as the Moon is from the Earth, but makes one revolution in its orbit (whose inclination and eccentricity are variable) in just under 16 days. Hyperion's mean distance from Saturn, 922,000 miles, is just about four times the Earth-Moon separation, but Hyperion travels round its very eccentric orbit in about three weeks.

Titan and Hyperion, with orbits both lying very nearly in the plane of the rings and relatively close together, form a true group of satellites and exemplify a particular solution of the problem of three bodies. Newcomb solved the problem of Hyperion's motion (as stated in Chapter 16), finding that the major axis of the orbit *regresses* $18°56$ per annum, and showing that this exception to the usual rule is caused by the extremely close commensurability of the mean motions of Titan and Hyperion:
$$3T - 4H = +0°051.$$
The best chance of seeing Hyperion, with a telescope of aperture not less than 6 inches, is to look for it when it is in conjunction with Titan; such conjunction can occur only when Hyperion is at aposaturnium and therefore also at its greatest distance from Titan, although they will appear close together in the line of sight. Hyperion is nearly seven magnitudes fainter than Titan and is a very much smaller body, with a diameter estimated at about 250 miles, while Titan's diameter is now reckoned at about 3,000 miles. In 1894 and 1895 Barnard made series of measures of the diameter of Titan with a micrometer in conjunction with the 36-inch aperture Lick refractor and a magnification of 1,000 (*M.N.*, vol. 55, p. 378, and vol. 56, p. 172). The result he obtained in 1894 was a diameter of 2,523 miles, which he said was smaller than had usually been assigned to Titan, and would indicate a mean density about 5·2 times that of Saturn. He raised this value to 2,720 miles on combining the two years' measures.

The outer satellites

Hyperion is separated by a gap of over a million miles from the outer group of satellites, Iapetus and Phoebe, which have little in common save their great distances from Saturn and their highly inclined orbits. The distances and periods of revolution of these two satellites and other particulars about Phoebe have been mentioned in Chapter 22, and a good deal of information about Iapetus in various earlier chapters: for example, the investigation of its light change by Cassini and Sir William Herschel, in Chapters 6 and 8; Chapters 16 and 18 have given particulars, illustrated by Figures 11a, b, of Iapetus's orbit and of Barnard's famous observation of the satellite's passage behind the crepe ring. Titan, five of whose revolution periods roughly equal one of Iapetus, has a long-period effect on the smaller satellite's motion. Iapetus is believed to have a diameter of about 750 miles, or three times that of Hyperion.

Phoebe's motion is retrograde and highly complex.

Statistics

The *N.A.* statistics of Saturn and the satellites which follow are from *B.A.A Hdbk.* 1960.

The dimensions of Saturn's rings are based on Bessel's observations but the factor for the inner ellipse of the crepe ring is founded on the observations of O. Struve, A. Hall, E. E. Barnard, and T. Lewis (Greenwich).

The statistics of the inner satellites and of Titan and Iapetus are based on the elements of O. Struve (1930 and 1933); those of Hyperion on J. Woltjer's elements (1928); those of Phoebe on the elements by F. E. Ross (1905).

The elements of the satellite orbits are all subject to considerable variations, especially in regard to inclination and eccentricity; the orbits of the outer satellites are not even approximately elliptic. The inclinations of the orbits are referred to the planet's equator. The masses of the satellites are in most cases unknown, and the diameters are mostly rough estimates.

The following statistics are *additional* to those of Saturn's orbit (in Chapter 1) and inclination (Chapter 5) and to the satellite statistics given earlier in this chapter.

(a) Saturn

Period of axial rotation (equatorial region)	$10^h 14^m$
Reciprocal mass (Sun = 1)	3501·6
Mean density (water = 1)	0·69
Escape velocity (miles per second)	22
Diameter (Earth = 1): equatorial	9·5
polar	8·5
Mass (Earth = 1)	95·2
Volume (Earth = 1)	763
Surface gravity (Earth = 1): equatorial	1·13
polar	1·15
(*Note.* Centrifugal force would reduce the equatorial, but not the polar, gravity by 16 per cent.)	
Albedo	0·42
Temperature	−150°C.

(b) Dimensions of Saturn's ring-system

Diameter		At unit Distance	At mean opposition Distance	Miles	Ratio
Ring A	{ outer	37″54	43″96	169,300	1·0000
	{ inner	330·4	38·69	149,000	0·8801
Ring B	{ outer	322·8	37·80	145,500	0·8599
	{ inner	249·6	29·24	112,600	0·6650
Ring C	inner	205·9	24·12	92,900	0·5486
Saturn	{ equat.	166·7	19·52	75,100	0·4440
	{ polar	149·1	17·46	67,200	0·3973

(c) Saturn's Satellites

Satellite	Mean distance from Primary		Sidereal period	Synodic period
	Astronomical units	Angular at mean opposition Distance		
		″	days	d h m s
I Mimas	0·001 240 1	30·0	0·942 422	22 37 12·4
II Enceladus	0·001 590 9	38·4	1·370 218	1 08 53 21·9
III Tethys	0·001 969 4	47·6	1·887 802	1 21 18 54·8
IV Dione	0·002 522 4	1 0·9	2·736 916	2 17 42 09·7
V Rhea	0·003 522 6	1 25·1	4·517 503	4 12 27 56·2
VI Titan	0·008 166 0	3 17·3	15·945 452	15 23 15 25
VII Hyperion	0·009 892 8	3 59·0	21·276 665	21 07 39 06
VIII Iapetus	0·023 797 6	9 34·9	79·330 82	79 22 04 56
IX Phoebe	0·086 575 2	34 56	550·45	523 16

Satellite	Orbit inclination	Orbit eccentricity	Satellite diameter (miles)	Reciprocal mass Saturn ÷ satellite
I	1°31′	0·0201	300	15 000 000
II	0 01·4	0·00444	400	8 000 000
III	1 05·6	0	600	870 000
IV	0 01·4	0·00221	600	550 000
V	0 21	0·00098	850	250 000
VI	0 20	0·0289	3000	4 150
VII	*	0·104	250	
VIII	14 43	0·02828	750	
IX	150 03	0·16326	150	

* Small and variable.

CHAPTER 24

ROTATION PERIOD OF NORTH TEMPERATE SPOTS, 1903

During the first decade of the present century the main attention of observers of Saturn seems to have been devoted to two occurrences: an outbreak of bright spots in the north temperate region in 1903 (the subject of this chapter), and the disappearance and reappearance of the rings in 1907–08 (to be related in Chapter 26).

Barnard's white spot and others (Denning, Comas Solá)

The history of these bright spots began with the detection of one of them on 15 June 1903 by Professor E. E. Barnard, who had never succeeded in seeing anything of the sort on Saturn during the previous decade and had always been very sceptical of the claims of observers with small telescopes to have discovered and followed up bright and dark spots (see Chapter 15). On this occasion Barnard was using an even larger aperture, the 40-inch refractor of Yerkes Observatory, and he announced the discovery after re-observing the spot on June 23 and accurately determining its position. This object, and the others subsequently seen, were all in a somewhat higher latitude than any previously well observed spots, namely 36° N., in the light north temperate zone, north of the double north equatorial belt.

Stanley Williams, in a very good short history of these observations (*M.N.*, vol. 64, p. 359), stated: 'Quite independently, and before he had heard of any spot having been observed elsewhere, Mr. W. F. Denning, on the night of July 1, detected another bright spot, which was followed by a dark mass, with his 10-inch reflector at Bristol. The bright spot was also observed on the same night by Señor José Comas Solá at Barcelona with a 6-inch refractor'.

Rotation periods and identification difficulties

Observations soon showed that there were a number of these white spots, all in approximately the same latitude, though the one first found seems to have been the most conspicuous. On comparing his observations of it Barnard was satisfied that the rotation period must be decidedly longer than $10^h 14^m$, but it was Dr. K. Graff of

Hamburg who first published a rotation period (*A.N.*, 3883). He compared the observations of Barnard's spot of June 23 and 26 and July 4, and derived a period of 10^h 39^m01, which seemed at first improbably long. Subsequent observations, however, showed this result to be substantially correct, and values in good agreement with it were soon published by several observers: Comas Solá from his own observations of the same spot found a period of 10^h 38^m4 (*A.N.*, 3894); Denning from a preliminary list of observations of this spot inferred a rotation period of 10^h 38^m, while from the mean of the results for seven other spots he obtained the value 10^h 39^m 21^s1; L. Brenner from his own data gave the period as exactly 10^h 38^m.

From a long list of observations of the spot first discovered Dr. H. C. Wilson derived a rotation period of 10^h 38^m 4^s8 (*Popular Astronomy*, 1903, p. 445). Professor G. W. Hough, of the Dearborn Observatory, from a few micrometrical measures of the position of the Barnard spot, found a variable motion, the rotation period being 10^h 38^m 18^s on June 27, but increasing by 0.1856 seconds per rotation after that date; three observations of another spot gave him a period of 10^h 38^m 30^s5 (*M.N.*, vol. 64, p. 124).

Hough asserted categorically in his paper that micrometrical measures of transits were of far greater value than eye estimates, an opinion which was contested by Denning, who knew what good results had been obtained for spots on Jupiter by eye estimations. Moreover, Denning in a paper (*M.N.*, vol. 64, p. 239) strongly questioned the correctness of some of Hough's identifications of the spots. With regard to this Stanley Williams stressed the very great importance of the correct identification of the observations in order to derive reliable rotation periods, and pointed out that in the case under consideration, identification was rendered difficult not only by the spots being quite numerous, but also by the fact that 'none of them seems to have possessed any permanent characteristic feature by which it might be recognised by its appearance'; furthermore, the spot originally found by Barnard was large, rather indefinite, and sometimes seemed to be double. Hence where only a few observations were available, mistakes in identification could easily be made.

On the other hand, Denning, in the paper just referred to, listed nearly fifty observations of the Barnard spot, the gaps between them being so slight that Williams considered that it would be difficult to suppose that any serious mistake in identification could have been made by Denning. He found from those observations a period of 10^h 38^m 3^s for Barnard's spot for the interval between late June and mid September, a result agreeing to within 2 seconds with that derived by Wilson. Denning stated that the mean period from about fifteen bright and dark spots from July to December was $10^h 37^m 56^s$.

Long afterwards, in 1918, Denning published (in *B.A.A.J.*, vol. 28, p. 161) a complete list of the rotation periods, based on some 170 observations by various observers, of 18 spots (11 white, 7 dark). The rotation periods of individual spots ranged from $10^h 37^m 48^s6$ to $10^h 38^m 2^s4$, an extreme range of 13·8 seconds, the mean rotation period for the 18 spots being $10^h 37^m 56^s4$.

In spite of differences of opinion about identification, there can be no doubt whatever about the most important result of these observations—the reality of the remarkably long rotation period of about $10^h 38^m$. The fact that Barnard's spot must have been extraordinarily brilliant during the first two or three weeks of its life makes this result all the more reliable.

Williams pointed out that previous rotation periods, all derived from observing spots in lower latitudes, were all round about the figure $10^h 14^m$. Denning had, however, stated (*Knowledge*, 1903, p. 271) that two observations by Dawes in 1858 of a bright spot in about 40° or 50° S. latitude (see Chapter 10) indicated a rotation period of about $10^h 24^m$, or at any rate a good deal longer than that of the equatorial spots.

The great equatorial current (Williams)

Stanley Williams drew the following conclusions from these observations: 'It seems clear, therefore, that we have on Saturn, as on Jupiter, a great equatorial current having an enormous velocity in the direction of the planet's rotation relative to the extraequatorial surface material. But the velocity of this current on Saturn is almost incomparably greater than it is on Jupiter, amounting as it does in miles per hour to between 800 and 900, as compared with the 250 miles an hour of the latter planet. The important bearing of the recent observations on our knowledge of the physical condition of Saturn can be judged from this. It may be inferred for one thing that the surface material of the planet is in a more mobile state than that of Jupiter, a circumstance that had been already foreshadowed by the earlier observations, and which is, perhaps, not unconnected with the lesser density of the planet'.

Further particulars of the observations (Williams, Denning)

It may well be asked, in view of the leading part played by Stanley Williams in the observation of spots and derivation of their rotation periods in the eighteen-nineties (see Chapter 15), why he did not fill a more prominent rôle on this occasion than that of historian and commentator. In fact he did make observations from July 24 to September 23 and obtained nineteen transits of the 'white' spots, which, he said, usually appeared to him to be yellowish. He published a list of his observations in November, 1903 (*M.N.*, vol. 64, p. 46), indicating the relative degree of accuracy of each, but without himself attempting to deduce a rotation period from them. Owing to being away in Ireland, he had not heard of Barnard's discovery until towards the end of July, and did not then know the probable position of the original spot, so he simply obtained transits, or estimated transits, of any spots that happened to be visible, using his $6\frac{1}{2}$-inch aperture reflector. As he was out of practice, the weather unfavourable and the altitude of the planet low, he did not obtain enough sufficiently accurate transits to justify the deduction of a rotation period from them alone, but he published them in the hope that other observers might find them useful in conjunction with their own. Denning, in fact, included seven of them in his list.

Without going into the details of Williams's observations it seems worth while to note some of his general remarks on the appearance of the spots. He found the north temperate zone extremely luminous in some parts, comparatively dull and lustreless in others, the spots constituting the brightest sectors of the zone. The spots were frequently indefinite, particularly in an east and west direction, so their exact position was difficult to locate, and it was hard to fix upon anything definite enough to observe its transit. They sometimes seemed double in the east-west direction, but it was difficult to tell how much of this impression was due to unsteady seeing, which was the chief obstacle to good observations. Similarly, the variable seeing conditions may have caused the selfsame spot to have been described as plain and bright on one night and faint and inconspicuous on another, without there necessarily having been any change in its brightness or appearance.

In general these north temperate spots had much resemblance to the white spots Williams had observed ten years before in the equatorial zone (see Chapter 15), but the latter were probably not quite so bright and luminous looking. On this occasion the equatorial zone looked very white, whiter than the north temperate zone, and there were distinct indications of brighter spots in it, but owing to the low position of Saturn no attempt was made at a systematic observation of equatorial spots. He saw one of those spots apparently divided into two by a narrow, faint, dark belt in the equatorial zone, probably the same that used to be known as the 'equatorial mottled belt' (see Chapter 20).

Hough had had the co-operation of Professor S. W. Burnham (better known for his stellar observations and star catalogue) in making micrometer measures of spots with the Yerkes 40-inch aperture refractor, but owing to much cloudy weather, poor seeing, and the low altitude of the planet, Hough was only able to get a total of about half a dozen transits of Barnard's spot and about eight of other spots, to form the basis of his rotation periods.

While admitting the accuracy of transits obtained with the use of a micrometer, the point that Denning made was that the manner of taking transits was less important than correct identification, and that mis-identification 'is scarcely avoidable unless observations are obtained at short intervals and a fairly numerous list of transits accumulated'. He himself made observations on 78 nights in the second half of the year, and in his long list of transits of Barnard's and two other spots he included those obtained by a dozen other observers as well as himself. He stated that the form, size and brilliancy of the luminous spots seemed very variable, and that there were changes in the rate of motion. Also several of these objects were compound, comprising two or three parts wholly or partly divided by dusky masses or wisps like those sometimes seen in the equatorial region of Jupiter. In Denning's opinion the rate of the spots merely represented that of an atmospheric current, and did not necessarily conform with the motion of the globe.

Rotation of the rings (Denning)

During the same period, in November 1903, Denning several times observed a large diffuse white spot on the western ansa, and by comparing observations deduced a very rough rotation period of about 14^h 24^m* for the ring (*B.A.A.J.*, vol. 15, p. 54). He compared this with previous determinations relating to the rotation of the rings: two periods attributed to Laplace, namely, 10^h 33^m 36^s (a theoretical period) and 10^h 29^m; and the period Sir William Herschel had deduced from observations of a 'lucid spot' in 1789, which was 10^h 32^m $15^s.4$ (see Chapter 8). He found the closest similarity to his period in one that Secchi had derived from the observation in 1854–56 of an ellipticity in the rings, which had approximated to the theoretical period of revolution of a satellite supposed to be situated at the distance of ring A, namely, 14^h 23^m 18^s.* It seems evident that the Laplace and Herschel periods would relate to the rotation of ring B rather than ring A.

* Cf. Campbell's period for outer edge for ring A (Chapter 15).

CHAPTER 25

VARIOUS OBSERVATIONS (DIMENSIONS – SPECTRUM – BRIGHTNESS), 1900–07

Dimensions of globe, rings and Titan (See)

During the 1900 opposition Professor Thomas J. J. See made a series of measurements of globe and rings with the Washington 26-inch aperture refractor. The rings were widely open and the seeing conditions were on the whole very good, while the low position of the planet in the sky was more than made up for by the use of colour screens. The main results were as follows:

	miles
Saturn's equatorial diameter	74,944
polar diameter	67,352
Ring A—outer diameter	173,115
Ring A—inner diameter	149,419
Diameter of centre of Encke's division	162,261
Ring B—outer diameter	145,828
Ring B—inner diameter	111,470
Ring C—inner diameter	88,405
Width of Encke's division	460
Width of Cassini's division	1,795
Width of Ring A	11,846
Width of Ring B	17,181
Width of Ring C	11,533
Width of space between globe and C	6,730
Diameter of Titan	2,092

Mean density of Saturn: 0·679 × water

These results (from *A.N.*, Nos. 3686–87, and *B.A.A.J.*, vol. 11, p. 219) are in good agreement with those of Barnard, Hall and Greenwich Observatory given in Chapter 19. The conversion to miles is based on a value for the solar parallax of 8″796. See's value for the diameter of Titan is rather surprisingly small, but in the autumn of 1901 he obtained a much larger figure for Titan of 3,140 miles from daylight observations, so as to avoid the effects of irradiation. He stated that, of the satellites, only Titan and Iapetus show measurable disks, but Titan's disk is rather obscure, and that of Iapetus even more so, only one side giving enough light for an observer to recognise a disk (*A.N.*, No. 3764 and *B.A.A.J.*, vol. 12, p. 192).

In the autumn of 1901 See also made some micrometrical measurements by daylight and by night of the planet and rings (*A.N.*, No. 3768 and *B.A.A.J.*, vol. 12, p. 192). He gave a few comparative results:

	By daylight miles	By night miles
Saturn's equatorial diameter	74,172	76,598
Ring A—outer diameter	171,948	
Ring C—outer diameter	111,349	

At Lick Observatory in 1903, W. J. Hussey measured the position angles and distances of each satellite from one of the others and gave a table of the details. From light values he concluded that Mimas is probably larger than Hyperion, a view that is held at present, and that the accepted diameter of Titan is too large; his estimate for the diameter of Titan was 2,500 miles (*Lick O.B.*, 34; *B.A.A.J.*. vol. 14, p. 43).

Bands in Saturn's spectrum (V. M. Slipher)

V. M. Slipher made a photographic study of the spectrum of Saturn in the autumn of 1905, comparing the photographs with some taken of the spectrum of the Moon at the same altitude. He found that certain absorption bands did not occur in the Moon's spectrum and were not due to the Earth's atmosphere; they must therefore be due to the atmosphere of Saturn. The wave-lengths (in Ångstrom units: $\lambda = 1$ Ångstrom $= 10^{-7}$ mm.) of these bands, in order of intensity, were: $\lambda\lambda6193, 5430, 6145, 6450, 5770$, and none of them occurred in the spectrum of the rings; the absence of the band $\lambda6183-\lambda6193$ from the spectrum of the rings had previously been observed by Keeler (1889), and Hale and Ellerman (1898). The indication was therefore that if the rings possess an atmosphere it must be very rare. Slipher found that Jupiter's spectrum was on the whole like Saturn's and those of Uranus and Neptune were similar— Neptune's spectrum had many strong absorption bands, much stronger than the corresponding ones of Uranus. No indications of water vapour were found: (*Ap. J.*, 26, no. 1, p. 59; *B.A.A.J.*, vol. 17, p. 413, and vol. 18, p. 98).

The nature of these mysterious bands was not known at the time, and it was not until some twenty-five years later that most of them were identified with methane and $\lambda6450$ with ammonia (see Plate XIV and Chapter 35). Secchi had discovered some of these bands in 1863 (see Chapter 14).

Attempt to glimpse globe through Cassini's division (Whitmell)

C. T. Whitmell published in April 1902 a statement (*B.A.A.J.*, vol. 12, p. 290) that at the opposition of 17 July 1902, the Earth and Sun would be equally elevated above the ring-plane, and that he computed that the effective opening of Cassini's division (at right angles to the line of sight, and in the direction of a meridian of the planet) would be 820 miles, or 0.''2 at the distance of the Earth. He considered that it might be possible for an observer to be able to glimpse through the gap in the rings a portion of the planet's surface lit up by sunlight, which passes through the rift and would be reflected back by the planet through the rift to the observer. Hence a small portion of the arc of the Cassini division across the globe should appear bright instead of dark. He pointed out that the observation would be very difficult, and asked that the planet be watched for a few days about the opposition date from observatories situated over a wide range. He published notices about the phenomenon in several astronomical journals and circularised public and private observatories (*B.A.A.J.*, vol. 13, p. 20).

This laudable attempt unfortunately did not succeed: Whitmell himself and some others in different places either could not trace the Cassini division across the globe at all, or, in some cases, saw it across the globe without any sign of a break. But it would appear that the observation was not an impossible one, because Dr. C. A. Young of Princeton Observatory stated in a letter to Whitmell that Lassell at Malta had noted a paleness of the division across the globe (see Chapter 12) on 29 October 1852, a date when the elevations of the Earth and Sun above the ring-plane were equal, and that Young himself with Asaph Hall had observed a similar phenomenon in late November 1883 (see Chapter 17), when they were observing with a $22\frac{1}{2}$-inch aperture refractor and the seeing conditions were magnificent, but according to Whitmell's calculations the elevations of Earth and Sun were not quite equal.

The geometrical conditions seem to have been very favourable in 1902, but the planet was low in the sky for observers in the northern hemisphere. Barnard tried to observe the phenomenon with the 40-inch aperture Yerkes refractor, but the weather conditions were unfortunately hopeless there, and none of the observers in England and Australia who were able to view Saturn on 17 July 1902 had an instrument with aperture more than ten inches.

D. G. McIntyre (*Translucency*, p. 5) drew attention to the fact that in 1852, 1883 and 1913, when astronomers appear to have 'succeeded in peering through the division', the part of the globe seen was in the northern hemisphere. Attempts made when the surface that might have been seen would have been in the southern hemisphere, had failed, even at the favourable opportunity of 1902. He put the question: 'Is this because the two hemispheres have different albedoes, or because the seeing was superior in the years of success?' Better seeing, and the use of apertures of 20 inches or more, as against ones of 10 inches or less, may have made the difference between success and failure.

A. C. D. Crommelin suggested three possible causes for the failure of the observations of 17 July 1902 (*B.A.A.J.*, vol. 13, p. 76): dusky penumbral fringes bordering the shadows of the edges of the division on the globe, but these would only reduce the width of the gap very slightly; or the albedo of the polar region of the globe (behind the division owing to the wide opening of the rings) being much lower than that of ring B; or optical illusion—the tendency of the eye to continue the Cassini division, seen in the ansae, across the globe.

Lunar occultation of Saturn viewed in Australia, 1906

Occultations of Saturn by the Moon had been observed before (see Chapters 6 and 10), but that of 27 October 1906 is interesting because the observation was one of the earliest of Saturn to be made in Australia, and also because the event happened at an hour when the Moon and Saturn were observable in Australia but not in western Europe, showing the advantage of having observers in other parts of the world.

The occultation was observed and timed at three stations in New South Wales by E. H. Beattie, Dr. R. D. Givin, and W. J. MacDonnell with E. W. Esdaile, all four being members of the New South Wales Branch of the B.A.A. (*B.A.A.J.*, vol. 17, p. 133). Beattie and Givin made drawings of the immersion and emersion of Saturn, and the former noted that on emersion planet and ring seemed faint and shadowy and of a strange greenish tint, by contrast with the brilliancy of the Moon's bright yellowish limb.

After the Moon had moved some little distance away but was still in the field, Saturn's faintness began to wear off and the northern belts, the shadow of globe on ring, the crepe ring's shadow on the globe, and two near satellites could be detected. Esdaile also noticed the dull greenish hue of Saturn on emersion, and stated that as the ring appeared to emerge almost at right angles to the bright limb of the Moon, the ring seemed at first to be 'projecting from the limb like a knitting needle in a ball of wool'. These observations were made with small telescopes.

Variability of satellites (Quignon)

For four years (1903–06) G. A. Quignon, a Belgian astronomer, kept several of the satellites under observation during each apparition in order to determine the variability of their apparent brightness, using a wire in the focus of the object glass to conceal the planet, and comparing the light of the satellites with that of certain stars of known magnitude (*B.A.A.J.*, vol. 17, p. 148). He found a regular variation of Titan from magnitude 8·0 to about 8·6, the light being greatest between western elongation and superior conjunction, and then decreasing rather suddenly. He found the mean magnitude of Iapetus to be 9·8. He considered that Rhea brightened from magnitude 10 at superior conjunction to 9 towards eastern elongation, then dropped again to about 10, again becoming 9 between western elongation and inferior conjunction. He also found a variation of about three-quarters of a magnitude for Dione and Tethys.

Theoretical variations in Saturn's brilliancy (Gaythorpe)

An interesting paper by S. B. Gaythorpe was read to the B.A.A. in January 1903 (*B.A.A.J.*, vol. 13, p. 167) in which an attempt was made to estimate theoretically the maximum variation in the brilliancy of Saturn at opposition, arising partly from the changing phases of the rings and partly from the varying distances of the planet from Sun and Earth. He first dealt with the variation as it would appear from the Sun: the ratio of maximum to minimum brilliancy would be equal to the ratio of (*a*) the total apparent area when the rings are most widely open to (*b*) the apparent area of the globe alone, assuming all parts of the system to have the same albedo. Hence there would be two maxima and two minima in each revolution, the maxima when the planet is in Gemini (near perihelion with rings wide open) and Sagittarius (near aphelion with rings wide open), and the minima when it is in Virgo and Pisces, near the ascending and descending nodes of the ring-plane respectively, with the rings edgewise.

He calculated the apparent area of the globe plus the two part ellipses of the ansae, deducting the areas of the dark spaces and the Cassini division, and using Dr. See's then recent measures of the Saturnian system, found that from the Sun, Saturn would appear nearly 2·3 times as bright when the rings are most widely open than when they are presented edgewise. He pointed out that the true ratio would be slightly greater than this because observation showed that the albedo of the globe is much lower than that of the rings.

To get the foregoing ratio he had assumed a circular orbit for

Saturn, but to find the results from the point of view of the Earth at oppositions of Saturn, he took into account the eccentricities of the orbits. The brilliancy would now vary directly as the apparent area presented and inversely as the product of the squares of Saturn's radius-vector (distance from Sun) and its distance from the Earth. The maximum brilliancy would occur when the rings are most widely open and the heliocentric longitude of Saturn is about 82°, within about 10° of perihelion; there is a secondary maximum when Saturn, in longitude 262°, is about the same distance from aphelion. The lower minimum brilliancy would be when the rings are edgewise and Saturn, in longitude 352°, is at rather more than mean distance from the Sun; the brilliancy at the other disappearance of the rings in longitude 172° would be somewhat greater.

If R, D are Saturn's radius-vector and distance from the Earth at the maximum brilliancy, and r, d their values at the minimum brilliancy, the ratio would be:

$$2 \cdot 29 \times \frac{r^2 \, d^2}{R^2 \, D^2}$$

From this formula he found a ratio between the higher maximum and the lower minimum of $2 \cdot 97$, equivalent to a difference of $1 \cdot 18$ stellar magnitudes. Similarly, between secondary maximum and lower minimum the ratio would be $1 \cdot 9$, or a difference in magnitude of $0 \cdot 68$. He pointed out, however, that the different conditions under which Saturn is seen at the maximum and minimum would tend to mask the effect of the variation.

A. C. D. Crommelin, commenting on this paper, said it was unfortunate that the varying albedo had not been allowed for, because that might have been done. C. T. Whitmell said that a rough estimate made by himself and Crommelin as to Saturn's difference in brightness as seen from the Sun, when the rings were fully opened as compared with when they were closed, was in agreement with Gaythorpe's figure of $2 \cdot 3$ (*B.A.A.J.*, vol. 13, p. 156).

From M. B. B. Heath's tables in Chapter 1 and Appendix I it will be seen that there was a secondary maximum at the 1958 opposition, and that a lower minimum will occur in 1966, and a higher maximum in 1973. Heath's calculated figures for the stellar magnitudes at these oppositions, based on Müller's formula, are: for 1973, $-0 \cdot 3$; for 1966, $+0 \cdot 8$; difference $1 \cdot 1$ magnitudes; for 1958, $+0 \cdot 2$; for 1966, $+0 \cdot 8$; difference $0 \cdot 6$ of a magnitude. With these differences those obtained by Gaythorpe are in fair agreement, the discrepancy being less than $0 \cdot 1$ magnitude in each case. Müller's formula (quoted in

Appendix I) enables the stellar magnitude at any opposition to be calculated, and not merely the difference in magnitude between a maximum and a minimum of brightness.

CHAPTER 26

RING DISAPPEARANCES AND REAPPEARANCES, 1907–08

Observations, July–November 1907 (Barnard)

Except for the 5° south declination of the planet, conditions for observing the phenomena connected with the disappearance of the rings were usually favourable in 1907, much more so than in October 1891, when Barnard with the 36-inch aperture Lick refractor was unable to see any sign of the rings until some days after the date of their reappearance (see Chapter 18). In his paper dealing with his observations of the reappearance in July 1907 and the disappearance in the following October (*M.N.*, vol. 68, p. 346), Barnard gave the following table of dates calculated by Hermann Struve:

1907	April	17	Disappearance	Earth in plane of rings.
	July	26	Reappearance	Sun in plane of rings.
	October	4	Disappearance	Earth in plane of rings.
1908	January	7	Reappearance	Earth in plane of rings.

The disappearance in April was invisible from the Earth owing to Saturn's position in the direction of the Sun. Bad weather prevented Barnard from observing in the early summer, and when at the beginning of July he was able to commence, using the Yerkes Observatory 40-inch aperture refractor, he found that the 'reappearance', so far from being a definite change that could be precisely timed when the Sun passed through the ring-plane on or about July 26, was 'a remarkably gradual phenomenon, and there was no possible means of telling when it occurred. The ring simply very slowly and gradually got brighter, and for several days it was impossible to tell that any change had taken place; and then it became bright and almost linear . . . . It was some time previous to the reappearance of the ring that the most important phenomena were visible'.

When Barnard examined Saturn on 2 July 1907 the 'entire surface of the ring was easily seen, though the Sun was not then shining on its visible surface'. The ring, where projected on the sky, looked a 'greyish hazy or nebulous strip', ill-defined under the best conditions, and about 1½ times as broad as its trace across the globe. The ring's sunlit edge could not be seen and must have been too thin to be visible. There were two ill-defined, nebulous, pale grey condensations on the ansae at each side of the planet, and brighter than the rest of the ring. The latter and the condensations were sufficiently pronounced to be strongly visible in oncoming daylight, only 36^m before sunrise, and were still visible though faint twelve minutes later. Barnard made rough measures with micrometer of the distances of the centres of the preceding condensations from the preceding limb; reduced to the mean distance of Saturn from the Sun the results were 7$''$6 and 2$''$8, 40$''$ of arc being the extreme diameter of the ring.

Enceladus was close to the outer condensation on the following side and appeared to be of the same brightness; the trace of the ring across the globe was not black, showed no irregularities, and on measurement was found to be nearly 1$''$ of arc nearer the north limb than the south limb.

Barnard continued his observations on July 5, 6, 12 and 23. The seeing was very poor, but the ring was faintly visible, and so were the condensations on July 5 and 6, the inner ones seeming brighter than the outer ones; there was no sign of the sunlit edge of the ring. On July 24 the ring was visible as 'a narrow, almost thread-like strip . . . faint and nebulous, without any irregularities'. He estimated it to be about one magnitude fainter than Enceladus, which was near the following ansa. The appearance on July 25, in rather better seeing, was similar: the ring was faint grey, palely illuminated, very narrow, with no irregularities, and quite distinct when best seen. On July 26, in bright moonlight, it was a little brighter, but perhaps half a magnitude fainter than Enceladus. The ring on the sky was broader than the trace. The illumination of the ansae did not look like direct sunlight when a power of 700 was applied. This was the date of the expected passage of the ring-plane through the Sun.

Two days later (July 28) the ring at first looked bright and the illumination like true sunlight, but later looked more of a pale nebulous light. Its projection on the sky was south of the trace on the globe and about twice as thick; it was discontinuous for a few seconds from the limbs, and the following ansa seemed a little thicker than the preceding one. In August and early September the ring when best seen was 'Clean cut like a bar sharply pointed at the ends'.

The next thing for Barnard to observe was the disappearance due to take place on October 4. On October 1, in bad seeing, he found the ring ashy in colour and comparatively faint, like 'a nebulous strip'. On the 4th at 8^h 30^m, it was entirely invisible with powers of 460 and 700, but ten minutes later he thought he glimpsed feeble traces of it following the planet; the seeing was poor and the presence of satellites masked the view of the ansae. A little later the same evening he was able, with an occulter in the field, to detect feeble traces of the ring following but not near the planet, and occasionally could see it feebly preceding Saturn.

On October 5, in rather better seeing, the ring was faintly visible with the occulter, decidedly brighter than the day before but very slender and faint, much fainter than Mimas near its following end. On October 6 the seeing was very bad but with the occulter occasional very feeble glimpses of the ansae were obtained. Later that evening, in moments of steadiness, he had fairly good glimpses of the ring, which was certainly brighter than on the 4th and perhaps rather brighter than on the 5th; the shadow on the globe could also be seen. On the 8th he could not detect the ring with Saturn uncovered in the field, but with the occulter he could see it on both sides of the planet very easily. The ring was faint, straight and very narrow with no irregularities, and easier than at previous observations since its 'disappearance'. The condensations had disappeared October 4–8 when they should have been best seen if they had been real masses on the ring system.

On October 13 Barnard could see the ansae with Saturn uncovered in the field. They were very thin but, when the occulter was used, quite conspicuous, and on each ansa there were two condensations (regions of greater brightness). He easily saw the ring and condensations on November 3; the centre of the trace was about 0″3 nearer the south limb than the north. It was hard to decide the relative brightness of the inner and outer condensations. The outer ones, 'like very small blurred stars', seemed more brightly condensed than the ring, which looked broken in each ansa between the condensations, though continuous up to the planet's limb. The inner condensations were about as bright as Tethys. As the Earth's elevation above the dark side of the rings increased, the brighter patches became more conspicuous.

In moments of steadiness on November 5 the outer condensation in the preceding ansa resembled a small ill-defined satellite, a small hazy spot like Enceladus, close to it, looking the same in size, form and brightness. The nearer condensation at times seemed double. On November 12 the outer condensations looked slightly brighter than Enceladus and twice as bright as Mimas, which was comparable with the fainter parts of the ring between the bright patches. Barnard found the following outer condensation was about 2″ of arc long and about 0″8 broad, and diffused rather abruptly on each side.

His measures of the distances of the condensations from the limbs of the planet made on July 2 and November 3, 5 and 12, gave the following means (reduced to the mean distance of Saturn):

preceding 2″68 and 7″46
following 2″74 and 7″42

Barnard therefore concluded that they were symmetrical with respect to the centre of Saturn, and as he had seen them only on dates when the Earth and the Sun were on opposite sides of the ring-plane, it seemed probable that they must be caused by sunlight filtering through from the illuminated to the unlit face of the rings. His final conclusion as to their situation in the ring system was only reached at the end of his observations in January 1908 and will be given later in this Chapter. He stated that he felt that the term 'condensation' which he had used for these luminous appearances was misleading, because although these bright places looked broader than the trace of the ring on the sky, this was probably due to contrast or irradiation, and he inferred that they were not really broader than the trace.

Observations, July–November 1907 (Aitken)

A summary (*B.A.A.J.*, vol. 18, p. 187, from *Lick O. B.* No. 127) gives the more important points· of the observations made by Professor R. G. Aitken. On the morning of July 25, the day before the expected reappearance of the rings, he saw them as a very bright line. On July 30 the space between the planet and the ring was seen clearly on both sides. The ansae were bright but very narrow, 0″3 or less, on September 10. On the critical date, October 4, Aitken could trace the ring to practically full distance, but only as a mere knife-edge line of light.

No unusual phenomena were to be seen until October 19, when two bright points or knots were observed on the line of the rings on each side of the planet. A third pair of knots nearer Saturn's limbs was detected on November 1. The knots (evidently the same objects that Barnard described as 'condensations') could easily be mistaken for satellites; all appeared to be tangential to the south edge of the line, so Aitken inferred that they were due to irregularities on the illuminated surface of the rings. The inner knots seemed much brighter than the outer ones. At several· more observations in early November no special change was noted in the aspect of the ring; the knots however, seemed to be bright condensations in the line with extensions towards the south, rather than lumps attached to the south edge. There was no extension beyond the north edge of the line. The knots appeared as sharp as satellites, the inner following two being judged brighter than Enceladus. The appearance of these objects and their development were considered to be similar to those described by W. C. Bond in 1848. Aitken's measures placed the eastern knots closer to the planet, so he did not find them to be distributed with perfect symmetry, and in this respect his results differed from those of Bond in 1848 and Barnard in 1907.

Other observations of the 'disappearance' in October 1907

Though Barnard and Aitken seem to have been the only observers of the reappearing rings in late July, there was no lack of corroborative evidence relating to the phenomena of the second disappearance which should have occurred on October 4, 1907. Schaer of Geneva saw a luminous line representing the ring on October 2 and occasionally and momentarily on October 4, with bands of a brownish tint sometimes visible on each side of the trace of the ring. Dr. Hassenstein of Königsberg found it distinctly perceptible on October 1, but on the 3rd it had disappeared, leaving only its shadow and a dark streak on the disk. Dr. Ristenpart stated that the whole ring had been perceived by Kirchhof on November 3rd and 4th with the 12-inch aperture refractor of the Urania Observatory at Berlin, and that he had seen it as an exceedingly fine line across the planet on November 5. On the next day Professor Hartwig of Bamberg detected the ring of a reddish-brown colour on both sides of Saturn, the parts nearest the globe being the brightest (*B.A.A.J.*, vol. 18, p. 97).

Meanwhile Professor Campbell had announced that during the week prior to October 28, four bright knots were visible on the ring, symmetrically placed, two east and two west. Professor Percival Lowell confirmed the observation of these condensations, which were repeatedly measured at his Observatory at Flagstaff, Arizona. He telegraphed on November 22: 'Ring-shadow on Saturn bisected, black medial line, phenomenon explicable by extraplane particles only' (*A.N.*, Nos. 4211, 4213, 4215, 4216.) As will be seen later in this Chapter, it was on the observation of this medial line across the shadow, and the bright knots on the ansae, that Lowell based his theory of 'tores'.

The October disappearance, and the final reappearance January 1908 (Innes, Phillips)

R. T. A. Innes, using a Grubb 9-inch aperture refractor at Johannesburg, made a series of observations of Saturn (the first of note* from South Africa) on 3 October 1907 (*M.N.*, vol. 68, p. 32). At 4^h 45^m p.m. G.M.T. the ring was distinctly seen, but seemed very faint and only about half its usual length. Later Innes noted the trace across the globe, just north of the equator, as sharper and darker than on preceding nights. By 7^h 30^m the ansae were still distinctly seen but ghost-like. At 9^h 45^m the ring was still glimpsed, but doubtfully, while at 10^h 30^m it was invisible. Therefore in good seeing the ring must have disappeared from sight in the 9-inch refractor soon after 9^h 45^m p.m. G.M.T. on October 3.

At various dates from October 16 to November 20, Innes saw whenever he observed Saturn, 'a ghost-like extension where the ring was', tapering off sharply, perhaps shorter than the ring, and on October 21, more distinct on the following side. On November 21 the ring-extension was seen to full length on each side; the trace on the planet was deep brown with just a suspicion of being divided into two lines. On November 25 and December 1 the discontinuous ghost-like extension was in line with the south side of the ring-shadow (*M.N.*, vol. 68, p. 209).

On December 25 Innes saw the ghost-ring and noted that the shadow of the ring was 'quite its own breadth N. of the equator and very black—no longer brown'. He continued to glimpse the ghost-ring till the end of the month, but on 3 January 1908, it was invisible to him, and he failed to see it on each of the next four days. On January 8, however, in daylight (5^h 35^m G.M.T.) he found it was visible, 'very fine and not unlike the ghost-ring' he had seen a month before. He estimated the time of reappearance at about six hours earlier, that is, about noon on January 8. The following ansa was the brighter. He stated that about 6^{h}0 G.M.T. on the same date another observer, H. E. Wood, saw it distinctly as 'a sharp well-defined line on either side' of the disk, and considered it brighter on the following side (*M.N.*, vol. 68, p. 372).

Another who, besides Barnard and Innes, reported observations at the epochs of the October disappearance and the January reappearance was the Rev. Theodore Phillips, a famous amateur astronomer who was for more than thirty years Director of the B.A.A. Jupiter Section and later for a time of the Saturn Section. He had been one of the observers of the north temperate spots on Saturn described in Chapter 24. Phillips stated that on October 4 the Earth crossed the plane of Saturn's rings, and from that date the ring-plane passed between Earth and Sun, so that the side turned slightly towards the Earth, the north side, should theoretically have been invisible. The

* Save Sir J. Herschel's.

purpose of his note on 'The Alleged Disappearance of Saturn's Rings' (*B.A.A.J.*, vol. 18, page 31) was, he wrote, to point out that the rings had never certainly disappeared in his 12¼-inch aperture Calver reflector, and that at the time of writing (late October 1907) the ring system was distinctly visible whenever seeing conditions were at all good. On the night of October 3 he saw the rings as 'an extremely thin and delicate line of light' of the usual length on each side of the globe. On the next night, the theoretical date of disappearance, seeing conditions were too bad either to detect the rings or to decide whether they were invisible. On several subsequent occasions he had the impression of a 'faint luminosity in the plane of the rings extending for a short distance on each side of the ball'. He suspected at first that it might be an illusion, but it was so persistent that he thought perhaps he was seeing the feeble light of the little scattered particles of the crepe ring. Subsequently in better conditions, notably on the nights of October 19, 23 and 25, he saw 'unmistakably an extremely thin though apparently continuous line of light extending on each side of the globe for nearly the whole distance of the rings from the planet'. He thought the aspect particularly beautiful about 9ʰ 40ᵐ G.M.T. on October 25 when for a few minutes the definition of the planet and the narrow line of light were very sharp and plain. On the p. side he saw what he took to be 'two satellites like threaded beads'. Afterwards he found that no satellites were in that position at the time, so he concluded that the objects briefly seen by himself and a friend were probably a first glimpse of 'the condensations which subsequently became so obvious with large apertures' (*B.A.A.J.*, vol. 18, p. 172).

Phillips noted particularly that 'the line connecting the two luminous arms extending from each side of the planet passed through the black band crossing the disc, though the exact position of so delicate an object was not easy to determine'. Apparently therefore he saw the shadow simply as a narrow dark line, collinear with the ansae.

Phillips's explanation of the visibility of the rings was that the narrow sharp line of light seemed likely to be due to the Sun shining through the Cassini division and illuminating the closely packed particles of the outer edge of ring B. He recalled that on several previous occasions when the dark side had been turned towards the Earth the ring had nevertheless been kept, at least partially, in view, for example, by Sir William Herschel, Bond, Dawes, and in 1861–62 (see Chapters 8, 10 and 14).

In his 'Note on the Reappearance . . .' (*B.A.A.J.*, vol. 18, p. 171)

Phillips wrote that the narrow line of light, which became visible on each side of the globe as the elevations of the Earth and Sun above the opposite sides of the ring increased, was not uniform in thickness, but as it came more into view it was seen to contain four bright condensations symmetrically placed, two near the extremities and two nearer the globe; these four objects were permanent in position, and did not appear to revolve with the ring. He had intended to watch Saturn carefully for all changes in visibility from late December till after the theoretical reappearance, but was prevented by bad weather. On December 25 he was able to see the ring distinctly though he considered that it was becoming more difficult.

On 3 January 1908, the ring was invisible though he looked for it at intervals during two and a half hours, and it was also invisible in poor seeing on January 4. The weather was very bad on the critical date, January 7, but he had one glimpse of the planet between clouds for about a minute and a half: there was no trace of the ring though the shadow on the globe was quite conspicuous. The next time he was able to see Saturn, January 9, the ring was quite prominent as a thread of light, not of uniform thickness, but seeming intensified towards the extremities. From this date the ring was perfectly obvious, even with small apertures.

A dark transit of Titan, 6 November 1907 (Cobham, Chauleur)

Several members of the New South Wales Branch of the B.A.A. saw Titan as a dark spot in transit over Saturn's disk on 6 November 1907 (*B.A.A.J.*, vol. 18, pp. 131, 172). This is of course by no means an exceptional phenomenon when the rings are almost edgewise; the interesting thing is that none of these observers was using a telescope of more than $4\frac{1}{2}$-inches aperture, and that A. B. Cobham and G. D. Hirst were also able to detect in glimpses the vestiges of the ring at an epoch when it was theoretically invisible, while Paul Chauleur, observing independently, was able to discern the ring with his $4\frac{1}{2}$-inch aperture frequently during that night 'on both sides of the planet as an extremely fine line of light, nearly on the limit of visibility', the eastern ansa seeming to him brighter than the western. Chauleur's aim was to observe the transit of Titan's shadow which had been predicted for that night, and he had never heard of a dark transit of the satellite, so the dark spot which he saw on the disk, and which was sharply defined only at intervals, he at first assumed to be the shadow. He followed it across the disk, losing it near the limb. Twenty minutes later he saw 'a small point of light, a little woolly, emerging from the limb not far from where the dark spot had vanished'. The supposed shadow was Titan itself! Nearly an hour before the egress of the satellite he noticed an indentation of the north-east limb, which soon became a dark spot on the disk. This spot (the shadow) was much larger, darker and easier to see than the first one, and he believed that it could have been seen with an aperture as small as $2\frac{1}{2}$ inches.

The final 'reappearance' (Barnard)

Barnard continued to observe, measure and make very detailed notes on the ring's aspect almost daily and from hour to hour throughout November and December 1907 (*M.N.*, vol. 68, p. 360), but no very noticeable change seems to have taken place until December 25, when the condensations and the ring in general seemed to him much thinner than they had looked a fortnight earlier. Though he could see the ring's full length it was narrow and faint at the ends. He estimated the thickness of its outer end at barely half that of the micrometer wire, i.e. half of $0''1$, and the thickness at the condensations $1\frac{1}{2}$ times the wire.

On 2 January (1908) the ring was very thin and hardly any trace of it could be seen on the sky without occultation. On January 5 at 5^h 40^m it was much more difficult than on January 2, no part being visible without the occulter; it was very thin throughout but possibly very slightly brighter at the condensations. On January 6, at 6^h, in very poor seeing he tried both telescopes (40– and 12-inch) with high and low powers, but even during the best moments no part of the ring was visible on either side of Saturn, and he doubted whether he could have detected it had the seeing been better.

At 5^h 15^m on 7 January 1908 he found the ring quite easily visible by occultation. It was linear, rather faint on the following side but well seen on the preceding. It must have been much brighter than on the day before because in even worse seeing he could now detect it: hazy and ill-defined, a straight bar with no trace of condensations and of uniform brightness throughout. 'It was of an ashy colour, rather thickish and fuzzy,' he noted.

Barnard concluded from these observations that the Earth must have passed the plane of the ring some time between 6^h on January 5 and 5^h 15^m on 7 January 1908, possibly at about the time of his observation on January 6, as the ring's distinctness on January 7 suggested that the plane had been passed some hours before. He considered that the thinness of the ring was demonstrated by the fact that at the many observations when the Earth and Sun were on opposite sides of the ring-plane, nothing could be seen of the edge though he had looked carefully for any sign of a thin rim of light.

Cause of condensations and of visibility of ring's dark side (Barnard)

Barnard discussed the possible causes of these phenomena (*M.N.*, vol. 68, pp. 355, 358, 365). Bond had supposed that what he had seen when the Earth was on the dark side of the ring in 1848, was its sunlit edge (see Chapter 10), but Barnard felt convinced that this easy explanation would not do, since his observations during the 1907 disappearance had shown that what was seen was not the edge but the very oblique unillumined surface. He considered that there were only two possible explanations: either that the rings are self-luminous, or that 'the sunlight sifts through among the particles composing them and thus makes them visible'. He felt that the temperature of particles could not be high enough to make them self-luminous. The thinness of the rings would also support the explanation that the sunlight sifts through, but his observation of the complete disappearance of Iapetus when in the shadow of the bright rings (see Chapter 18) showed that though the rings A and B might be translucent they were not transparent. He excluded another possible alternative—that the illumination could be due to reflection of sunlight from the globe of Saturn—because the entire surface of the ring was visible, which he felt could not be the case by reflection from the planet.

The bright condensations or knots seen on the ring during its theoretical 'invisibility' in 1848 had been explained by Bond as the edges of rings A and B seen through the Cassini division. In Barnard's opinion, this ingenious suggestion was refuted by the thinness of the ring and the largeness of the condensations. Barnard himself had at first concluded from his measurements that the centres of the outer condensations fell in the middle of the outer part of ring B, and those of the inner ones in the midst of the crepe ring, and he was puzzled as to why similar phenomena should be produced by the most dissimilar parts—the brightest and the faintest—of the ring system. But at the conclusion of his observations in January 1908 he compared the positions of the outer edges of the condensations with his previous measures of the system, as follows:

Outer edge of outer condensation from centre of system					17″·48
Outer edge of Cassini's division	,,	,,	,,	,,	17·52
Outer edge of inner condensation	,,	,,	,,	,,	12·90
Outer edge of ring C	,,	,,	,,	,,	12·82

The measures therefore seemed to connect the condensations directly with the Cassini division and the crepe ring (see Plate X). The condensations could therefore be explained by sunlight shining through and illuminating the particles in these parts of the ring system, because it could be assumed that the Cassini division is not entirely devoid of particles. The drawback to the explanation is that as the inner and outer condensations were essentially of the same intensity, it would require the particles to be as closely clustered in the division as in the crepe ring, which seems unlikely. Barnard also pointed out that the inner condensations always looked much brighter than the crepe ring had ever appeared to him.

Commenting on Barnard's explanation, D. G. McIntyre (*Translucency* . . ., p. 10) suggested that in the case of the outer condensations the effect may not be entirely due to sunlight sifting through but also to reflection of light from Saturn's globe to the surface of the rings. Professor Russell calculated that the light reflected by Saturn at the position of the outer condensations would be about 160 times full moonlight on Earth. 'We may therefore conclude,' Russell wrote, 'that the outer condensations, and the general visibility of the surface of the rings, may be accounted for by their illumination by light reflected from Saturn, while the inner condensations are due to sunlight transmitted through the practically transparent Crape Ring.' McIntyre therefore suggested that possibly both the phenomenon of the outer condensations and the general visibility of the 'dark' side of the rings might be a combined effect of light penetrating and light reflected. If so the effect might be enhanced, McIntyre felt, by fine, widely-scattered dust which, Dr. L. Bell suggested, may envelop the ring system. This dust becomes 'more and more tenuous away from the ring-plane', according to Bell's theory, and is everywhere 'so thin a cloud that it reflects no visible light save when seen in great thickness edgewise'.

Having regard to all the observations of the rings during 1907–08, there will probably be a feeling that the terms 'disappearance' and 'reappearance' are not happily chosen to represent what actually takes place during epochs when the Earth and the Sun in turn cross the ring-plane. In theory the ring should be invisible for the whole of each period during which the Earth and the Sun are on opposite sides of it. In actual practice this does not seem to happen, and it certainly did not, even with smallish telescopes, during the relevant periods in 1907, when the only true disappearance of the ring occurred during the few hours on each occasion when the Earth was actually passing through the plane. Hence the paradoxical situation arose in January 1908 that when the 'reappearance' was due, the first thing that happened was that the ring became temporarily invisible!

Lowell's theory of 'tores'

Percival Lowell, the founder of Flagstaff Observatory, famous for his studies of the 'canals' of Mars, had, like Aitken but unlike Barnard, gained the impression from his observations of Saturn in 1907 that the bright condensations projected considerably from the line of the ansae, and also Lowell had observed 'a dark medial core' running longitudinally across the whole length of the ring shadow on the globe.

These observations caused Lowell to enunciate a new theory to explain the condensations, namely, that they are caused by elevations or ridges on the surface of the rings which he called 'tores', a term not hitherto used in astronomy and derived from the Latin word 'torus', used in botany for the raised receptacle of the whorls of a flower (*B.A.A.J.*, vol. 18, p. 299). Lowell concluded that rings B and C are tores, not flat rings, because he assumed that the black core (observed only at Flagstaff), running through the middle of the shadow band, was the shadow of the flat ring A bordered by the particles of B and C above and below its plane. He also felt that this theory accounted better than any other for the condensations seen by many observers. He claimed in addition that observational results showed that the rings were disintegrating (*B.A.A.J.*, vol. 18, p. 392, from *Lowell O. B.* 32).

McIntyre (*Translucency*, p. 9) has pointed out that Barnard's observations of the invisibility of the rings when exactly edgewise runs counter to this supposed existence of thickened 'tores', and that Dr. (now Sir) Harold Jeffreys in a subsequent investigation (1916) of the dynamics of the rings showed the probability that collisions between the ring particles would be so frequent 'that motion perpendicular to the plane of the ring must be annulled in a year or two, so that the thickness of the ring must be too small to be observed when seen edge-on'.

CHAPTER 27

PASSAGES OF THE EARTH THROUGH THE RING-PLANE

Dates of passages, 1848–1966

From Barnard's list (*M.N.*, vol. 68, p. 357) and Porter's (*B.A.A.J.*, vol. 70, p. 52) the following complete catalogue of passages of Earth and Sun through the ring-plane from 1848 to 1966 is derived, the letters *D*, *R* denoting respectively the theoretical disappearances and reappearances of the ring:

Date	D or R	Passage by	Observations given in
1848 April	D	Earth going South	
Sept. 3	R	Sun ,, S.	
12/13	D	Earth ,, North	Chapter 10
1849 Jan. 19	R	Earth ,, S.	
1861 Nov. 22	D	Earth ,, N.	
1862 Jan. 31	R	Earth ,, S.	
May 17	D	Sun ,, N.	Chapter 14
Aug. 12	R	Earth ,, N.	
1878 Feb. 6	D	Sun ,, S.	
Mar. 1	R	Earth ,, S.	Chapter 14
1891 Sept. 22	D	Earth ,, N.	
Oct. 30	R	Sun ,, N.	Chapter 18
1907 Apr. 13	D	Earth ,, S.	
July 27	R	Sun ,, S.	
Oct. 3	D	Earth ,, N.	Chapter 26
1908 Jan. 7	R	Earth ,, S.	

(The *N.A.* dates for 1907 are given here in preference to Barnard's quoted in Chapter 26.)

Date	D or R	Passage by	Observations given in
1920 Nov. 7	D	Earth going N.	
1921 Feb. 22	R	Earth ,, S.	Chapter 32
Apr. 9	D	Sun ,, N.	
Aug. 3	R	Earth ,, N.	
1936 June 28	–	(Earth did not cross plane of rings; minimum saturnicentric lat. of Earth 0°0001 North	Chapter 36
Dec. 28	D	Sun going S.	
1937 Feb. 21	R	Earth ,, S.	
1950 Sept. 14	D	Earth ,, N.	
21	R	Sun ,, N.	
1966 Apr. 2	D	Earth going S.	
June 15	R	Sun ,, S.	
Oct. 29	D	Earth ,, N.	
Dec. 18	R	Earth ,, S.	

These dates are given by J. Meeus (*Ciel et Terre*, vol. 73, p. 55, 1957) and E. W. Woolard, Director of Nautical Almanac Office, Washington (*Sky and Telescope*, 1961 Dec., p. 355).

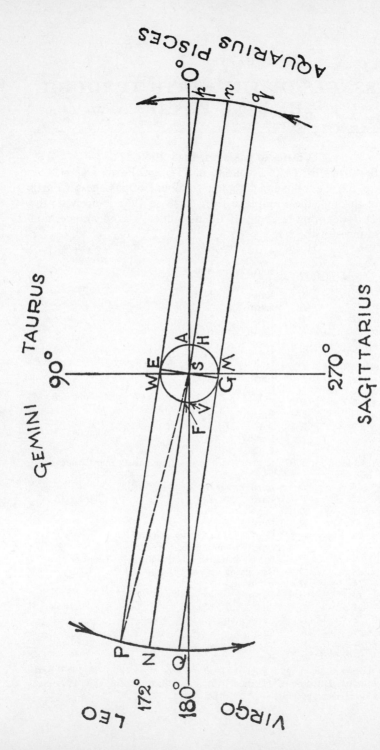

FIGURE 15. General diagram of passages of Earth through ring-plane. Heliocentric longitude of n : 352°.

Porter states that the phenomena in 1965–66 should be quite as favourable for observation as those in 1907–08: conjunction and opposition are calculated to fall at 1966 March 10 and September 19 respectively; in 1907 they were on March 8 and September 17 respectively.

	Conjunction	Opposition
1920–21	1920 Sept. 8	1921 Mar. 12
1936	1936 Mar. 16	1936 Sept. 25
1950–51	1950 Sept. 16	1951 Mar. 20

It will be noticed that, as the result of each of these groups of events, both Sun and Earth pass from one side of the ring-plane to the other, remaining on that side until the next group of passages occurs; also that, whereas at each epoch the Sun passes once only through the plane, the Earth sometimes makes one passage and at other times three. The object of this chapter is to explain with the aid of some simple diagrams (see Figures 15 to 18b) why the number of passages by the Earth through the ring-plane of Saturn varies in this way. It is hoped that this rather full explanation will remove all possible misunderstanding of what is mathematically a straightforward case of the combination of simple harmonic motions which, as is shown later in the chapter, can be illustrated by the superposition of two sine curves.

Proctor's explanation (*Saturn*, chap. 4) has three drawbacks: a faint and confusing diagram, a complication of the issue by discussing ring disappearances and reappearances along with the Earth's passages through the ring-plane, and an unfortunate miscalculation of the time taken by the ring-plane to traverse the Earth's orbit. Sir John Herschel (*Outlines of Astronomy*, 1859, pp. 343–46) gave a clearer diagram but an explanation that is too superficial and not entirely correct. The one that will now be given follows closely that worked out, with illustrative examples and diagrams, by M. B. B. Heath.

General explanation (Heath)

For simplicity the orbits of the two planets may be assumed to be circular, concentric and co-planar; though none of these assumptions is strictly true, this does not vitiate the argument in a general discussion. In Figure 15 the point S represents the Sun, the small circle $AEWFVGMH$ the Earth's orbit, and the arcs PNQ, qnp small parts of Saturn's orbit (approximately to scale) with a radius 9·54 times that of the Earth's orbit. The points A, W, V, M, represent the Earth's positions on its orbit at the autumnal equinox, winter solstice, vernal equinox and summer solstice respectively; the heliocentric longitudes of these directions and of N, n (Saturn's positions corresponding respectively to the ascending and descending nodes of the ring-plane) are shown, and also the zodiacal constellations in these directions.

For example, at the winter solstice the Earth will be at W, the Sun will be in the direction of Sagittarius, and Gemini and Taurus will be in the south at midnight. The diagrams show the time of year roughly, according to the Earth's position on the small circle: from late September to late December the Earth will be moving from A to W, from late December to late March from W to V, and so on.

In Figure 15 let PEp, QGq be two tangents to the Earth's orbit parallel to Saturn's ring-plane. Then for the Earth to pass through the ring-plane the line joining the Earth to Saturn must be parallel to PEp or QGq, and Saturn must be in the portion PQ or the portion qp of its orbit; since Saturn's orbit is more than $9\frac{1}{2}$ times the size of the Earth's these opportunities are confined to limited periods recurring at long intervals.

Through S draw NSn parallel to PEp, QGq and meeting the Earth's orbit at F and H. It will then be clear that if, when Saturn reaches N, the Earth happens to be at F, Saturn will be in opposition and readily seen at the time of the Earth's passage through the ring-plane; but if the Earth should be at or near H when Saturn reaches N, Saturn will be in conjunction and invisible. Join ES and SP.

Then, since Pp and Nn are parallel, the angles NSP and SPE are equal. SP is 9.54 times SE; hence sin $\widehat{SPE}$ is $1/9.54$, whence angle NSP is $6° 1'$, and the arc PQ, and also the arc pq, subtends an angle of $12° 2'$, at S. As the heliocentric longitude of Saturn's perihelion is about $92°$, the arcs PNQ and pnq are not at all near either the perihelion or the aphelion of the planet's orbit, so Saturn can be assumed to traverse these arcs at its mean angular motion, so that it will take 359.63 days to traverse either arc, that is, 5.6 days less than a year. Hence the ring-plane will take approximately 5.6 days *less* than a year to cross the Earth's orbit from the line PEp to QGq, or from qGQ to pEP.

Obviously during any one of its crossings of the Earth's orbit, the ring-plane is bound to pass once, and only once, through the Sun, and is also bound to pass at least once through the Earth wherever the Earth may be on its orbit. It also follows that, at the present time, these phenomena can occur only when Saturn's heliocentric longitude lies between about $166°$ and $178°$, or between about $346°$ and $358°$.

Whenever the ring-plane passes through the Earth, the line joining the Earth and Saturn must be parallel to the direction of the line of the nodes. This is the crux of the matter. The various cases can now be considered:

(1) If when Saturn reaches P and the ring-plane the line PEp, the Earth happens to be somewhere in the near side of its orbit, between W and F or F and G, the Earth's motion, being more than three times as fast as Saturn's, will keep it ahead of the advancing plane, and the line Earth–Saturn will not become parallel to the direction of the line of nodes until after the Earth has passed the point G, but somewhere in the semi-circle GHE the line Earth–Saturn will come into parallelism with the line PEp and the Earth will meet and pass through the ring-plane. It will not be able to overtake the ring-plane before the latter has passed off the Earth's orbit at QGq, so there will only be one passage of the ring-plane through the Earth.

(2) If when Saturn reaches P the Earth is on the far side of the orbit, somewhere in the arc MHE, it will meet and pass through the ring-plane before reaching E, will overtake and pass through it again somewhere in the semi-circle EFG, and will meet and pass through it a third time between G and H shortly before the ring-plane passes off the Earth's orbit.

(3) If the Earth is at, or a little on the near side of, E when Saturn reaches P, the Earth will for the moment be moving almost directly towards Saturn, and the ring-plane will overtake and envelop the Earth, but before the latter passes through, the Earth's motion will carry it clear and ahead of the plane. Hence in this unusual case there will be (in Proctor's opinion) a double passage, the Earth seeming to hang in the ring-plane. The Earth will meet and pass through it properly somewhere in the arc GMH.

(4) If the Earth is at or a little beyond G when Saturn reaches P, it will meet and pass through the ring-plane somewhere in the arc HAE, and will again almost pass through the plane by overtaking it shortly before it reaches QGq, but as the Earth will then be moving almost directly away from Saturn, the ring-plane will as it were shake off the Earth and leave it behind. This was also regarded as a double passage by Proctor, but it seems more satisfactory to look upon the 'double passages' mentioned under (3) and (4) as not being passages at all, since in neither would the Earth really pass through the plane, and in both the Earth emerges on the same side of the plane as it entered. (3) and (4) are really limiting cases of (1), each producing only one true passage through the plane.

In all the foregoing cases Saturn has been assumed to be near the

307

direction of the ascending node of the ring-plane, but precisely similar circumstances can of course occur when Saturn is at the other side of the orbit, near the direction of the descending node, i.e. in the part of its orbit between q and p as in 1907–08 and 1936–37.

Examples from 1907–08, 1920–21, 1936–37 and 1950 (Heath)

These examples are shown in Figures 16 and 17, based on diagrams worked out by M. B. B. Heath, and they should be clear from the diagrams and need only a brief explanation. To avoid any confusion the same system of lettering has been used throughout, as in the general diagram (Figure 15) and with the addition of the letter X for the Earth's position when the ring-plane reached the Earth's orbit, Y for the Earth's place when the plane was passing through the Sun, and B, C, D for positions of the Earth when it passed through the ring-plane, the corresponding positions of Saturn being b, c, d, and the lines Bb, Cc and Dd being of course parallel to PEp, QGq.

In 1907–08 the ring-plane began to cross the Earth's orbit in late January 1907, Saturn being at q and the Earth at X. As the two planets were moving in opposite directions, the Earth soon (in April) met and passed through the plane at B; the Sun passed through in July when the Earth was at Y, and the Earth overtook and passed the plane again at C in early October, and met it and passed through a third time at D in early January 1908.

In 1920 also the Earth was on the far side of its orbit at X when at mid-October Saturn entered the arc PQ at P and the ring-plane began to cross the Earth's orbit. Hence there were again three passages of the Earth through the ring-plane (at B, C, D): in 1920 November and 1921 February and August, from south to north on the first and third occasions and from north to south on the second. The Sun passed from south to north in 1921 April.

While 1907–08 and 1920–21 were typical of case (2) in the above general description, the phenomenon in 1936–37 provided an example of the unusual happening referred to as case (3). Saturn, approaching the descending node of the ring-plane, entered the arc qp at q towards the end of June 1936, the Earth being at X, not far from G. As the Earth's motion was almost directly towards Saturn, the ring-plane quickly overtook the Earth at B, the latter hanging in the plane for two days but without actually passing through to the southern side, because the Earth's motion then enabled it to get clear and ahead of the plane, so that it was still on the northern side of that plane as it had been since 1921. The ring-plane passed

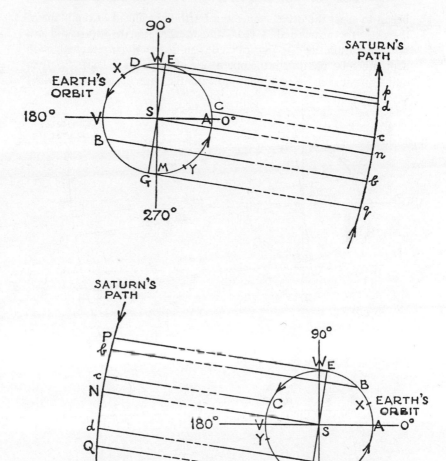

FIGURES 16a, 16b. Diagrams of passages of Earth through ring-plane:
(a) 1907–08, (b) 1920–21

through the Sun late in 1936 December, the Earth being at *Y*. In February the Earth met the plane and passed through it at *C*, only about three weeks before conjunction, so the position of Saturn was very unsuitable for observation.

In 1950 there was provided a normal example of case (1), the Earth being on the near side of its orbit near *V* when the ring-plane began to cross the orbit. Hence the Earth kept ahead of the plane of the rings for nearly half a year, and the line Earth–Saturn did not become parallel to the line of nodes until the Earth reached *B* on September 14, passing to the north side of the rings. Saturn was then quite invisible as its conjunction with the Sun occurred two days

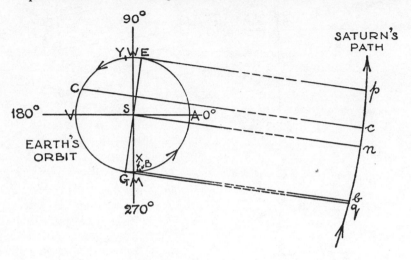

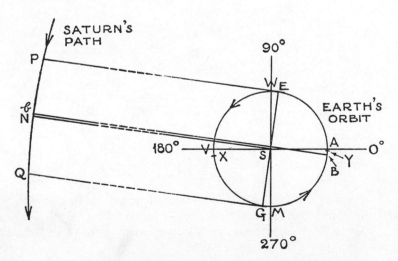

FIGURES 17a, 17b. Diagrams of passages of Earth through ring-plane:
(a) 1936–37, (b) 1950

later; on September 21, when the Earth was at Y, the ring-plane passed through the Sun. This passage was not observable either, and so the phenomena of 1950 were of purely academic interest. The ring-plane passed off the Earth's orbit at QG long before the Earth could reach G, so there was only one passage of the Earth through the plane.

The intervals between successive passages of the Sun through the plane of the rings are alternately about $13\frac{3}{4}$ years (1848–62, 1878–91, 1907–21, 1936–50) and about $15\frac{3}{4}$ years (1862–78, 1891–1907, 1921–36). This is because of the ellipticity of Saturn's orbit, as explained in Chapter 5.

Condition for passage of Earth through ring-plane (Turner)

In a paper entitled 'Note on the condition for the passage of the Earth through the plane of Saturn's Ring' (*M.N.*, vol. 68, p. 460), Professor H. H. Turner, starting with the same assumptions as Mr. Heath of circular, concentric and co-planar orbits for the two planets, gave an equally interesting but more mathematical explanation. Turner showed that the passages of the Earth through the ring-plane can be found from the points of intersection of two sine curves (see Figure 18a), one of small amplitude ($y = \sin t$) representing the Earth's motion, and one of large amplitude representing that of Saturn:

$$y = n^2 \sin (t - a)/n^3,$$

where the radius of the Earth's orbit is unity, and that of Saturn is n^2, which therefore equals $9 \cdot 54$, whence n is $3 \cdot 088$ and n^3 is $29 \cdot 46$, the number of years of Saturn's revolution; t is the time, the unit of time being $\dfrac{1}{2\pi}$ year; and at the origin of time Saturn's angular distance behind the Earth is a/n^3.

Some of the results of Turner's discussion are as follows:

(1) The Earth's curve alternates much more rapidly while Saturn's is a broad sweep; there may be one or three intersections (there must be an odd number) where the curves cross.

(2) If there is only one intersection the slopes of the two curves are opposite (i.e. one upward, the other downward), so the planets are on opposite sides of the Sun; hence the single passage will occur near Saturn's conjunction with the Sun and will not be so readily observable.

(3) If there are three intersections, the middle one is near opposition and the other two near quadratures.

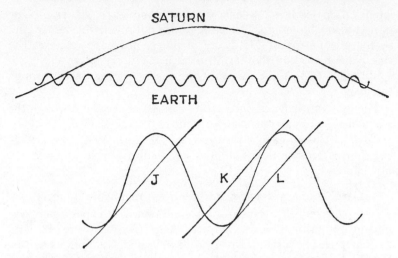

FIGURES 18a, 18b. H. H. Turner's sine curves of condition for passage of Earth through ring-plane. (Redrawn after diagrams in *M.N.*, vol. 68, p. 461).

(4) There is an occasional case where the curves cross at one point and touch at another; the latter would represent a case of the Earth hanging in the plane without passing through.

(5) Turner calculated that, in a long series of years, the chance of three intersections is about 53 per cent, that of one about 47 per cent.

(6) He found two separate series of values for α referring to the two nodes, and pointed out that if the common difference in the value of α were really 162° the numbers would repeat after twenty terms, but since it is about 162° or 163°, under actual conditions the series will slowly diverge from the following calculated values:

First Series (*Descending Node*)

Year:	1789	1819	1848	1878	1907	1937	1966	1996	2025	2055	2084
α	25°	187°	349°	151°	313°	115°	279°	79°	241°	43°	205° etc.
Crossings	3	1	3	1	3	1	3	3	1	3	1

Second Series (*Ascending Node*)

Year	1803	1832	1862	1891	1921	1950	1980	2009
α	101°	263°	65°	227°	29°	191°	83°	245° etc.
Crossings	1	3	3	1	3	1	3	1

An interesting point for observers arising from these calculations is that there seems to be a probability of the Earth making three passages through the ring-plane in 1966, 1980 and 1996, which are the next three epochs of the edgewise ring, and as Turner pointed out, the conditions for observation where there are three passages should be much more favourable than when there is only a single passage.

Figure 18b shows a 'close-up' of the intersections of the two sine curves, illustrating all the various possibilities. If the curves cross between positions J and K there is a single passage only; if between K and L there are three passages. The limiting cases are J (or L) and K: at J there is (as occurred in 1936) a 'double-passage' (or appulse), the Earth shaking off the ring-plane, followed later by a single passage; at K the single passage precedes the appulse, in which the plane shakes off the Earth.

CHAPTER 28

THE LIMITS OF THE RING

(A) OBSERVATIONS OF A SUSPECTED OUTER DUSKY RING

Discovery (G. Fournier, Schaer)

The edgewise ring epoch of 1907–08 produced one more surprising problem. Georges Fournier, a leading French planetary observer of the early twentieth century, was then assisting at Jarry Desloges's observatory which Fournier had himself installed on Mont Revard in Savoy (altitude 1550 metres—about 5,100 feet). Observing Saturn's ring with a refractor of aperture about 11 inches on 1907 September 5 and 7, he noticed a peculiar appearance surrounding the outer bright ring. Fournier's description of what he saw on September 7 is as follows:

'10^h 25^m Grossissement 320. Très belles images, absolument parfaites par moments. Tout autour de l'anneau dans les meilleurs moments de calme on aperçoit une zone lumineuse très pâle, surtout bien visible à gauche et au-dessous de l'anneau, ainsi qu'à ses extrémités. Cette remarque avait déjà été faite le 5 septembre et prise pour une erreur de mise au point; mais la présence réelle de cette zone n'est pas douteuse . . . on aperçoit dans la zone lumineuse de petits corpuscules lumineux qui se meuvent très vite. . . .' (see Jarry-Desloges; *Observations des Surfaces Planétaires*, vol. 1, 1907, p. 106; also references in *B.A.A.J.*, vols. 18, p. 223, and 19, p. 152, and a useful note by G. H. Lepper in vol. 31, p. 125).

It is important to note that Fournier described what he had seen as 'a luminous zone' containing swiftly moving luminous particles, and that his description did not refer to the phenomenon as an outer faint or dusky ring. It was Jarry Desloges who referred to it as '. . . un anneau extérieur, lumineux et transparent . . .'; as they failed to detect it on September 11, in spite of good seeing, Jarry Desloges concluded that it could only be perceived in quite exceptional conditions of atmospheric calm. He suggested that high latitude observatories such as Arequipa and Flagstaff would seem to be most favourably situated for studying this very difficult object, and added that the numerous minute bodies seen on the ring seemed to be composed of denser matter reflecting light differently from the rest of the ring. They seemed to vary rapidly in shape and luminosity, and perhaps also to break up into many fragments.

Fournier's surprising observation, though prominently reported at the time, seems to have passed unnoticed for nearly a year. Then, in 1908 October, came a telegram from E. Schaer of the Geneva Observatory, announcing that a new dusky ring had been discovered surrounding the bright rings of Saturn (*B.A.A.J.*, vol. 19, p. 55 and *A.N.*, No. 4276). It would seem that the earlier observation of Fournier had not come to the notice of the Swiss astronomer, and that his discovery was an independent one; moreover, while Fournier had noted a pale, luminous zone, Schaer described and drew his newly discovered ring as brown, and doubts were expressed by some commentators as to whether they were referring to the same object (see *Bulletin S.A.F.*, 1909, e.g. p. 233).

Schaer was using for the observation of Saturn, 1908 September to 1909 January, a Cassegrain reflector of 16-inch aperture which he had just completed, and powers from 250 to 660. His record of these observations (*Bulletin S.A.F.*, vol. 23, p. 232) shows that prior to 1908 October 5 he had noticed that the bright ring was bordered in front of the planet by a 'bande brune ou sombre' about 1 second of arc wide, which he had assumed to be the shadow of the rings. But on October 5 he realised that the dark edge was not due entirely to the shadow because he could distinctly make out part of it to right and left beyond the planet's globe, following the line of the rings as far as the limits of Cassini's division in the ansae. On the next day he could detect a strip of the bright disk less than 1″ wide separating the brown border from the shadow of the rings, so he then supposed it might be merely a dusky belt. However, on October 7 he again traced the dark outer ring, as he was now convinced it was, in the ansae as far as the limits of Cassini's division and momentarily as far as the extremities of the bright rings in both ansae. He then sent a telegram: 'Un nouvel anneau brun entoure les anneaux blancs de Saturne'. At subsequent observations he felt even more certain of its reality; on October 9 he described it as darker than the sky background.

Schaer also noted in 1908 October, that the part of the rings covered by the globe's shadow was nevertheless faintly but decidedly visible, and that a darker tinge was present in the central part of ring B but no definite division there (*B.A.A.J.*, vol. 19, p. 414, and *A.N.*, No. 4331).

A paper by R. M. Baum (*B.A.A.J.*, vol. 64, p. 192) gives a useful outline of the observations of Fournier and Schaer and of those of the Greenwich observers and Barnard (related in the next two sections of this chapter) and discusses his own more recent observations and some of the problems involved, but does not refer to Antoniadi's important contribution.

Following up the discovery (Bowyer, Lewis, Eddington)

In consequence of the Geneva announcement, a careful examination of the ring system was made at the Royal Observatory, Greenwich, during 1908 October with the 28-inch aperture refractor. On October 10, in bright moonlight but fairly good definition, the ring, all round, appeared to W. Bowyer to be dusky on the outer edge; duskiness was also noticed on the following limb of the planet. T. Lewis, who then made a briefer observation, found the crepe ring, Cassini's division and the satellites Enceladus and Tethys very distinct, and remarked: 'Two features seem different from what I remember of Saturn in 1895—(1) the north preceding edge of the outer ring had a hazy border; (2) the north following edge is fairly sharp and dark,—the impression being that it is the thickness of the ring one is looking at. The south edge, preceding and following, is quite normal. The varying effects may possibly be due to the bright moonlight. Two white dots on the rings near the position of Encke's division, one on each ansa, are very bright.'

On October 11, with power 550, Lewis could not for an hour see anything unusual about the edge of the ring, but he then noticed that the north following edge 'appeared indented, while a faint fringe extended hence to the end of the ansa'. A few minutes later this fringe was seen both preceding and following. Bowyer considered that the north following edge had a different appearance from the south edge; 'there was a trace of a faint, fuzzy ring', which soon after was suspected on the north preceding edge also.

On October 12 Lewis suspected a fringe to the north preceding edge of ring A; Bowyer twice found the north edge of ring A bordered by a faint dusky ring—at $10^h 38^m$ and $10^h 50^m$, but it does not appear to have been seen continually.

On October 15 Bowyer again saw the other features well, including the bright spots on ring A; the preceding and following edges of that ring on the north side seemed sharper than on the 12th, and 'in short spells of good definition it appeared bordered by an outer dusky ring . . . fairly well outlined'. Eddington, also observing, with a power of 670, reported: 'At the moments of good definition there appeared to be a narrow dusky ring surrounding the bright ring, visible on the N. edge. I could not detect it on the S. edge, nor where it crossed the disk of the planet. The north and south edges differ in colour (owing to atmospheric dispersion), and this renders a comparison of their appearances somewhat uncertain.' (This observer was later, as Sir Arthur Eddington, to become famous in connection with relativity and stellar evolution.)

Bowyer made further observations on October 22, 27 and 30, and on each occasion when the seeing was favourable was able to see traces of an outer dusky ring on the north preceding and following edges of the bright ring (*M.N.*, vol. 69, p. 39).

Negative evidence of Barnard

Although the evidence related so far seems fairly strong for the existence of a dusky fringe outside parts of the edge of ring A, it does not seem to have established the presence of a complete outer crepe ring. Had Barnard been able to get a good view of the object with the 40-inch aperture Yerkes refractor, the evidence would have been conclusive, but he had seen nothing of it during a number of careful observations during 1908, though he admitted that it might have escaped his notice through his not being at that time aware of its suspected existence, and that in November and December when Saturn was well placed for observation he had been engaged on other work.

In January 1909, however, when Schaer claimed that the new dusky ring was more easily seen than before, Barnard searched for it very carefully on two occasions, but without success. On January 12, in satisfactory seeing, he closely examined the outside edge of the ring, especially at the ansae, using an occulter to reduce the glare of Saturn, but could see '. . . nothing abnormal anywhere'. On January 19, again with powers of 460 and 700, and trying a reduced aperture of 35 inches, another careful search produced no evidence of an exterior ring, although the reduced aperture gave appreciably better definition (*M.N.*, vol. 69, p. 621).

Antoniadi suggests extraplanar particles

In an interesting paper entitled 'Corpuscules en dehors du plan de l'anneau de Saturne' (*Bulletin S.A.F.*, vol. 23, p. 448) Antoniadi referred to the negative findings of Barnard. Although he recalled that Barnard was not an observer of delicate planetary details, such as faint spots, yet this was a case, not of a planetary detail in the ordinary sense, but of a nebulous fringe of very feeble luminosity, and Barnard had the discovery of many faint objects to his credit such as nebulae, comets and the fifth satellite of Jupiter. Moreover, the great Yerkes telescope with its powerful light-grasp was the ideal instrument for the detection of faint nebulous objects.

Earlier in his paper Antoniadi wrote that his first glance at Fournier's drawing had convinced him that the appearance observed at Mont Revard could not be explained by the hypothesis of an exterior nebulous ring in the plane of the other rings. He then proceeded to show by a diagram that, because of the very narrow opening of the rings at the time of the observation, perspective would not allow an outer ring to be visible before and behind the narrow part of the rings near the globe unless it extended in the ansae for a great distance beyond the extremities of ring A. Fournier had shown it extending for only a short distance at the extremities of the ansae, but also bordering both edges of ring A to the limbs of the globe. Antoniadi therefore concluded that the only possible explanation of the drawing was that it showed not only an exterior dark ring but also particles outside the plane of the rings, above and below their surfaces, and he felt that the Greenwich observations seemed to establish the existence of these extraplanar particles. Moreover, Barnard's observations, though they seemed to dispose of the ring, would not negative the objectivity of such particles, because the rings, more open when he was observing than they had been in 1907 at Mont Revard, could have hidden them. He concluded that the existence of these particles seemed highly probable and that Fournier deserved great credit for discovering them.

A further communication from Antoniadi (*Bulletin S.A.F.*, vol. 23, p. 506) stated that he had received, following the publication of his paper, a letter from Fournier accepting his conclusion about the existence of particles outside the plane of the rings and the non-existence of a nebulous exterior ring; Fournier said that he had always interpreted his observation of 7 September 1907 as 'une sorte de gaine nébuleuse, corpusculaire, enveloppant l'anneau sur ses deux faces' and had so described it in a letter to Nice Observatory at the end of 1907. Antoniadi recalled the observation of extraplanar particles by Wray in 1861 (see Chapter 14) and said that Fournier had reminded him of Lowell's telegram mentioning 'extraplane particles' in November 1907 (see Chapter 26). Hence Antoniadi concluded that the existence of extraplanar particles could no longer be doubted, but that 'il n'y a pas d'anneau crépusculaire extérieur de Saturne'.

The negative testimony of Barnard using one of the world's largest telescopes, combined with the arguments of Antoniadi and their acceptance by Fournier, seem to weigh heavily against the other not very conclusive evidence for the existence of an outer crepe ring. In recent years, as related in Baum's paper, various observers with small telescopes have claimed to have detected this object, and no doubt it will continue from time to time to attract the efforts of those who regard it as one of the unsolved mysteries of Saturn, a sort of 'Loch Ness monster' of Saturn in which some believe, but of whose reality most astronomers are very sceptical.

(B) INVESTIGATIONS OF THE LUMINOSITY AND THINNESS OF THE RINGS

Illumination of the dark side and thinness (Russell)

Professor Henry N. Russell of Princeton University Observatory, famous for his work on the luminosities and spectral types of stars, wrote a paper in 1908 (*Ap. J.*, vol. 27, p. 230) suggesting, as mentioned in Chapter 26, that light reflected from Saturn caused the general visibility of the dark side of the rings and the outer condensations observed by Barnard in 1907–08. From available photometric data of the Sun and planets Russell calculated that the intensity of the light reflected from Saturn at the distance of the outer condensations is 560 times moonlight on Earth and 1/1500 of terrestrial sunlight, but since, as Bond had once said, Saturn would appear from the condensations 'as a half-moon half set', the actual intensity of the reflected light at the condensations should, Russell stated, be estimated at 2/7 of the figures for full illumination, or 160 times terrestrial moonlight. From the reflecting power of the ring particles and the apparent area of the condensations, he calculated that their apparent magnitude would be between 10 and 11 magnitudes fainter than Saturn, which would make them decidedly brighter than Mimas and Enceladus and perhaps brighter than Tethys, which accorded with Barnard's observations. Russell agreed with Barnard that the inner condensations are due to sunlight filtering through the crepe ring, and showed that according to calculation the inner condensations should have about the same brightness as the outer, which tallied with observation. The most important conclusion came at the end of the paper, based on the fact that Barnard could at no time see any evidence of a bright rim to the faintly illuminated surface of the rings. The surface brightness of the edge, illuminated by full sunlight, would be about 160 times that of the condensations, and it would be as bright as they, if its width were 1/160 of 0″.5, that is, 0″.003. Hence, Russell concluded that the thickness of the rings must be much less than this, which, at Saturn's mean distance, corresponds to 13 miles (21 km.). Therefore Russell gave an upper limit to the thickness of the rings which was less than one-third of Bond's estimate of sixty years before.

Barnard, commenting on Russell's paper (in *Ap. J.*, vol. 27, p. 234), though admitting that reflected light from Saturn may have had something to do with the visibility of the dark side of the rings, considered the illumination to be chiefly caused by sunlight percolating through between the particles. He also mentioned that the inner condensations were brighter in general than the crepe ring had ever appeared to be, and that he as well as Professor Aitken had on some occasions seen two inner as well as an outer condensation on each ansa.

Degeneration, flatness and thinness (Jeffreys)

The investigation of the dynamics of the rings by Dr. Jeffreys, mentioned at the end of Chapter 26, formed part of a mathematical paper dealing with the distribution of meteoric bodies in the solar system (*M.N.*, vol. 77, p. 84, December 1916), and he amplified some of the arguments about the ring particles four years later in a letter (*B.A.A.J.*, vol. 30, p. 294). Jeffreys calculated that, owing to the loss of relative motion at every collision between particles, the rings must long ago have reached a state in which the particles are moving in very accurate circles all in one plane. He considered that Barnard's observation of Iapetus in 1889, visible behind the crepe ring, but quite invisible when it passed into the shadow of ring B (see Chapter 18), showed that the particles were numerous enough to prevent the light of a faint object from penetrating the denser parts, and by reflection, absorption and scattering, to allow very little light to get through. This was also confirmed by the high albedo of the bright rings, but the fact that the two main rings are not equally bright showed that the number of particles did not greatly exceed a density sufficient to block most of the incident light. He considered that the ring particles must have come to be distributed in narrow circular strips, each strip being practically a broken solid ring, perhaps undergoing something not unlike friction from its neighbours on either side, moving at speeds differing only slightly from its own. Frequent collisions must still be going on, though with less violence than in the past. This must have reduced much of their substance to powder, each of the new reflecting surfaces adding to the apparent albedo. The bright rings, he thought, were probably not absolutely opaque; hence they would be expected to be faintly visible when Sun and Earth were on opposite sides of them. With reference to Russell's upper limit for their thickness, Jeffreys considered it very unlikely that any measurable thickness would ever be found.

Albedo and thinness (Bell)

In his second paper on the 'physical interpretation of albedo' Dr. Louis Bell in 1919 discussed the albedo of Saturn's rings, and deduced an upper limit for their thickness (*Ap. J.*, vol. 50, p. 9). Having regard to the invisibility, apart from the condensations, of the edgewise rings in all but the very largest telescopes, and their doubtful visibility in those, and to the fact that, in theory, they should be at least as bright edgewise as at any other angle with a deeper mass of reflecting matter in the line of sight, Bell deduced from their disappearance that their thickness is 'too small at their effective albedo, to render them visible as linear bright objects'. He quoted experiments showing that a bright line eight or ten times its breadth can be seen as easily at a distance as a round bright spot of the same area, and that a bright line on a dark background is more easily visible than a black one on a white background.

Mars's satellites have diameters of the order of 10 km.; hence a bright line should be visible at the distance of Mars down to a thickness of 1 km., equivalent to not over 15 km. at Saturn's distance. After quoting other experiments on the visibility of thin lines to the naked eye and through the telescope leading to similar results, Bell concluded that it was highly probable that the substantial layer of Saturn's rings does not exceed 15 km. in thickness. He pointed out that this value, 0″0025 of arc, agrees closely with Russell's suggestion of 0″003 made on a different basis.

CHAPTER 29

OBSERVATIONS AND STUDIES OF THE OPENING RINGS, GLOBE AND SATELLITES, AND PHOTOGRAPHY OF SATURN, 1908–14

Oblique aspect of the rings, 1908 (Barnard)

Barnard, using the Yerkes 40-inch aperture refractor, made a series of observations from July to October 1908 of the very narrowly opened rings, only a few months after the last passage of the Earth through the ring-plane, in the hope of finding something that would explain the luminous condensations seen by him on the dark side of the rings in 1907 (see Chapter 26). His report (*M.N.*, vol. 69, p. 621) confessed that these later observations failed to provide any important clue to the cause of the condensations, but he expressed himself as being quite satisfied that the inner ones were due to the sunlight sifting through and being reflected from the particles of ring C, while the outer ones were connected in some way with the Cassini division.

In these later observations Barnard noted, as Carpenter had in 1863 (see Chapter 14), the unusual brightness of the crepe ring. He explained this by the fact that the ring was seen very obliquely; hence the sparsely distributed particles were compressed by perspective into a much smaller space, and so looked brighter. He remarked also that when the rings are seen at a low angle, ring A appears brighter than B for similar reasons. It followed from this that the particles are sparser in A than in B. In Barnard's view: 'these observations alone could almost be taken as a proof of the meteoric constitution of the rings'. He expressed the opinion that the visibility of the dark side of the rings in 1907–08 was due to sunlight passing through among the particles of the rings and being diffused and reflected from the dark side, and added: 'I am now inclined to attribute most of this . . . effect to the outer bright ring' . . . , where the particles are more thinly scattered. He then made the following interesting deduction: 'I believe that if the surface of the rings was turned at right angles to us, the outer ring would be comparatively dark, and the crepe ring would be perhaps invisible', because of the apparent widening out of the spaces between the particles.

Barnard evidently disagreed most emphatically with Lowell's theory of 'tores' (see Chapter 26), for he stated in this report: 'It is needless to say that the condensations were not due to thicker places on the rings'. He considered that this was decisively disproved by his observations in both July and October 1907 and in January 1908, when the rings were edgewise; at these times the condensations, if due to greater thickness, should have been at their most conspicuous, whereas 'in reality, under these conditions, they entirely disappeared'. He also pointed out that no trace whatever of the condensations was seen when the bright side of the rings was presented at a similar low angle to that of the dark side when they were visible. This also showed 'that they were due in some manner to the transmission of the sunlight through the rings (through the crepe ring and Cassini division, perhaps). Barnard also noted that the south pole of Saturn appeared dark or darkish during these latest observations.

Titan's atmosphere (Comas Solá)

The Spanish astronomer, J. Comas Solá, who had taken a prominent part in the observation of bright north temperate spots in 1903 (see Chapter 24), made a series of observations of Saturn's satellites at the Fabra Observatory, Barcelona, in 1908, and reported (*A.N.*, No. 4290; *B.A.A.J.*, vol. 19, p. 151) that his observations of Titan appeared to indicate the existence on that satellite of a more or less dense atmosphere, the limbs being dark and difficult to see, while towards the centre of the disk lighter patches were visible. This striking discovery was confirmed spectroscopically 36 years later by Dr. G. P. Kuiper (see Chapter 38).

Wisps, spots and streaks, 1909–10 (Lowell, Maggini)

Professor Percival Lowell noticed, in September 1909, when observing Saturn at the Flagstaff Observatory, Arizona, what seemed to be faint lacings diagonally crossing the bright equatorial zone, so faint that he at first doubted their objective reality. On mentioning them to E. C. Slipher he found that the latter had suspected them ten days before, on September 9. Further observations convinced Lowell of their real existence and on November 4 he managed to record them photographically (*Lowell O.B.*, No. 44; *B.A.A.J.*, vol. 20, p. 278).

On 29 September 1910 Signor M. Maggini, of Florence Observatory, noticed, near a large bright spot at the extremity of the south equatorial belt, a bright projection on the west limb of Saturn, against the background of the globe's shadow on the ring. His drawing also showed round spots on the equatorial belts accompanied by straight streaks across the equator (*A.N.*, No. 4445; *L'Astronomie*, March 1911; *B.A.A.J.*, vol. 21, pp. 65, 286).

Irradiating spots, diagonal wisps and straight streaks are more familiar features on Jupiter than on Saturn; the bright projection may have been a contrast effect caused by the contiguity of the bright spot and the dark shadow.

Photographs with colour screens, 1909–15
(Belopolsky; Barnard; Wood)

In 1909 Belopolsky photographed Saturn with the 30-inch aperture refractor of Pulkovo Observatory, using two colour screens, one transmitting 'indigo violet' (390–450$\mu\mu$), the other 'yellow green' (495–620$\mu\mu$). (1$\mu\mu$ = 10λ = 10^{-6}mm.) He found that the difference in intensity between the limbs and the centre of the disk was greatest in the red and disappeared in the violet, and that the equatorial zone was most brilliant in the red and darkest in the violet. The behaviour of the rings was opposite to that of the disk, but the edges of the disk and adjoining parts of the rings were equally intense in all radiations. Observations with the spectroscope confirmed the explanation that there is an atmosphere about the disk but none about the rings, and the similarity of transmission seemed to indicate that the separate particles in the rings are of about the same size as those forming the atmosphere, and in the mean less than the wave-length of light (*Nat.*, November 1911; *B.A.A.J.*, vol. 22, p. 113).

Dr. A. C. D. Crommelin made some interesting comments in June 1912 on a set of very fine photographs of Saturn taken by Barnard in 1911 with the 60-inch aperture reflector of Mount Wilson and yellow colour screen—(Plate XII, Fig. 1 shows four of them). Crommelin drew attention to the extreme width of the Cassini division on the photographs—much wider than it appears with small telescopes. He said that Antoniadi had remarked on how wide the Cassini division looked with the Meudon 33-inch aperture refractor, and said that the unduly narrow appearance in small instruments was due to part of the division being encroached on and covered by the diffraction fringes of the bright ring on each side. Crommelin also pointed out in the photographs the great difference of brightness between different regions of the planet and rings: for example, between rings B and A, which is so much less bright that in photographs of shorter exposure its extremities are nearly lost; the planet's poles are also of feeble luminosity. The ordinary exposure time was 10 to 12 seconds, but one of the photographs had been given 40 seconds, too long for the globe and bright rings, in order to bring out the crepe ring, which showed faintly on the print and, accoruing to Barnard, could easily be seen on the negative (*B.A.A.J.*, vol. 22, p. 421).

Barnard (in *Ap. J.*, vol. 40, p. 259) gave the results of measures he had made on the negative and a larger glass positive of these 1911 photographs, taking the diameter of the centre of the Cassini division as unity, and expressing the other dimensions in relation to that standard; he also worked out for comparison his visual measures related to the same standard; from comparison of the preceding and following measures he could find no evidence of eccentricity of the rings. The main results were:

	Visual actual	Visual ratio	Negative ratio	Enlarged positive ratio
Diam. outer A	40″186	1·164	1·141	1·139
inner A	35·034	1·015	1·019	1·018
mid Cas. div.	34·517	1·000	1·000	1·000
outer B	34·000	0·985	0·981	0·982
inner B	25·626	0·742	0·761	0·758
Equatl. diam.	17·798	0·516	0·500	0·501
Polar diam.	16·246	0·471	0·451	0·447
Width Cas. div.	0·517	0·015	0·019	0·018
Diam. inner C	20·528	0·595	0·620	—

He stated that the better photographs of 10 seconds exposure showed everything except the crepe ring that could be seen with the eye: any object, especially an elongated marking, as large as 0″1 of arc was readily visible. The Cassini division was easily to be traced all round; in front of the globe it was very thin and less distinct as if the globe could be seen through it. Ring A across the globe was also lighter than elsewhere. The belts were of similar tone to ring A, the polar regions much darker. The shadow on the rings, seemingly visible on both sides of the globe, seemed to bend out at the Cassini division, just as in visual observations near opposition.

In 1915 R. W. Wood of the Johns Hopkins University photographed Saturn using the Mount Wilson 60-inch aperture reflector and colour screens transmitting infra-red, yellow, violet and ultra-violet light respectively. Describing the results (*Ap. J.*, vol. 43, p. 310) he stated that in infra-red light the globe appeared almost devoid of surface markings, showing only the merest traces of the usual belts, and much brighter than the brightest part of the ring; this suggested the presence of mist or dust in Saturn's atmosphere. In yellow light the usual visual appearance was shown, the narrow belts being distinct. The violet photograph showed, as Belopolsky had found in

1909, the usually bright equatorial zone covered by a very broad dark belt; also in the violet there was a dark polar cap of considerable size. The aspect was much the same in the ultra-violet, but the dark central belt was not quite so wide and the bright region between it and the polar cap looked distinctly broader (cf. Wright's photographs: Plate XII, fig. 2).

Wood suggested two possible explanations of the darkness of the equatorial zone in the violet: (a) fine mist or dust extending from the crepe ring to the globe—the idea that the dark spaces between the crepe ring and the globe were filled with some material reflecting short wave-lengths slightly was supported by the slightly denser appearance of those dark spaces in the violet and ultra-violet photographs as compared with the sky outside the rings; (b) the existence in Saturn's atmosphere of some substance, such as a pale yellow gas or fine mist or dust, able to absorb violet and ultra-violet light. Wood also noted a decrease in contrast between rings A and B as the wave-length of the light decreased.

He suggested that if trichromatic colour photography were applied to the pictures made through yellow, violet and ultra-violet filters, a coloured picture of Saturn would be obtained showing the planet as it would appear if it were receding fast enough to make ultra-violet appear violet, or if the eye's sensitiveness were shifted slightly towards the ultra-violet. The result would be that the globe would look yellowish with a broad orange-red equatorial zone, dark red polar cap and narrow blue belts; ring A (more luminous in ultra-violet) would look bluish-green, while ring B would appear white.

Saturn in 1910–11 (Phillips)

Although the B.A.A. Saturn Section had shown little activity during the first decade of the present century, one member, the Rev. Theodore Phillips, though much occupied as Director of the Jupiter Section and also with observing Mars, had contributed very usefully to the observations of Saturn related in Chapters 24 and 26. Employing a 12¼-inch aperture Calver reflector with powers from 230 to 400 and occasionally 550, he secured an excellent series of Saturn observations in 1910–11 (B.A.A.J., vol. 21, p. 323), using occasional very good nights amidst much unfavourable weather.

The Globe. In August and September 1910 Phillips found that the darkest region of the disk was a well-defined south polar cap of diameter nearly 30°; next to this was a bright but narrow zone, then a narrow dark belt, then a broad lightish zone, then a broad double belt (presumably the south equatorial belt) with a widish lighter space between the two components. The belts appeared mottled, and rather indefinite irregularities at the edges were suspected, as were vague lighter patches from time to time on the equator. He felt more certain of bright spots on the south temperate zone and obtained some transits, but owing to unfavourable weather was unable to follow these spots systematically enough to deduce a rotation period.

The light zone bordering the polar cap appeared particularly brilliant at each limb, white spots seeming to stand out slightly by irradiation and to distort the shape of the cap. He thought at first that they were genuine, but as he found them constant in position night after night when the seeing was good enough to reveal them, 'their illusory character was revealed'. He felt that phenomena of this kind must have caused the 'square-shouldered' aspect first noticed by Sir W. Herschel.

In these observations he noticed one striking difference between Saturn and Jupiter: on the latter, white spots appear more brilliant near the central meridian, fading or disappearing through atmospheric absorption near the limbs, whereas on Saturn he found that (perhaps partly through contrast) the light zone next to the cap was far more intense at the limb and terminator than centrally, and also the white spots on the south temperate zone, sometimes seen only with difficulty near the central meridian, became so bright towards the limb that one of them on November 14 looked almost like a satellite passing off the disk.

Phillips noted a great change in the globe features from late September onwards, the details, especially the bright zone by the cap and the south temperate zone, fading, so that by January 1911 the whole region from pole to equatorial zone became more or less uniformly dusky.

The Rings. When the seeing was really good Encke's division on ring A was clearly observed on several occasions at the ansae, and became fairly easy after September. He considered it like a delicate pencil shading, less dark than Cassini's division. He remarked: 'I do not think this feature is illusory, though it is evidently not a real division and is probably inconstant'. He saw a delicate shading towards the inner edge of ring B, 'but nothing that could be called a division'. Its inner edge was fairly sharply defined. Ring C was somewhat conspicuous, rather bright and clear at the inner edge; he detected no division between rings B and C, and he could not be sure that the planet's limb was visible through ring C. The crepe ring did not seem to extend quite half way from the inner edge of B to the planet's limb. 'Nothing whatever was seen of the new exterior ring, although it was looked for several times.' He found the Cassini division very dark and conspicuous throughout the apparition. In late October–early November the elevation of Earth and Sun above the ring-plane was about the same, but bad weather prevented him from trying to glimpse a strip of the globe through the division. He noted a sharp bend at the Cassini division in the globe's shadow across the rings on January 15 and subsequently, although previously the shadow had appeared perfectly regular. This 'peaked shadow' appearance, with a notch or peak at the Cassini division, became a favourite subject for observation and discussion by the B.A.A. Saturn Section during the following few years.

Revival of B.A.A. Saturn Section

After an energetic start in the eighteen-nineties, the Section seems to have become much less active mainly owing to the retirement of the planet for one of its long periods of years in southern declination, making observation from northern latitudes difficult and unsatisfactory. There was also the loss, through death, of the active observing leadership of the Rev. A. Freeman, and the fact that the exceptional frequency and persistence of observable spots in the nineties did not continue, except in 1903, to provide a regular incentive for observation such as similar objects do on Jupiter. It may also have been realised that, while important discoveries were still being made with smallish telescopes in the middle of the nineteenth century, by the end of that century there seemed little remaining to be found out about Saturn that was within the scope of small apertures, and nearly all the plums of observation and discovery were by that time falling to professional astronomers at great observatories with telescopes of 25 to 40 inches' aperture at their disposal.

At the beginning of the second decade of the present century, with the return of Saturn to northern declination and the observations of Phillips and Harold Thomson (a few years later to become Director of the Mars Section), the activity of the Saturn Section was already beginning to revive, and this was greatly helped by the appointment, in the summer of 1912, of Patrick H. Hepburn as Director. Under his inspiring leadership the efforts of observers were guided into what seemed the most promising channels, the writing of papers was encouraged, and interesting discussions were initiated at meetings, so that a greater awareness of Saturn and its current problems was aroused among the members of the Association. In June 1912 Hepburn pointed out, in a message to the Section as its new Director (*B.A.A.J.*, vol. 22, p. 430), that the planet's situation in the next three

PLATE XI.—Nine Lowell Observatory photographs of Saturn's ring phases 1909–1921, taken by P. Lowell and E. C. Slipher.

Fig. 1.—Four photographs of Saturn, 1911 November 19, taken by Barnard with the 60-inch reflector of Mount Wilson Observatory.

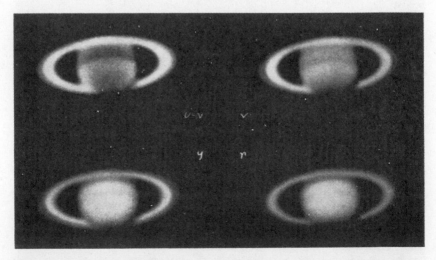

Fig. 2.—Four photographs of Saturn, 1927 July 13 and 14, taken by W. H. Wright at Lick Observatory in ultra-violet, violet, yellow and red light.

Plate XII

or four years would be most favourable for northern observers, adding: 'If, during the next few years, Saturn could be observed by members of the Association as assiduously as Jupiter and Mars have been in the past, it is not at all unlikely that some considerable addition may be made to our knowledge of the planet.' These words turned out to be prophetic because it was less than five years later that two members proved by observation the translucency of ring A.

Saturn in 1911–12 and 1912–13 (Hepburn, Phillips, Thomson)

The report of the Saturn Section on the apparition of 1911–12 (*B.A.A.J.*, vol. 22, p. 453) was based on the observations of the Rev. T. E. R. Phillips and H. Thomson, who both saw independently a white area in the centre of the south polar cap, a feature also seen and remarked on by Jarry-Desloges. Otherwise the planet's appearance seems to have been very similar to that of the previous apparition; the peaked shadow and bright spots on the limb near the polar cap were noted as before, and Phillips invited attention to the disagreement between the apparent dimensions seen by him and Barnard's measures, especially as to the width of ring C, and he suggested that there might be an inner extension of the crepe ring visible only in the largest telescopes.

The question of the true dimensions of the Saturnian system was a particularly thorny one at this time, since there were two distinct sets of measures in current use: those of Meyer, 1881, used in the *N.A.*, and those of Barnard, 1894–95 (see Chapter 19). Hepburn thought of a rather ingenious way in which the observers of the Saturn Section could test visually the relative merits of these measures so far as the polar diameter of the globe and the external diameter of ring A were concerned. Accordingly he supplied members of the Section in 1912 with particulars of the phenomena for Saturn worked out according to both sets of measures. As he explained in a paper (*B.A.A.J.*, vol. 24, p. 315), the first test lay in observing at what date the shadow of the planet first began to fall wholly within the ring system, through the opening out of the rings. Members watched for the first contact of the edge of the globe's shadow with the exterior edge of ring A, and noted the date. The results of this observation appeared to be wholly inconsistent with the date based on Meyer's measures, but quite consistent with that calculated from those of Barnard.

Hepburn then explained that the maximum opening of the rings during that apparition was due about 31 August 1912; the rings would close to a minimum (for the apparition) about 12 January

1913, and afterwards reopen. According to calculation based on Barnard's measures, at the time of minimum opening the globe should project very slightly (0″06 by computation) beyond the outer edge of ring A. This constituted the second test. Some of the Section's observers thought there was a slight projection, others that the globe just failed to project. Barnard himself, with whom Hepburn was in communication, was of opinion from his own observations (*M.N.*, vol. 73, p. 603) that the globe fell within the ring's edge by something like 0″1. It followed from this investigation that Barnard's measures of the exterior diameter of A and of the globe's polar diameter gave results for this computation that are accurate within 0″1 or 0″2. This was a case of a useful job done by observers with smallish telescopes and an admirable example of co-operation between amateur and professional astronomers.

Hepburn's reports of the Section's work in 1912–13 (*B.A.A.J.*, vol. 23, pp. 78, 451) also noted that the white area in the centre of the south polar cap had not been seen during this apparition, and no ring divisions had been seen except Cassini's and Encke's. Thomson and Phillips had made observations and drawings of the peaked shadow on the rings, showing a shadow on both sides 'giving a square appearance to the dusky limit of the planet'. This dark peak must have been visible on both sides around opposition date, when no true shadow should have been seen at all; moreover, photographs also showed the shadow on both sides. Not long afterwards, two members found a satisfactory solution of this difficult problem, as will appear later in this Chapter.

The Section report also stated that attempts were being made, in co-operation with the Variable Star Section, to determine the variability of the brightness of the satellites, especially Iapetus. Hepburn stated that the photometric observations made at Harvard in 1877–78 had been inconclusive on whether any of the satellites except Iapetus were variable in brightness. For Iapetus a consistent curve of variability had resulted from 303 Harvard observations, but the whole series had covered less than four revolutions of the satellite, its presentation in 1878 was quite different from that of 1913, the limits found at Harvard—from magnitude 11·4 down to 12·7—were narrower than those deduced from visual observation, and there was some evidence suggesting occasional sudden and very considerable irregular fluctuations; the matter therefore needed investigation.

Saturn's mass and Le Verrier's Tables (Gaillot)

In Chapter 16 Le Verrier's Tables of Saturn were referred to, and particulars were given of the values obtained by him and other astronomers at earlier and later dates for the reciprocal mass of Saturn. By 1913, Gaillot, who had been working at the Paris Observatory in Le Verrier's lifetime, completed the revision of the latter's Tables, and also corrected Le Verrier's figure for the reciprocal mass, based on the perturbations of Jupiter's orbit caused by the attraction of Saturn, to a value of 3499·8, which is in good agreement with the following values obtained from the elongations of Titan or Iapetus: 1833 Bessel: 3501·6; 1888 H. Struve: 3498; 1889 A. Hall: 3500·5 (*Comptes Rendus*, vol. 157, p. 191: *B.A.A.J.*, vol. 24, p. 54).

New subdivisions on the rings (Maggini)

Mentore Maggini, of the Royal Observatory, Arcetri, near Florence, in October 1913 during a few instants of perfect seeing noticed a new subdivision in ring B at the side of Cassini's. The delicate line had been seen in November 1910 and again in 1912, but was then considered an illusion; it runs around ring B, and the space between the two lines is always dusky. On 19 October 1913 Cassini's division seemed double throughout the northern part of the ring. On October 25 Encke's division, instead of a single line, showed two lines separated by a band much brighter than the rest of ring A. A similar observation had been made by G. Fournier on 4 September 1909 (*L'Astronomie*, Feb. 1914; *B.A.A.J.*, vol. 24, p. 228).

Globe again glimpsed through Cassini's division, 1913
(G. and V. Fournier)

It was reported (in *Bulletin S.A.F.*, vol. 27, p. 546) that in observations made between 20 October and 3 December 1913, Georges Fournier and V. Fournier saw easily through the Cassini division a strip of Saturn's disk (and also several subdivisions in rings A and B). These observations were made at Sétif, Algeria, at an altitude of 3,660 feet, using a 19·7-inch aperture refractor, the conditions on some of the nights being so good that high magnifying powers on the whole aperture could be employed. The only previous occasions on which a similar observation was successfully made were in 1852 and 1883 (see Chapters 12 and 17 and also p. 337).

C. T. Whitmell, who had tried to organise a similar observation in 1902 (see Chapter 25), calculated that the elevations of the Earth and Sun above the ring-plane must have been equal on 23 October 1913, and that the meridian altitude of the planet at Sétif was 75°; as on the other two successful occasions, the strip of the planet's surface seen was in the northern hemisphere. He pointed out (*B.A.A.J.*, vol. 28, p. 24) that the best conditions for such an observation are: (1) that the saturnicentric declinations of Earth and Sun should be equal, and (2) as large as possible, and that at the same time (3) the saturnicentric R.A.s of Sun and Earth should be equal. In 1902 conditions (1) and (3) only were fulfilled; in 1913, condition (1), and to a large extent (2), but not (3), would have been realised. In any case the observation must always be a delicate and difficult one.

Brightness of Rhea (Steavenson)

In March 1914 W. H. Steavenson, now Gresham Professor in Astronomy but at that time a new member of the B.A.A. and its Saturn Section, read a short paper (*B.A.A.J.*, vol. 24, p. 314) on the variable brightness of Rhea. Looking over records of past observations made with his 3-inch aperture refractor, he noticed that on occasions when he had recorded the satellite as being visible it had almost invariably been in the eastern half of its orbit. As he thought this might possibly indicate a variation in brightness, he observed Rhea on about forty nights in the winter of 1913-14, and found the satellite, as he had suspected, consistently brighter when in the eastern than when in the western part of its orbit, and possibly faintest of all when in the northern quadrant of the western half of the orbit. Owing to the smallness of the aperture he was unable to estimate the variations in terms of exact magnitudes, but as Rhea is near the limit of visibility of a 3-inch, he found it quite easy to classify his observations as 'very faint', 'faint', 'average', etc. As a result he estimated the total range of brightness to be about half a magnitude, which seems more reasonable than Quignon's estimate of one magnitude (see Chapter 25). Quignon had apparently found Rhea faintest near conjunctions, which rather agrees with Cassini's statement that it was invisible in his telescope at conjunctions (see Chapter 6). Steavenson, however, took into account the state of the atmosphere and the apparent distance of Rhea from Saturn at each observation.

In the discussion of Steavenson's paper (*B.A.A.J.*, vol. 24, p. 284) Phillips said that Thomson, observing with a 12¼-inch reflector, had noticed that the light of Rhea seemed inconstant; and also that Lowell had telegraphed to the *A.N.* that observers at his Observatory had found that not only Iapetus and Rhea but also Dione and Tethys varied systematically in light, and were brighter in one part of the orbit than another, the amount of variation of the two last-named being about a quarter of a magnitude. Quignon's estimate in these cases was three-quarters of a magnitude.

Double shadow and peaked shadow explained (Bartrum, Steavenson)

In April 1914 a paper was read by C. O. Bartrum on 'the appearance of Saturn at opposition' (*B.A.A.J.*, vol. 24, p. 359) which gave very satisfactory explanations of the puzzles of the globe's shadow on the rings. Regarding the double shadow he wrote: 'The so-called false shadow is believed . . . to be no shadow, but to be the limb of the planet, sufficiently dark at opposition phase when projected on the bright ring to cause the appearance seen'. He deduced this complete explanation from the law that the 'illumination of a surface varies with the cosine of the angle of incidence of the light', and showed that at opposition the illumination of the planet's limbs as seen from the Earth must be negligible. As at the then recent opposition the angle between the directions of Earth and Sun from Saturn changed at the rate of $0°12$ per day, and the dark edge would be very narrow, the false shadow would only be apparent on the following side for a few days before opposition, and continue on the preceding side for a few days after that phase.

Bartrum accounted for the notches or peaks by the assumption that the image of the Saturnian system is somewhat blurred, even under the best conditions of observation. He pointed out that even in the beautiful photographs taken by Barnard and Lowell within a few days of the 1911 opposition, the bright parts of the image were diffuse at the edges. He suggested that this want of exact definition may be caused by lack of atmospheric homogeneity, irradiation on the retina at visual observations, and halation on the plate in photographs; diffraction might also contribute. Each of these effects would expand every point of light into a small circle. He demonstrated that the effect could be imitated by viewing an object through a lens or telescope slightly out of focus, or by projecting a picture on to a screen similarly out of focus. He made a model to represent Saturn and its rings, illuminated it and photographed it with different deviations from accurate focusing. The resulting photographs showed a remarkable representation of the peaked shadow. (For a striking photograph of the peaked shadow see Plate XVI, Fig. 2).

During the discussion of this paper W. H. Steavenson said that visual observation of Saturn and experiments with drawings and photographs of it had brought him to the same conclusion, namely that the peak was purely an illusion, in some cases optical, in others atmospheric in origin. Although, like other observers, he nearly always saw a peak in the shadow even under good seeing conditions, yet under particularly favourable conditions with a 10-inch aperture refractor he had found the contour of the shadow an unbroken curve. He said that he had been able to produce the peak by photographing Proctor's fine engraving very slightly out of focus, though the image provided by the engraving was sharper than any photograph of Saturn yet obtained. Bartrum added the suggestion that the peak was produced, not by a widening of the shadow or dark limb, but by its apparent obliteration at other points by the brightness of the rings (*B.A.A.J.*, vol. 24, p. 341).

Ring particles—Kirkwood's gaps (Stromeyer, Crommelin)

In the course of a paper about Saturn's ring divisions, C. E. Stromeyer worked out that particles in Encke's division would have a period of about 13 hours, approximately 1/5 that of Dione, 2/7 that of Tethys, 2/5 of Enceladus and 4/7 of Mimas, and that the mean radius of the perturbed orbits of particles would practically coincide with that of Encke's division. Similarly, a period of about 11 hours, approximately 1/6 that of Dione, 1/4 Tethys, 1/3 Enceladus, 1/2 Mimas, would give a mean radius of perturbed orbits approximating closely to the mean radius of Cassini's division. (*B.A.A.J.*, vol. 25, p. 135.)

In the discussion of the paper Dr. A. C. D. Crommelin pointed out that the perturbative action of a more massive body would make the period of less massive bodies undergo oscillations if their period was a simple fraction of that of the perturbing body. In some cases they would move inwards, in others outwards. Any oscillating body was stayed much longer near the limits of its oscillation than in the middle of its range. Hence, owing to the action of Mimas, there would at any moment be very few particles in the Cassini division; most of the time they were sensibly inside or outside that position. Though Crommelin mainly used as his illustration Jupiter and the asteroids he said the same argument would explain the Cassini division due to the action of Mimas (*B.A.A.J.*, vol. 25, p. 114, December 1914).

In the course of another discussion about a year later Crommelin made the interesting observation that such tiny bodies as the ring particles could not be expected to have a truly spherical form (*B.A.A.J.*, vol. 26, p. 63). As will appear in Chapter 40 this opinion seems to be borne out by the results of recent polarimetric investigations by Dr. Audouin Dollfus.

Measures of rings and subdivisions and Saturn's rotation (Lowell)

Professor Percival Lowell (*Lowell O.B.*, 68, summarised in *B.A.A.J.*, vol. 26, p. 130) gave measures, especially of the rings, made by himself and E. C. Slipher in February to May 1915, continuing similar work done in 1913–14. In 1914 several subdivisions had been noticed and these were confirmed and extended in 1915. It was found possible to measure the positions of many of those in ring B, and most proved to be due to perturbations by Mimas. A filar micrometer was used under different conditions of illumination so as to discover the irradiation correction for each, and hence the true as well as the apparent dimensions of the several parts of the system. The 24-inch aperture refractor at full aperture with a power of 392 was commonly used and an iris diaphragm outside the eyepiece.

The following means of micrometer measures, for a mean distance of Saturn assumed to be 9·53885 A.U., are taken from Lowell's summary tables based on many pages of detailed observations given in *Lowell O.B.*, 66 and 68:

	1913 *Dec.* to 1914 *Mar.*	1915 *Feb. to May* Lowell	*E. C. Slipher*
Ring A outer diameter	39″71	39″67	39″71
inner „	35·18	35·27	35·11
Cassini division width	·49	·51	·51
Ring B outer diameter	33·64	33·71	33·78
inner „	26·33	26·42	26·44
width		3·72	3·72
Ring C inner diameter		21·93	21·59
width (to right)		2·08	2·00
width (to left)		1·94	1·91
Diameter of Ball	17·26	17·21	17·20
Inner edge B to Ball (to right)		4·85	4·15
(to left)		4·37	4·00
Divisions in Ring B:			
3/7 period of Mimas		30·64	
2/5 „ „ „		29·48	29·48
3/8 „ „ ,		28·01	28·15

Lowell also gave in Bulletin 66 the following determinations for Titan:

Diameter at mean distance	0″568
Observed magnitude	9·6
Mass (from three measures)	1/4700 of Saturn's mass
Resulting albedo	0·177
Resulting density	0·691 × Earth's density

In Bulletin 68 he calculated the oblateness of Saturn to be 1/9·18.

In Bulletin 68 Lowell also gave the following means of measures made on drawings of the radii of various minor divisions observed by himself and E. C. Slipher on the southern face of rings A and B:

	Lowell	*E. C. Slipher*	
Ring A			
(inner edge of ring	17"64)		
Minor division 1	18·63	18"51	(Encke's inner comp.)
2	18·97	19·00	(Encke's outer comp.)
3	19·62	19·59	
(outer edge of ring	19·84)		
Ring B			
(inner edge of ring	13·21)		
Minor division 1	13·39		
2	13·95	14·01	(3/8 period Mimas)
3	14·77	14·71	(2/5 period Mimas)
4	15·06	14·97	
5	15·32	15·32	(3/7 period Mimas)
6	15·69		
7	15·95	15·89	
(outer edge of ring	16·86)		

(Particulars in brackets have been inserted here, derived from micrometrical measures given elsewhere in the Bulletins.)

Lowell also stated that, more than once, glimpses had been obtained of a division 'as a black band' between the inner edge of ring B and the outer edge of ring C.

Lowell's conclusions were: that the crepe ring is not of equal width on both sides of the system; that there is no evidence of a change in the width of the rings such as Otto Struve thought he had detected (see Chapter 13); that ring B seems to be elliptic not circular, and that it is probable that all the rings are elliptic.

A remarkable theory about the rotation of Saturn was suggested by Lowell (in *Lowell Observatory Mem.* I, ii, given briefly in *B.A.A.J.*, vol. 26, p. 172), namely that the planet actually rotates in layers with different velocities, the inside one turning the faster. If these layers were two only, or substantially two, this arrangement would result in Saturn's being composed of a very oblate kernel, surrounded by a less oblate husk of cloud.

Plate XI shows a fine series of photographs by Lowell and Slipher illustrating the ring phases during nearly half a cycle.

Variability and rotation of Mimas and Enceladus (Lowell)

The observations of Lowell and E. C. Slipher on the brilliancy of Mimas and Enceladus in 1913 December and 1914 January to March (*Lowell O.B.*, 64) showed those satellites to have a very different albedo in different parts of their apparent orbits, the variations recurring regularly. They concluded from this that both satellites turn always the same face towards Saturn. They found that the differences in brightness are greater than those of any other moons of Saturn except Iapetus. Both Mimas and Enceladus appeared brightest near western elongation and faintest near eastern, the variation in apparent magnitude being:

Mimas from 12·9 to 13·33—range 0·43 magnitude
Enceladus from 12·33 to 12·67—range 0·34 magnitude.

Saturn in 1913–14 (Fournier)

Observing Saturn in Algeria on 20 dates in 1913–14, G. and V. Fournier made eight good drawings showing the peaked shadow and other anomalies. They found: the number and positions of subdivisions on rings A and B variable and often different in the two ansae, those on ring B being less distinct than in 1911–12; irregularities in Cassini's division, especially at the edges; ring C granular and of varying intensity with dusky knots and lighter areas; on the globe a small dusky south polar cap, a dusky ring round the polar region, and up to six belts, some occasionally double; the south equatorial belt vague at the south edge and irregular on the north, that edge being joined by wisps to a faint equatorial band. They considered the Cassini division not to be devoid of particles, and the ring system to be very unstable. (Jarry-Desloges: *Observations des Surfaces Planétaires*, vol. 4, p. 199–215.)

CHAPTER 30

TRANSLUCENCY OF RING A, 1917

Measurements on Barnard's photographs, 1914 (Hepburn, Hollis)
In Hepburn's paper on the 'Dimensions of Saturn's Rings' (*B.A.A.J.*, vol. 24, p. 315) he drew attention to two problems which were very much in the minds of the B.A.A. Saturn Section in 1914. One of these was whether in that apparition, when the maximum opening of the rings was due to occur about May 27, the inner edge of the far part of ring A would, or would not, clear the polar limb of the globe. Theoretically, according to Barnard's and other authoritative measures, the Cassini division should not be seen clear of the polar limb at maximum opening, yet for 60 years at least the best observers had been unanimous in showing it clear on such occasions, and the majority of the members of the Section were of opinion that at least the upper edge of the division was already clear of the limb at that time (March 1914). The other problem was that the Rev. T. E. R. Phillips had made a set of measures of the planet in January 1913, which, while agreeing with Barnard's in the overall dimensions of the system, did not accord in detail with his measures of the rings, and that a year previously Hepburn had independently made eye estimates which were found to agree very well with Phillips's measures.

As the R.A.S. had obtained a glass positive of the set of 12 photographs of Saturn, taken by Barnard on 19 November 1911 (ten days after opposition) with the 60-inch aperture reflector of Mount Wilson, and as these photographs (see Plate XII, Fig. 1 and Chapter 29) showed the images beautifully defined, Hepburn and H. P. Hollis made detailed measurements of the rings on them with a special instrument. The results agreed closely with micrometrical measures of the same photographs made by Graff of Bergedorf Observatory, Hamburg, and were in substantial agreement also with the measures of Phillips and the eye estimates of Hepburn mentioned above, but they made rings A and B considerably narrower than did the earlier direct measures of the planet by Hall, Barnard, Greenwich Observatory, and See (given in Chapters 19 and 25). Hollis's measurements appeared to confirm that at maximum opening the inner edge of the far part of ring A should not appear to be clear of the polar limb.

Apparent transparency of ring A on Barnard's photographs
(Hepburn; Maxwell, Thomson)

The minute inspection of Barnard's photographs made by P. H. Hepburn for the purpose of the measurements, caused him to read another paper in June 1914 giving particulars of the details shown in the photographs (*B.A.A.J.*, vol. 24, p. 479). Among other matters Hepburn called attention to the non-uniform brightness of ring A, the inner portion being the brighter, and the narrower in the ansae, and the line of demarcation seeming to spring from the inner edge in two alternate quadrants, tapering towards the outer edge in the others like a two-branched spiral. He also found ring A wider in the preceding than in the following ansa, and looking much brighter where it crossed the globe than ring C, although ring A was projected on its own (presumably darker) shadow, and on a part of the planet probably darker than the region lying under ring C. But Hepburn's most startling pronouncement about these photographs was: 'Ring A, as appearing on the photograph is transparent, the globe of the planet is clearly visible through it, both limbs being fairly sharply defined'. He added that the region where ring A was superposed on the disk was brighter than the adjacent parts of the ring, and that the limbs of the disk could be traced across the grey band of the Cassini division up to the edge of ring B. He cited Trouvelot's suspicions of 1880–81 of the possible occasional transparency of ring A (see Chapter 17).

Hepburn's assertion of the transparency of ring A gave rise to further papers and a series of long discussions at B.A.A. meetings during the rest of that year and the following one (*B.A.A.J.*, vol. 25, pp. 331, 343, 350, 363, 386; vol. 26, pp. 63, 96). The controversy makes interesting reading but is now of purely academic interest, since whatever might be the true cause of the photographic effect, the more important question of the translucency of ring A was decided by the observation of M. A. Ainslie and J. Knight in 1917. The chief opponent of ring A's transparency on the photographs, J. E. Maxwell, based his arguments on the black shadow cast by ring A, which when projected on its shadow should, if transparent, look dark, not light, and he attributed the appearance on the photographs to atmospheric conditions and trembling of the image. H. Thomson, with a photographed model, suggested specular reflection by the particles of A rather than transparency.

Hepburn's interpretation of the photographic appearance of ring A found ready acceptance from quite a number of astronomers, among them Hollis and Barnard himself. These had both noticed the unusual brightness of ring A but had not detected the limbs of the globe showing through the ring until Hepburn drew attention to this. Barnard's conclusions from his 1908 observations, mentioned at the beginning of Chapter 29, had strongly suggested the probable translucency of ring A, but its usual aspect and the darkness of its shadow made some astronomers reluctant to class ring A with ring C rather than with ring B. Moreover, the demonstrations of Bartrum and Steavenson (related in Chapter 29) must have made many of them illusion-conscious regarding the photographic as well as the visual aspect of Saturn. Even those most sceptical of the photographic evidence, such as Maxwell, were prepared to admit that ring A might very possibly be flimsy and semi-transparent. Ainslie said he would have preferred a visual proof: his wish was soon to be fulfilled.

Ainslie and Knight prove the translucency of ring A by visual observation, 1917 February 9

The historic observation on the evening of 1917 February 9 of the passage of the seventh magnitude star B.D. +21° 1714 behind ring A and the Cassini division by Naval-Instructor Maurice A. Ainslie at Blackheath, and John Knight at Rye, was one of the greatest triumphs ever achieved by observers of Saturn in Great Britain. The observations were unpremeditated, as the occultation had not been predicted, and they were quite independent of each other. Also, as it happened, they were complementary, since Knight watched the first part of the occultation, and Ainslie the later stage. They also showed that great achievements in the observation of Saturn were still occasionally possible with fairly small telescopes, since Ainslie used a 9-inch aperture reflector with powers of 180 and 270, while Knight's instrument was a 5-inch aperture refractor with powers 100, 180, 250. On this occasion the final proof of Cassini's division being a true gap which had been suggested by the elder Herschel was attained, and Trouvelot's dream of proving the translucency of ring A was realised.

In the accounts of the observation (*M.N.*, vol. 77, p. 456; *B.A.A.J.*, vol. 27, pp. 7, 212) Ainslie stated that he first noticed the star, of a golden-yellow colour, north of and preceding the north pole of Saturn at about $8^h 45^m$ G.M.T on that evening, but that circumstances prevented his continuing the observation until 10^h. He gave the apparent position of the star for 1917 February $9^d 10^h$ G.M.T. as:

R.A. $7^h 49^m 23^s 163$, Decl. $+21° 19' 15''43$.

The star would therefore be in Gemini and near the border of Cancer, a little north of 85 Gem. and within the triangle 82 Gem.– 85 Gem.–μ Cancri.

Fortunately it happened that Knight noticed the star at about 9 p.m., it appearing to be projected on the disk between the limb and the outer edge of the rings in the north polar region, 'like a satellite of Jupiter in transit', but really (he assumed) skirting the north-east limb. Knight then 'saw it apparently eat its way into the outer ring, and apart from isolated moments when the air was particularly unsteady, it never seemed wholly to disappear. 'Sometimes it was exceedingly faint and I was only conscious of a "something extra" being present in the rings.' A little before 10 p.m. it was plainly discernible in the Cassini division and it was traced easily through this till about 10.25. When in the division, perhaps through indifferent seeing, he thought the star's image looked elongated. After 10.25 the seeing deteriorated, and Knight could not be certain that he saw it again till it emerged a little after 11 p.m. His conclusions were that the star was not quite occulted by the globe; that its furthest penetration behind the ring system was to the inner edge of the Cassini division; that while passing behind ring A, apart from momentary obscurations (perhaps due to unsteady air), it was seen with varying degrees of brightness all the time; and that it was never so bright during the whole phenomenon as when clear of the rings after occultation.

It so happened that Ainslie was able to pick up the star about 10.15 p.m., shortly before the other observer lost sight of it. 'It appeared as a very conspicuous cream-white spot, very small, and apparently projected (through irradiation) on the extreme edge of ring B; it passed into the division and travelled along it, remaining. . ., as nearly as I could judge, as bright as when clear of the planet. It does not appear to have passed behind any portion of ring B, as its brightness was so little affected. At about 10.30 it was seen on the outer edge of the division, and (owing to bad seeing) disappeared for a few minutes; at about 10.35 it was very easily seen, greatly reduced in brightness, through ring A. The star remained visible until the end of the observation, traversing the ring with little or no variation in brightness, except as follows:

(i) 'at $11^h 3^m$ it increased very rapidly, not quite instantaneously, to perhaps double intensity, at which it remained for from 10 to 15 seconds, fading again as rapidly as it had brightened;'

(ii) 'a second brightening was observed at $11^h 8^m$, this time for not more than about 5 seconds.'

The star finally appeared from behind ring A at $11^h 10^m$, and five minutes later was seen separated from the ring by a distinct black interval. During the passage of the ring and Cassini division over the star its yellow colour was much less pronounced than when it was clear of Saturn, and it looked more creamy white. The image was 'stellar' throughout, especially in the division; the image may have appeared slightly enlarged when seen through the ring. The seeing was good during the occultation. Ainslie estimated the time taken for the alteration of brightness on the two occasions behind the ring to have been about half a second, 'and the final brightening when clearing the ring took about the same time'. He roughly estimated that, if the star's original brightness is taken as 100, its brightness in the Cassini division was almost if not quite 100; behind most of ring A about 25; at the two places in the ring where it temporarily brightened, about 50.

Ainslie concluded, on comparing his observation with Knight's, 'that the star was visible through ring A; that while behind ring A it was seen, though not with anything like its full brilliancy, through Encke's division, and through another exterior to the same; that it did not actually pass behind the ball, although it must have made a very close approach thereto; and that, as far as can be judged from the present observation, the measures of the rings, on which the data given in the N.A. are based, would appear to be somewhat too small.'

Figure 19 reproduces Ainslie's diagram based on data from the *Nautical Almanac* for the globe and rings, the polar semidiameter being corrected for tilt of the axis and the position of the planet for parallax. The diagram is to scale and the lower straight line shows the path of the star from positions of the planet and star supplied to him by Dr. Crommelin; it shows the computed positions at 8^h, 10^h and 12^h. The upper straight line seemed to Ainslie better to represent his observations, though it would have involved the disappearance of the star behind the globe which does not seem to have taken place.

Ainslie expressed the opinion that the ring particles must be very small compared with the apparent diameter of the star; otherwise the ring particles streaming past at 16 miles per second would have caused a flickering of the star's image during the occultation. In the discussion, Dr. Crommelin gave the opinion that the apparent diameter of the star would be only 1/10,000 of a second of arc, or less than half a mile at Saturn's distance. The particles must be much smaller than this, probably at most 200 yards.

Another member said that it would appear that there were only two divisions in ring A and not six or eight as Lowell represented in his drawing. The Rev. T. E. R. Phillips pointed out that there

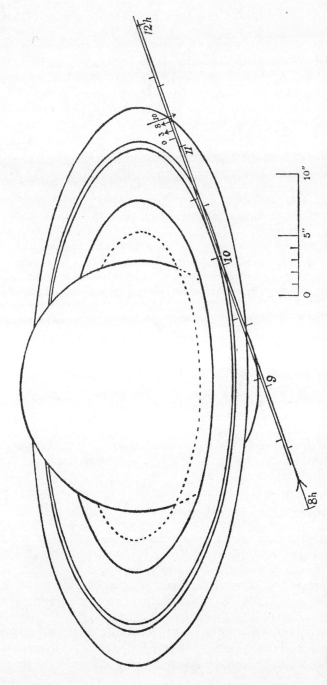

FIGURE 19. M. A. Ainslie: diagram of passage of star BD +21° 1714 behind ring A, 1917 February 9. (*From M.N.*, vol. 77 p. 458).

might possibly have been only two divisions on the night of the occultation of the star; might not perturbations in the streams of minute particles cause divisions in the rings to appear from time to time and close up again? He also remarked that Encke's division had often seemed to him decidedly nearer the outer edge than the middle of ring A; a week or two before he had seen it as a narrow sharp line distinctly near the outer edge of ring A.

Saturn in 1917 (Steavenson)

Dr. W. H. Steavenson, directing the B.A.A. Saturn Section in the absence of P. H. Hepburn on military service, reported (*B.A.A.J.*, vol. 27, p. 278) the famous observation of Ainslie and Knight, and some features of Saturn observed during 1917 by J. E. Prior, R. L. Waterfield and himself. These included: white spots and dark condensations in the south equatorial belt; the reduction of the dark antarctic belt (so conspicuous two years before) to a mere southern edge of a south temperate shading; and the visibility of Encke's division as a sharply defined line, often eccentrically placed in ring A, to several observers using large apertures. Also Phillips, Prior, Waterfield and Steavenson had detected unmistakable variations in the relative brightness of Tethys and Dione; and Prior, in a series of photometric observations of Rhea, had found some evidence of variability.

Eclipse of Iapetus by the rings, 1919 (W. F. A. Ellison)

An eclipse of Iapetus by the rings, similar to that observed by Barnard in 1889 (see Chapter 18) was predicted for 1919 February 28, but owing to bad weather the only one to observe it even partly seems to have been the Rev. W. F. A. Ellison, Director of Armagh Observatory (McIntyre: *Translucency*, p. 20). Ellison, using a 10-inch aperture Grubb refractor, first saw it at $7^h 17^m$ when the satellite had emerged from the shadow of ring B and was passing behind the Cassini division. At maximum brightness at 7.20, it had commenced to fade five minutes later through entering the shadow of ring A, and 'disappeared or nearly so' at 7.32. Very unfortunately the observer at this point was called away and could not continue the observation until 8.22, when he found Iapetus 'visible but difficult' still in ring A's shadow. Although the predicted time of emergence from the shadow of A was 8.29, Ellison found the satellite gradually increasing in brightness from 8.22 until it became conspicuous at 8.42.

Ellison's observation therefore certainly confirms the translucency of ring A. McIntyre, though admitting that the predicted time of emergence must be considered as approximate owing to uncertainty about the exact dimensions of the rings, suggested that Ellison's observation seemed to show that the ring system extends further than is generally supposed and gradually dwindles into space, rather than having a definite outer limit. McIntyre, in fact, considered that the observation lent some support to the suspected outer dusky ring.

CHAPTER 31

TRANSLUCENCY OF RING B, 1920

Predicted occultation of star Lalande 20654 (Comrie)

In January 1920 L. J. Comrie published (*B.A.A.J.*, vol. 30, p. 130) full particulars with a diagram of his prediction of the occultation by Saturn on 1920 March 14 of the star Lalande 20654 (No. 1460 in *Cape General Catalogue*, 1900) of magnitude 7·3. His forecast was that, as seen from Greenwich, the occultation would last 1ʰ 35ᵐ, but the star would not cross the rings. Later in the same evening there would be a close approach to the star by Titan, which might appear from some position on the Earth to occult it for a maximum duration of about five minutes. The observations of several widely scattered observers might lead to a reliable result for Titan's diameter. He gave the position of the star as:

R.A. 10ʰ 38ᵐ 6ˢ·84 Dec. N. 10° 46′ 10″·3.

This would place it roughly 5ᵐ preceding the star 53 Leonis, the nearest really bright star being ρ Leonis. Major P. H. Hepburn, now once more in charge of the Saturn Section, called special attention to this prediction and asked for reports of any observations (*B.A.A.J.*, vol. 30, p. 147).

The reports from various observers in England and other parts of Europe (*B.A.A.J.* vol. 30, pp. 189, 230) were in depressing agreement as to the very unfavourable weather, and it would seem that the only case in which the disappearance of the star was observed in this part of the world was at Antibes by Raymond with a 12-inch aperture reflector; his report was that the disappearance had taken place behind the ring about half an hour before the predicted time, that no special phenomena were observed, but that the star 'flickered distinctly' and then disappeared. Maynard at Hampstead and J. Knight at Rye, alone among English observers, managed to glimpse, between clouds the star and Titan looking like a close double star, so that the approach must have been very close, but no reliable estimate of distance could be made. Dr. Bernewitz at Berlin-Babelsberg Observatory observed that Titan did not occult the star but passed 1″ or 2″ of arc to the north of it. At Berlin-Babelsberg, Smichow, Leipzig and Jena Observatories the emersion was observed and timed, though generally in poor seeing conditions, some help however being obtained from the reddish colour of the star.

Successful observation of the occultation, 1920 March 14, at Rondebosch, South Africa (W. Reid, McIntyre, Dutton and H. Reid)

The only place where the immersion of the star behind the rings and the planet was observed under ideal conditions was at William Reid's observatory at Rondebosch, near Cape Town, and the observation was made by him in company with three other amateur astronomers, D. G. McIntyre, C. L. O'B. Dutton and H. Reid. The instrument used was a 6-inch aperture Cooke photo-visual refractor, the power giving best definition being 216. Reid felt so certain that the occultation would be observed by many other astronomers that in the first instance he merely sent a report on it, as a matter of local interest, to the *Cape Times*, and unfortunately the account was rather incomplete and as printed contained an error of one hour in the time of immersion.

This report (*B.A.A.J.*, vol. 30, p. 230) stated that the night was the finest of the whole summer; Saturn looked like a copper-plate engraving; the thin line of the ring with a thin dark line of shadow on one side and the crepe ring on the other could be clearly seen across the planet; the belts were plainly visible and the Cassini division was conspicuous. The bright orange colour of the star helped, but the acute angle between the star's path and the ring rendered difficult the determination of the time of immersion. 'While the star was behind the ring its light fluctuated considerably and once gave a momentary flicker. When disappearing, its light seemed to gradually die out until only a slight orange speck could be seen, the speck going out very suddenly.'

A constant watch was kept for the reappearance. 'At 10.36 the small orange speck was detected at the spot where we expected it to emerge, but inside the planet. It was seen by all ... observers clearly inside the limb. It gradually brightened up until it emerged.' They then watched it going straight towards Titan, but clouds came up, so they could not see whether it was occulted by Titan.

Discussion of the Rondebosch observation (Hepburn, Ainslie, Crommelin, Thomson)

Major Hepburn drew the special attention of the B.A.A. to the importance of the observation of the South African astronomers, and on the strength of it read a very interesting paper on 'The

Transparency of Saturn's Rings and a Suggestion as to their Constitution' (*B.A.A.J.*, vol. 30, p. 246). He gave a résumé of previous evidence of the transparency of the rings, and pointed out that in the observation in 1917, Knight had seen the star B.D. +21°1714 apparently projected within the limb, and Ainslie had seen it apparently projected on the extreme edge of ring B, while Ainslie had felt that his observations would be best represented by a path for the star taking it behind the globe (see Chapter 30). Although the apparent projections of that star on the limb and on ring B had been assumed to be due to irradiation, Hepburn suggested that, in the light of the Rondebosch observation, those appearances of projection might have been real and not optical, 'implying in that case that both ring B and the limb were transparent'. He then affirmed that the observations of Reid and his colleagues 'constitute the first direct evidence for the transparency of ring B'. Hepburn also discussed the tenuity of the rings, and pointed to their high intrinsic brightness as a difficulty in the way of the acceptance of the meteoric theory for their constitution. He stated that conspicuously white substances are usually found to 'consist of finely divided transparent matter, often crystalline in structure', and he therefore offered the suggestion that Saturn's rings might be of the nature of thin sheets of cirrus cloud, the particles being ice crystals. Though he admitted there were difficulties, Hepburn's idea that the ring particles may be of the nature of ice crystals seems to be favoured by astronomers at the present time (see Chapter 38).

During the discussion (*B.A.A.J.*, vol. 30, p. 235) Instructor-Commander Ainslie expressed surprise that the account of the Rondebosch observation seemed to indicate no diminution of light and no alteration of colour of the star when it was behind the rings, although it must have been seen through a much greater depth of ring-substance than the star observed by him in 1917, when the rings were more open.

Dr. Crommelin considered that the transparency of the rings seemed to be substantiated by the evidence presented by Major Hepburn, but that there did not appear to be enough evidence for the visibility of the star through the ball of the planet. The apparent semi-diameter of the Moon's disk obtained from meridian observations was 1″ of arc larger than that deduced from occultations of stars during lunar eclipses; this apparent enlargement is due to diffraction and irradiation. Stars had often been seen projected on the Moon's disk although it was known not to be transparent. Enlargement of

Saturn's semi-diameter by a second of arc would be amply sufficient to explain the star being seen apparently through the ball. Crommelin also drew attention to the difference between the times of disappearance and reappearance in the *Cape Times* report being an hour too short. He said that parallax would account for the star possibly going behind the ring in South Africa whereas it had been predicted not to do so in Europe.

Hepburn said he had written to South Africa to clear up what was evidently an error or misprint regarding the times; the observation was certainly remarkable, but it was circumstantially described, and made by an observer of repute. The President (H. Thomson) said he did not believe that a ring dense enough to cast a dark shadow could be so transparent that a star would not be dimmed when passing behind it, nor did he believe that a star could be seen through the limb of the planet.

Several of the difficulties raised during the foregoing discussion were cleared up by the report on the occultation which, at the request of Hepburn, was drawn up by W. Reid and confirmed by his fellow observers, Dutton and McIntyre (*B.A.A.J.*, vol. 31, p. 37)

Translucency of ring B proved by the Rondebosch observation

In his official report, made at the request of Major Hepburn, Reid stated that there was an obvious error in the times given in the newspaper report for contact with ring and disappearance behind planet, an hour too late in both cases, probably a printer's error or a slip of the pen on his part. His report continued:

'The time, 8.46, which we decided upon as the moment when the star was in contact with the ring is the result of a compromise. It is the time when we were all positively certain it was behind the ring. Personally I think it touched the ring three minutes earlier, and at the time given was on the edge of Ring B. At first there was very little diminution of light, but as soon as it touched ring B the light gradually faded for about half a magnitude. It remained like this for a few seconds, when it again fell a little further, and almost immediately the flicker took place—that is, it suddenly almost went out, but not quite, (then) it rose again fairly suddenly; after this its light fluctuated very considerably, but never reached more than a magnitude less than its original brightness.

'When it reached what we considered the limb of the planet, it was about two-thirds across Ring B. Instead of disappearing it continued to traverse the ring until it was almost on the inner edge of Ring B. From the time it touched the limb of the planet its light did not fluctuate; it gradually got dimmer until it was two and a half magnitudes below its original brightness. The disappearance took place well within the limb of the planet, and, although it was sudden, it was quite unlike the disappearance of a star in a lunar occultation. *It did not go out with a snap.*

'Regarding change of colour. Before and after occultation the star was a beautiful bright orange; during the passage of the ring it faded to a dull orange, and before disappearance and at reappearance was a brownish orange.

'We had no difficulty in following the star during the whole time, except when the flicker took place, but even then I do not think we lost it. I was fortunate to observe the star at the moment of reappearance. It was then about two magnitudes below its original value, and brownish orange in colour; it gradually brightened up until it emerged, and the time occupied until we saw it just clear of the disk was almost three minutes. Cassini's Division was not visible at the spot where the star crossed, but it was clearly visible further out.'

Reid called special attention to certain points: that the seeing was perfect; that the instrument (6-inch Cooke O.G.) gives beautiful definition; that the times given: 8.46 when star was touching ring, 8.54 disappearance behind planet, and 10.36 reappearance from behind planet, were taken from McIntyre's watch, considered the nearest to South African Standard Time; the interval of time when the star was behind the planet (8.54–10.36) is correct to within a few seconds; the bright orange colour of the star contrasted well with the planet: 'In all probability we would have failed to see the star continuously when behind the ring, if it had been the same colour as the planet'.

This detailed statement from William Reid and his colleagues corrected the error in the newspaper report of the observation and resolved most of the doubts to which the incompleteness of that report had given rise. It became clear that the observation had proved the translucency of ring B, and had even made out a very fair case for the star having been seen through the limbs of the globe, the outer layers of which must consist of a deep atmosphere. Even those who are reluctant to accept this interpretation of the appearance of the star at the limbs must recognise that the South African astronomers had made a very great contribution to the knowledge of Saturn. By proving the translucency of ring B they had dealt the final blow to the seventeenth century idea of a single, solid, opaque ring.

350

In 1934 D. G. McIntyre, one of those who had taken part in the epoch-making observation, gave a presidential address to the Astronomical Society of South Africa on 'The Translucency of Saturn's Rings', in the course of which he complained (p. 16) of the rather marked manner in which all reference to the Rondebosch observation had been omitted by various astronomers and writers of astronomical textbooks, who had, either explicitly or by implication, continued to assert the opacity of ring B. He discussed possible causes for this refusal to accept the observation as proof of the translucency of ring B: the obliquity of the rings at the time, so that the loss of light from the star might have been expected to be greater than Reid's estimates; the high albedo and dark shadow of ring B. He showed that such objections could be met by the extreme thinness of the rings, and the probability, from theoretical considerations and from photometric observations, that the ring particles are very finely divided. He also referred to the imperfections of the original report as a possible cause for the neglect. But it seems probable that, at least in some instances, the omission may have been due to pure ignorance: that, as in the cases of various other observations of Saturn mentioned in earlier chapters, for some reason or other the Rondebosch observation did not, in spite of its importance, become widely known.

Reid's description is easier to follow if it is realised that the ring system was within a few months of the edgewise phase, and that the star crossed behind ring B obliquely close to where the rings are hidden behind the limb of the globe.

CHAPTER 32

RING DISAPPEARANCES AND REAPPEARANCES, 1920–21

Passages through ring-plane and satellite ephemeris
(Hepburn, Comrie)

To prepare the B.A.A. Saturn Section for the observation of Saturn during the passages of Earth and Sun through the ring-plane in 1920–21 Hepburn in two papers (*B.A.A.J.*, vol. 30, pp. 158 and 240) gave a clear and detailed explanation of the circumstances, marred only by the unfortunate repetition of Proctor's error that the ring-plane takes more than a year to pass across the Earth's orbit. Hepburn used as illustrations two plates of Barnard's drawings of 1907 (one of which is reproduced here as Plate X), and diagrams which he had prepared, the chief one being shown here as Figure 20.

On this diagram the vertical straight line AA represents the ring-plane, to the left of it being South, to the right North. The curve BB and line $B'B'$ are the graphs of the columns in the *N.A.* giving respectively the saturnicentric latitudes of Earth and Sun referred to the ring-plane. At each date therefore the horizontal distance from AA to $B'B'$ or BB shows the elevation (to scale) of Sun or Earth with respect to the ring-plane at that date. Whereas the Sun's southerly elevation steadily decreases until it crosses the ring-plane at S, and thereafter the Sun steadily increases its northerly elevation, the Earth's elevation is much more variable, the curve BB crossing the line AA three times, at E_1, E_2, E_3, representing three passages of the Earth through the plane. The shaded intercepts between AA and BB indicate that the unillumined face of the rings is presented to the Earth. The sketches of Saturn on the right show the varying tilt of the rings as it would appear from the Earth at the respective dates. Figure 16b in Chapter 27 represents in a different way the Earth's passages through the ring-plane in 1920–21.

Hepburn also wanted to obtain for his Section information about forthcoming satellite phenomena. Marth had provided such ephemeris for many years in the later nineteenth century, and H. Struve had continued them (for eclipses of satellites) in the early years of the present century. By 1920 such help was lacking, so (at Melotte's suggestion) Hepburn appealed to L. J. Comrie to provide

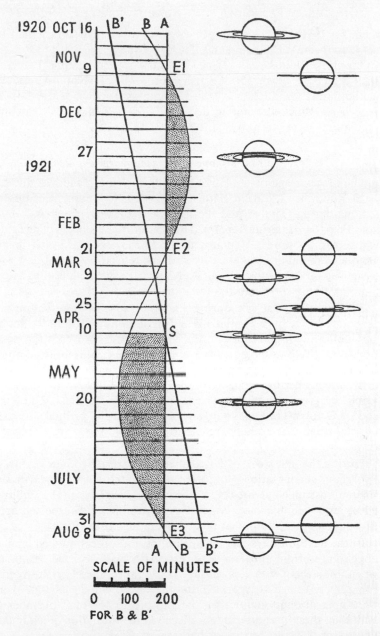

FIGURE 20. P. H. Hepburn: diagram of varying elevations of Earth and Sun in relation to ring-plane, and of ring phases, 1920–21. (From *B.A.A.J.*, vol. 30, p. 241).

353

them for 1920–21. The work proved so onerous that Comrie had to get help from A. E. Levin and others with the result that the B.A.A. Computing Section was formed, Comrie being its first Director (*B.A.A.J.*, vol. 70, p. 55). A further proposal of Hepburn resulted in the annual production by the Computing Section of the *B.A.A. Handbook*, which was started in 1922. (Comrie's first set of satellite predictions were given in *B.A.A.J.*, vol. 30, p. 292.)

Before the first disappearance (Hepburn, Phillips)

The first passage of the Earth through the ring-plane was due on 1920 November 7, but the Saturn Section made a number of observations during October and the first week of November (*B.A.A.J.*, vol. 31, p. 64—Interim Report). They tested the visibility of the ring with small apertures: Hepburn found it easy with a 3-inch up to October 27; Phillips saw it easily with his reflector stopped down to $3\frac{1}{2}$ inches on November 3; H. R. Maynard saw the ansae as 'very fine lines' with a $4\frac{1}{2}$-inch refractor on November 5.

Typical of the aspect during October was an observation by Hepburn on October 11: the ansae were continuous up to the limbs and uniformly bright; they appeared to taper from about the position of ring B towards the limb, but he considered this an irradiation effect; the ring was too narrow for its actual shape to be seen with his $12\frac{1}{2}$-inch aperture reflector, but across ring B the effective breadth would be greatest and the surface intrinsically brightest. Until the end of October the only change noted was a very gradual decrease of total brightness and apparent breadth of the line representing the rings.

During the first week of November there were two new features: apparent 'condensations' of brightness in the ansae, and (in the smaller instruments) breaks in the ansae near the limb. Phillips, using an 8-inch aperture refractor on November 1, 'suspected condensations on each ansa' at places in the region of ring B.

In the next few days Hepburn, Phillips and others noticed similar appearances from time to time, especially with averted vision or when the seeing was less good; to Hepburn these bright points sometimes seemed to sparkle. But they were convinced that there were no real abnormalities on the ring up to 18^h on November 6, and that these pseudo-condensations were irradiation phenomena due to insufficient aperture or bad seeing. Breaks in the ansae were not seen at this period by any observer using an aperture of 12 inches or more. There were many reports of the preceding ansa appearing brighter than the other, but Hepburn considered this an optical, not a real, effect.

On November 6, the day before the Earth was due to pass through the ring-plane, Phillips, Thomson and Waterfield found the ring easy to detect with apertures of 18, 12¼ and 8 inches, under indifferent seeing conditions. In the early morning of November 8, they found, with the same instruments, that the ring was absolutely invisible, though the seeing was on the whole good. Ainslie, observing with the 28-inch aperture refractor of the Royal Observatory, Greenwich, was unable to see any sign of the ring even when the planet was concealed by an occulting bar. These observations seem to have been made about 11½ to 13 hours after the predicted time of the Earth's passage through the ring-plane; Barnard, 17 hours after passage, using the Yerkes 40-inch aperture refractor, could see no sign of the ring (*M.N.*, vol. 82, p. 275).

First views of ring after disappearance on 1920 November 7
(Barnard, Steavenson, Hepburn, Ainslie)

After the ring's disappearance on November 7, Hepburn and Phillips failed to find any definite trace of it in good seeing on November 11 and 15, nor did Ainslie and Waterfield fare any better with the Greenwich 28-inch aperture refractor in hazy conditions on November 13. It appears, however, that the Yerkes 40-inch revealed the ansae, though feebly, to Barnard on November 13.

The first observer in England to see the ring again was Dr. W. H. Steavenson, using the 28-inch aperture refractor at Greenwich from 14^h 30^m to about 18^h on November 16 (*M.N.*, vol. 81, p. 127; *B.A.A.J.*, vol. 31, pp. 67, 116). After seeing nothing of the ring during the first hour, he glimpsed it repeatedly during the following half hour, and from 16^h onwards held it pretty steadily. He found it continuous but not uniform in brightness. The brightest portions were estimated to be (i) in the position of ring C and (ii) just internal to the Cassini division. The 'condensations' he saw at those points were objective, best seen when the seeing was best: it had been just the converse with the illusory ones noted before disappearance. There was no bulging of the brighter portions, which were of considerable length and quite unlike satellites. The ring appeared to taper at both ends to fine points. The whole ring was a most delicate object, much more difficult to see than Enceladus, and probably at this date invisible in any instrument that would not show Enceladus with ease. 'The quality of the light suggested that of a nebula.' Across the ball Steavenson could see no trace of the ring but saw its shadow, concave towards the north and broader at the ends than in the centre (see Plate XIII, Fig. 1). It is a good illustration of the difference in light-grasp between apertures of 28 and 12½ inches that Hepburn, observing the same day with his 12½-inch reflector, could record no sign of the ring and only a doubtful suspicion of Enceladus for which he looked closely.

On November 20 Hepburn and Ainslie observed with the 28-inch at Greenwich from 17^h 20^m to just after 19^h, the power used being 450, as in the other observations with this instrument mentioned above. The seeing was very fair and the ring was seen at first glance; Enceladus near eastern elongation, Mimas near western, provided standards for estimating its brightness. The complete elliptical out-line of the ring was seen, with condensations in the positions described by Steavenson. The following ansa and its condensations were more easily seen than the preceding. The uncondensed parts of the ansae looked nebulous, though with sharp outlines. Irradiation effect, if existing, was small. Evidently the ring had brightened considerably since November 16, because it was certainly seen by Hepburn at 19^h 1^m, 15 minutes after Mimas and 12 minutes after Enceladus had disappeared, and when Saturn was invisible to the naked eye, less than half an hour before sunrise. (For full details of this observation see *M.N.*, vol. 81, p. 127; throughout these reports G.M.A.T., starting at noon, is used.)

The ansae, as seen with the 28-inch refractor on November 20, 'did not appear as lines, but as very elongated areas tapering towards their extremities, with fairly well-defined outlines. The condensations were but little brighter than the rest of the ring, and bulged little, if at all, as a result of irradiation'. Steavenson was inclined to think that the apparent difference in brightness of the ansae was objective, because, though he found their brightness equal on November 16, he and Waterfield found the preceding one quite distinctly the brighter on November 23, when observing under similar conditions.

On November 21 Hepburn was able to see the ansae, at least in part, though broken between the condensations, with his 12½-inch reflector, and from that date onwards the ansae were visible with instruments of moderate apertures. Few features were to be seen on the disk at this time: the south equatorial belt was well marked, and there was a fainter belt further south; the north equatorial belt was ill-defined; the equatorial zone seemed 'pinkish' to Steavenson on November 16; Waterfield described the rings as 'copper-coloured' on November 6.

Dark side of rings (Phillips, Hepburn, Steavenson, Waterfield)

The second interim report of the Saturn Section (*B.A.A.J.*, vol. 31, p. 222) deals with observations from mid-November 1920 until 1921 February 22, the period during which the Earth and Sun were on opposite sides of the ring-plane, and also with the 'reappearance' of the rings after the Earth had recrossed the plane.

Phillips first glimpsed the ring on November 20, using apertures of 12¼ and 8 inches; from early December till late January he found it very easy; from the beginning of February it became more difficult, only the parts of the ansae near the limbs being seen or glimpsed; after February 14 until February 22 it was totally invisible, but quite easily seen after 14^h on that night with the 8-inch aperture refractor, and very conspicuous after February 22. He could not make out any definite condensations, but was impressed with the intense brightness of the part of the equatorial zone south of the equator.

Hepburn and others also found that the condensations were very little in evidence after the end of November. During December and January Hepburn sometimes found the ansae broken, and the brightest parts near the limbs.

On 1921 February 1 Steavenson made another observation with the 28-inch aperture refractor at Greenwich, and saw both condensations in each ansa, but the space between them was not so dark as shown in his drawing of November 16. He had frequent glimpses of a minute star-like point of light in the centre of the outer condensation of the p. ansa, rather fainter and smaller than Enceladus. The inner portions of both ansae were, as before, brighter and thicker than the outer parts. Waterfield, observing on the same occasion, reported similar stellar points on the outer condensations. Steavenson also noted that the shadow across the disk was nearly straight and added: 'I sometimes suspected that there was a darker central core to the shadow'. Waterfield reported: 'Shadow concave downwards, upper border more nearly straight. . . . In best seeing I suspect a central darker line. The ring appeared in same plane as upper part of shadow.' Both these observers expressed doubt as to the objective reality of the appearance of a dark medial core in the shadow, though a similar appearance had been observed on several occasions in 1907 by Lowell (see Chapter 26). W. H. Pickering, observing with an 11-inch aperture refractor in Jamaica, also recorded a similar appearance in 1920 December.

Hepburn stated in his second interim report that the black line seen across the equatorial region of the planet is the synthesis of shadow and projection of the ring. He never completely lost it with his 12½-inch reflector, but others, using smaller apertures, lost it on certain dates between February 14 and 23. On February 22 the projection of the ring vanished, and the black line seen about that date was the shadow only, 0″1 to 0″15 broad. After February 22 the black line became easier, and about March 8 was caused by the projection only, the ring being at that date superposed on its shadow; 'the darkness of the projection at that date is interesting, as we were then looking at the illuminated side of the ring'.

Reappearance of ring, 1921 February 22
(Rambaut, Graff, E. C. Slipher)

Dr. A. A. Rambaut, Radcliffe Observer at Oxford, made an interesting report (*M.N.*, vol. 81, p. 391) of the observations of himself and his colleagues on 1921 February 21 and 22 with an 18-inch aperture refractor, power 290. During the evening of February 21, though they could see many markings on the disk, and Tethys and all the brighter satellites, and one of them thought he could glimpse Hyperion, they could see no sign of the ring. The seeing was good enough, they considered, to show an object two magnitudes fainter than Tethys. It had been predicted in the *N.A.* that the Earth would pass to the sunlit side of the ring-plane at about 9^h G.M.T. on the evening of 1921 February 22. At 8^h 38^m the 18-inch refractor was again set on the planet, and 'at once an exceedingly fine streak of light on either side of it could be seen', although both ansae had been quite invisible only 20 hours previously. The eastern ansa was rather more distinct than the western, which 'thinned off towards its extremity'; each ansa was estimated to extend from the limb about 7/10 of the planet's diameter.

For a week before February 22 no trace of the ring could be seen by K. Graff (Bergedorf, 24-inch aperture refractor) or other observers using as large or larger apertures under favourable conditions, but he found the rings fairly easily at 8^h on February 22, and he and others noticed that they brightened steadily as the evening advanced (*M.N.*, vol. 82, p. 275). E. C. Slipher (Lowell Observatory, 24-inch refractor) was, however, able to *photograph* the rings at 21^h on the evenings of February 20 and 21, and found them to have considerably brightened by the latter date; the full outer extension of the ansae is shown in these remarkable photographs, but the ansae near the limbs are obliterated by the over-exposed image of the planet.

It seems that none of the observers of the B.A.A. Saturn Section were able to see any trace of the ring on February 21, but at about

10^h on February 22, in spite of rather poor conditions, J. Bougon (10-inch aperture reflector) and J. Knight (5-inch refractor) were able to glimpse it now and then. Four hours later on the same night Steavenson and Phillips saw it easily with an 8-inch aperture refractor as a steady continuous line of light. On February 23 the ring was visible in all apertures: Steavenson, using 2·8-inches' aperture, found it faintly but unmistakably visible. Apparently while its unilluminated side was presented to the Earth no one saw it with any aperture less than 4¼ inches.

The apparent profile of the ansae after the reappearance was found to be much the same as that before disappearance, the most pronounced bulge lying about at the projection of the brightest zone of ring B, thence tapering off gradually inwards and fairly abruptly outwards. When the illuminated side is on view the parts of the ansae next the limbs appear faintest, while the contrary is the case when the dark side is presented.

Fading of rings through Sun's passage, 1921 April 9 (Hepburn)

P. H. Hepburn made brief reports (*M.N.*, vol. 82, p. 277; *B.A.A.J.*, vol. 31, p. 252) on observations made by the Section on the occasion of the Sun's passage through the ring-plane 1921 April 9, which involved another theoretical 'disappearance' of the rings, which were then a little more open. The sole effect observed was the gradual fading in brightness and the reappearance of the condensations. The rings remained visible in instruments of moderate aperture, and disappeared temporarily only in telescopes of 4 inches and under. From observations made in twilight in May and June when the unilluminated rings attained their greatest opening at this epoch, Hepburn found their visibility intermediate between those of Titan and Rhea.

Although W. H. Pickering claimed that measurements made with his 11-inch aperture refractor in Jamaica in 1920 November and 1921 April indicated a thickness for the crepe ring of something between 400 and 1,000 miles, Hepburn agreed with Barnard's conclusion from his 1920 November observations that the rings must be incredibly thin, and that Russell's estimate of a thickness of about 21 kilometres seemed reasonable. Two condensations had been seen in each ansa when the unilluminated side of the rings was presented, and Barnard, measuring their positions in 1920 November, had found them to agree closely with those he had determined in similar circumstances in 1907 (see Chapter 26). Hepburn concluded that the observations of 1920–21, made under very favourable circumstances 'gave no support to the view that the rings are other than exceedingly thin and of perfectly uniform contour'.

For the final passage of the Earth through the ring-plane to the same side as the Sun, due on 1921 August 3, Saturn's position would allow of observations being made only in the Earth's southern hemisphere.

Eclipse of Rhea by shadow of Titan, 1921 April 8
(Comrie, Hepburn, Levin)

The same report by Hepburn said: 'The satellite phenomena were disappointing. In England no transits, shadow transits, or eclipse disappearances of Titan were visible, and but three eclipse reappearances, all under unfavourable conditions'. R. T. A. Innes (in South Africa) observed eclipses of Titan and Rhea, and Professor W. H. Pickering (in Jamaica) watched mutual occultations on 1921 February 20 of Titan–Rhea and Dione–Rhea. Pickering, who was using an 11-inch aperture refractor, was unable· to see the shadow of Rhea though he looked for it several times under the most favourable conditions (*B.A.A.J.*, vol. 31, p. 240).

The one really successful satellite observation carried out in England was that of the eclipse of Rhea by the shadow of Titan, predicted by L. J. Comrie and Major A. E. Levin for 1921 April 8 (*B.A.A.J.*, vol. 31, p. 161). Observations of the eclipse of one satellite by the shadow of another are very rare, and this was probably the first instance ever recorded of such an event for two of Saturn's satellites having been both predicted and actually watched. The following is a summary of the observations (from *M.N.*, vol. 81, p. 487; for full details see *B.A.A.J.*, vol. 31, p. 271):

P. H. Hepburn and A. E. Levin, observing together with a 12-inch aperture reflector at Hampstead (powers 160 and 310), obtained a good view of both immersion and emersion, and from estimates of brightness at both phases deduced the time of mid-eclipse as $10^{\text{h}} 47^{\text{m}}4 \pm 0^{\text{m}}5$; Rhea was completely invisible for over half an hour. L. J. Comrie observed at Cambridge with a 12-inch aperture refractor, power 200, watched both phases, and estimated the time of mid-eclipse as $10^{\text{h}} 47^{\text{m}} \pm 1^{\text{m}}$; Rhea was invisible for 38 minutes. These observations were in excellent agreement. E. A. L. Attkins and F. Burnerd, using $8\frac{1}{2}$-inch aperture reflectors at different places, were unable to get a complete observation: the latter could only see the phenomenon through breaks in clouds, the former, while obtaining good observations of the immersion, was prevented by clouds from witnessing the emersion. C. J. Spencer at Halifax, with a $4\frac{1}{2}$-inch aperture refractor, observed both phases, but his estimated time for immersion was earlier and for emersion later than those of the other observers, probably owing to the small aperture he was using.

Although these last three observations were only partial, Comrie and Levin found the data particularly useful in determining when the fading of Rhea was first detected. They concluded from all the records that the time of mid-eclipse was approximately $10^h47^m3 \pm 0^m5$, and the detected duration at least 50^m, i.e. from 10^h 22^m to 11^h 12^m. Both disappearance and reappearance were fairly sudden, for Rhea was invisible in the 12-inch telescopes for more than two-thirds of the total duration of the eclipse. These records were expected to afford a good check on the tables for Rhea and Titan, and suggested a diameter for Rhea nearer $0''4$ than the assumed value of $0''3$.

CHAPTER 33

ASTROPHYSICAL STUDIES IN THE 1920s OF THE GLOBE, RINGS AND SATELLITES

Visual observations (from France only), 1924–30 (Antoniadi)
Had it not been for some observing at Meudon and Juvisy, the break in visual study of Saturn during the decade 1922–31 would have been as complete as that of 60 years earlier (see Chapter 14), when the planet had also spent some years in southern declination. A similar hiatus in the later 1890s had been avoided partly through opportune outbreaks of spots on Saturn, and partly through the observing zeal of Barnard, H. Struve, Antoniadi, See and others (see Chapters 15, 19–22, 24 and 25). On the other hand, valuable progress was made during the 1920s, especially in England and America, with mathematical and other astrophysical investigations of the Saturnian system.

The visual observations in France were made chiefly by Antoniadi, using the great 33-inch aperture refractor of Meudon (*L'Astronomie*, vols. 38, p. 302; 40, p. 339; 43, pp. 371, 375; 44, pp. 6, 7, 9, 160–1).

In 1924 June, Bernard Lyot, later to become famous for his planetary and solar observations, observed a rose tint on the north equatorial belt; Antoniadi, confirming this, noticed also a slight rose tinge on the equatorial zone, and remarked that Trouvelot had recorded a similar tint on the same belt in 1874. Variations in size, tint and general appearance of the north polar cap were noted by Antoniadi 1924–29; in early 1926 he found the cap slate grey in colour, surrounded by a clear zone, to the south of which were two brownish belts, one of them—the north equatorial belt— showing dusky spots like those found by Stanley Williams in 1891 (see Chapter 15).

In the later 1920s Antoniadi also noted whitish complex patches and irregular filaments on the equatorial zone, a narrow discontinuous belt on the equator, and considerable variations in the tint and visibility of the north equatorial belt, which once in 1928 was almost invisible in the great telescope. Unusual appearances on the ring were recorded by F. Quénisset (from Juvisy) and by Graff, Baldet and Lyot, who noted its obscurity towards the outer edge in

1928 and 1929: Encke's division seemed a mere dusky border to the outer darker part of ring A. In 1929 Mme Camille Flammarion and Antoniadi, using the 9½-inch aperture refractor at Juvisy, found the ring clearly separated from the globe by a wider space on the east than on the west, an appearance the reality of which is open to doubt—see the beginning of Chapter 34, and especially Stroobant's conclusions in Chapter 35.

Influence of satellites on the form of the rings
(Goldsbrough, Greaves)

In 1921 Dr. G. R. Goldsbrough in a paper (*Phil. Trans.*, vol. 222, p. 101; summarised in *B.A.A.J.*, vol. 33, p. 124) examined mathematically the perturbing influence of the satellites on the ring particles and of the particles on one another. He considered the small oscillations of each particle about its position of relative equilibrium in the ring to a first approximation, and deduced the existence of zones of instability. He first assumed concentric rings of particles surrounding the planet, performing approximately circular orbits when unperturbed. He considered the effect of one satellite, moving in an unperturbed circular orbit, on the particles of one ring, subject to their mutual attraction as well as to those of Saturn and the satellite. If the particles in the rings are all equal, then in certain places large perturbations will occur, collisions with adjacent rings will ensue, and a division will be formed. He then showed that if the particles are unequal, the divisions must be more extended. If some of the particles are indefinitely small, Cassini's division is obtained. Using the dimensions of this division to set an upper limit to the magnitude of the particles in any ring, he obtained the following results: Mimas should produce a clearance of particles from radius 20″2 (about the outer edge of ring A) up to itself; it should also produce a division from radius 16″9 to 17″64 (Cassini's division is from about 16″87 to 17″64). Dione should cause a clearance of particles from the region of the planet's surface up to radius 9″34 (inner edge of crepe ring is 10″83 from planet's centre). Rhea should produce a clearance up to radius 13″07 (inner edge of ring B is at about 13″21). He concluded that in time the crepe ring would be dispersed by Rhea, and the whole of the ring particles by Titan.

In a further paper (*Proc. Roy. Soc.*, Series A, vol. 101, p. 280; summarised in *B.A.A.J.*, vol. 33, p. 45) Goldsbrough allowed for the fact that the orbits of the satellites are not precisely co-planar with the ring system. As the result of strict mathematical analysis he concluded that the effect of a satellite moving in an orbit inclined to the ring-plane is to produce a new division. Mimas would cause a narrow division at a distance 19".05 from Saturn, which corresponds closely with the observed position of Encke's division. Enceladus should produce a division at 16".5, but no observation* of one seems to have been recorded at this place (near the outer edge of ring B); in any case a division there would be expected to be faint, because of the minute inclination of the orbit of Enceladus, and its distance from that place.

W. M. H. Greaves (who later became Astronomer Royal for Scotland) discussed Goldsbrough's first paper mathematically (in *M.N.*, vol. 82, p. 356) and suggested that if it were possible to take into account the terms arising from higher approximations than the first, Goldsbrough's conclusions regarding the existence of zones of instability might prove to be illusory. The instability found by Goldsbrough is of an extremely slow nature, and so at a remote date the effect of the neglected terms might be appreciable. Greaves expressed the opinion that in the divisions in Saturn's rings there is a greater tendency to collision and a smaller density of particles than elsewhere in the rings, so that it is probable that particles do exist in the divisions but very much more sparsely than in the rings themselves.

In a further paper (*M.N.*, vol. 82, p. 360) Greaves, after a full mathematical discussion, concluded that the perturbations due to Mimas are not large enough to ensure a small body initially in the Cassini division being trapped in one of the bright rings, though he admitted that he had not taken into account perturbations by other satellites or by the particles in the rings or the other particles in the division. He pointed out that instability of the rings within the regions of clearance develops slowly and that it is by no means certain that dissipation has been completed: in fact, in one case it certainly has not, for one of the areas of clearance corresponding to Titan includes the whole ring system! He concluded that there would appear to be no objection mathematically to Barnard's belief (see Chapter 26) that matter is to be found within the Cassini division.

Possible ellipticity of the rings (Goldsbrough)

In 1924 Dr. Goldsbrough investigated mathematically the question, suggested by certain past observations, whether the rings might be elliptical instead of circular (*Proc. Roy. Soc.*, Series A, vol. 106, p. 526; summarised in *B.A.A.J.*, vol. 35, p. 41). Of the

* Except by Maggini 1910, 1912, 1913 (see Chapter 29).

various solutions he obtained for a stable elliptical ring he found one to be valid over the whole range of values of the ratio of the masses. Although in his solution only a single ring had been considered, he was of opinion that the results form a first approximation to a composite ring made of confocal ellipses. He pointed out that measurements of the observed diameter of such a ring made at different times should show discrepancies (of a periodical nature) according to the relative positions of Saturn and the Earth, as well as others arising from the rotation of the ring. He suggested an investigation of observations of the ring diameter to test this theory.

In the course of a lecture in 1928 (*B.A.A.J.*, vol. 39, p. 64) Professor Goldsbrough showed that the mass of the rings is not more than 1/27,000 of the planet's mass, and he explained that the satellites are entirely responsible for all the divisions, and that certain thickenings of the rings named 'tores' near the divisions could be explained by oscillation, subject to satellite attraction, taking place at right angles to as well as in the ring-plane.

Satellite diameters and densities (Hepburn, Levin)

Major P. H. Hepburn read a very interesting paper in March 1923 (*B.A.A.J.*, vol. 33, p. 244) in which he attempted to derive values for the diameters and densities of the six inner satellites from their masses, as determined by H. Struve, their stellar magnitudes, as corrected by Guthnick from observations 1905–08, and their albedo. He could not deal with Iapetus and Hyperion because no satisfactory determinations of their masses appeared to have been made. Taking first Struve's values for the reciprocal masses relative to that of Saturn, and assuming densities similar to those of Saturn, the Moon and the Earth respectively, he obtained three alternative sets of values for the diameters:

Satellite	Reciprocal of Mass Satellite/Saturn × 1,000	Density		
		as Saturn (0·125)	as Moon (0·605)	as Earth (1)
		Diameter (miles)		
Mimas	16340	285	168	143
Enceladus	4000	456	270	228
Tethys	921·5	744	440	372
Dione	536	891	526	445
Rhea	250	1149	678	574
Titan	4·7	4319	2552	2160

Since the diameter of Titan as measured by Barnard was 2,720 miles, and by Lowell 2,440 miles, the mean of those values (2,600) strongly suggested that Titan's density is similar to that of the Moon. From Guthnick's values for the stellar magnitudes:

Mi. 12·07   En. 11·69   Te. 10·56   Di. 10·71   Rh. 9·98   Ti. 8·28

he then worked out the relative intensities for the three alternative densities, and on the basis of Professor H. N. Russell's value of 0·33 as the geometric albedo of Titan, Hepburn deduced from the masses and magnitudes the geometric albedo on each of the three assumptions for density. (Russell's definition of geometric albedo was 'the ratio of the actual brightness of the planet at full phase to that of a self-luminous body of the same size and position which radiates as much light from each unit of its surface as the planet receives from the Sun under normal illumination'.) Hepburn's calculations gave impossible results for geometric albedo for all these satellites except Titan if their densities were assumed to be similar to those of Earth or Moon. Hence the correct assumption for their densities was evidently that of densities similar to Saturn's which would make their diameters range from 400–500 miles for Enceladus up to rather over 1,000 miles for Rhea. Even that assumption, however, gave an inadmissible albedo for Mimas, for which he had to assume a density only 0·45 times that of Saturn, or only one-third that of water, which would give a diameter for Mimas of about 370 miles.

Hepburn's conclusions from these results are interesting. He had found Titan to be very similar to the Moon in size and density, but with more than three times the reflective intensity of the full Moon. He went on to say: 'Since it is difficult to see how Titan can have an atmosphere, this points to the surface of Titan being composed of, or covered with, very white material.' (Presumably it was not known at that time that low temperature due to remoteness from the Sun would enable a body of Titan's size and mass to retain an atmosphere, and evidence of an atmosphere for Titan was not found until 1944, by Kuiper from its spectrum—see Chapter 38.) As Hepburn believed that Saturn's low density was explained by 'great heat and gaseous condition', he naturally found it difficult to explain how such small bodies as Rhea and the others, which must be cold and solid, could have densities similar to or lower than that of water. He pointed out that his result for Mimas implied a density not more than one-third that of water and 'a brightness comparable with that of new-fallen snow'; he therefore suggested that Mimas might actually be a sort of gigantic snowball.

Major A. E. Levin gave a short account (*B.A.A.J.*, vol. 33, p. 214) of the derivation of values for the diameters of Titan and Rhea from the observations of the eclipse of Rhea by the shadow of Titan (see end of Chapter 32). He pointed out that the penumbral fringe of the shadow must have produced a certain diminution of light before it became sensible to the eye, and some residual light when the satellite was no longer visible: these sources of uncertainty would, he showed, affect the result for Rhea much more than that for Titan. There was also an uncertainty as to the fraction of a magnitude by which the light had diminished by the time it became noticeable to the observers. He drew up a table of values to cover these uncertainties which gave possible values for Titan's diameter ranging from 3,260 to 3,680 miles, and a much greater range for Rhea from 510 to 1,150 miles. He considered that the most probable of these values would be a pair in the middle of the table giving a diameter of 3,550 miles for Titan and about 700 miles for Rhea.

In a note (*B.A.A.J.*, vol. 33, p. 284) Hepburn showed that Levin's diameter of Titan would give a density about twice that of Saturn and a geometric albedo twice that of the Moon, both of which seemed reasonable enough; the value for Rhea would, however, give an impossible albedo.

Micrometer measures of rings and globe (Phillips)

In 1924 the Rev. T. E. R. Phillips read a paper (*B.A.A.J.*, vol. 34, p. 185) giving the results of a large number of micrometer measures of the rings and globe made by him, chiefly in 1912,–19,–21, and comparing the means with measures by Asaph Hall, Barnard, Lewis and Dyson (Greenwich), and See as follows:

	1 Phillips	2 Hall	3 Barnard	4 Lewis & Dyson	5 See
	(1912–21)	(1885–87)	(1894–95)	(1895)	(1900)
Outer diam. of Ring A	40″176 ± ·31	40″450	40″186	40″590	40″274
Diam. centre of Cassini div.	34·775 ± ·20	34·530	34·517	34·349	34·339
Diam. inner edge Ring B	26·552 ± ·16	25·750	25·626	25·647	25·932
Equat. diam. of Saturn	17·910 ± ·20	17·720	17·798	17·754	17·804
Polar diam. of Saturn	16·345 ± ·16	—	16·246	16·793	16·005

Phillips pointed out that the accordance on the whole was very satisfactory, the agreement of his and Barnard's figures being particularly close, but the figures for the inner diameter of ring B were a striking exception, those of the other observers being all smaller than his, by about 0″8 in the mean. All his measures were made with an 8-inch aperture Cooke refractor, and when seen with that instrument under good definition, the ring had always looked to him further away from the planet than it appeared to be on Barnard's drawings, or than the other measures would indicate.

Phillips also compared his own measures with the mean of the earlier visual ones and with the mean of the measures made by Hollis and Hepburn on Barnard's set of photographs (see beginning of Chapter 30), which were reduced to seconds of arc by equating the external diameter of ring A in Hollis's measures to Barnard's value of 40″186 and scaling the others accordingly:

	Mean of photograph measures by Hollis and Hepburn	Mean of earlier visual measures	Phillips (means)
Outer diam. of A	40″195	40″375	40″176
Centre Cas. div.	35·393	34·434	34·775
Inner edge of B	27·180	25·739	26·552
Breadth of A and B with Cas. div.	6·558	7·318	6·812
Inner edge of B to Ball	4·818	3·986	4·321
Equatorial diam. of Ball	17·443	17·769	17·910

In reply to a point raised by Dr. Crommelin, Phillips said that his measures of the polar diameter were made in 1921 when the equator was not appreciably tilted, and therefore the polar diameter was seen at its true length (*B.A.A.J.*, vol. 34, p. 172).

Constitution and temperature of the outer planets (Jeffreys)

In April 1923 a very important and revolutionary paper was presented to the R.A.S. by Dr. Harold Jeffreys on the constitution of the planets Jupiter, Saturn, Uranus and Neptune (*M.N.*, vol. 83, p. 350). He referred to the belief then widely held that the four great planets are very hot and largely gaseous, and suggested that the evidence hitherto put forward to support that belief was inadequate; in fact, the data could be more easily reconciled with the theory that those planets are cold and solid, but composed of materials low in density compared with terrestrial rocks. He first investigated the heat loss by radiation during the existence of Jupiter, which could be assumed to be at least as old as the Earth, and found that on any reasonable assumption of original temperature, there had been ample time to solidify the planet, the surface cooling down till it became much cooler than the interior, from which heat could then only be slowly conducted out. The same would hold good for the other three large planets. He then considered the surface temperature

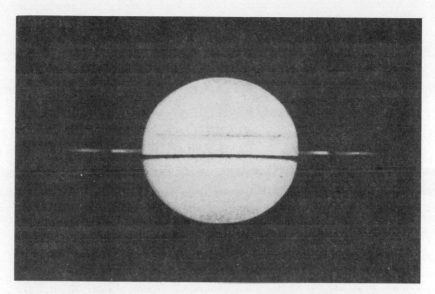

Fig. 1.—W. H. Steavenson's drawing of Saturn showing condensations on the ring's unillumined face, 1920 November 16 (28-inch refractor of the Royal Observatory, Greenwich)—(from *B.A.A.J.* vol. 31, plate 4).

Fig. 2.—W. T. Hay's discovery drawing of the equatorial white spot, 1933 August 3, 6-inch refractor (from *B.A.A.J.* vol. 44, frontispiece).

PLATE XIII

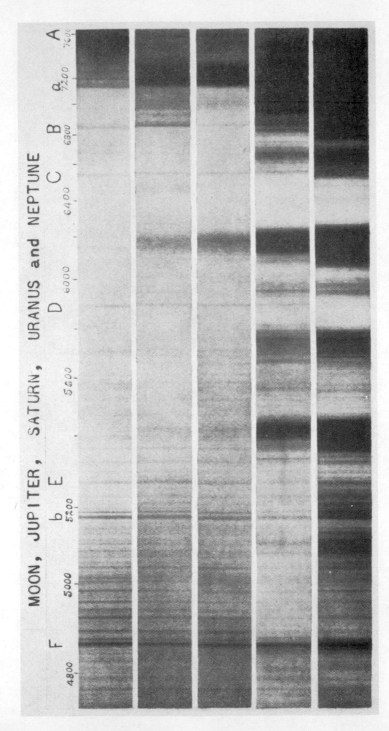

MOON, JUPITER, SATURN, URANUS and NEPTUNE

F   b E   D   C   B   a A

4800   5000   5200   5600   6000   6400   6800   7200   7600

PLATE XIV.—Photograph of spectra of the major planets showing bands of methane gas (from Sir Harold Spencer Jones: *Life on Other Worlds*, English Universities Press, 1955, plate 9, facing p. 97). For explanation see pp. 386–7 of text.

in connection with heat supplied by slow conduction from the interior and by solar radiation, and he found that even at the distance of Neptune the amount of heat received from the Sun would exceed that conducted from the interior.

Even if the outer planets had more radio-active matter than the Earth in their surface layers, the amount would have to be 10,000 times as great as on the Earth to produce heat comparable with the radiation received on the Earth from the Sun. It seemed very probable, therefore, that the surface temperature on the outer planets would be lower than on the Earth.

He then considered the atmospheres and wind velocities on Jupiter and Saturn, as indicated by the rotation periods derived from the observation of spots at various latitudes, and concluded that the depth of Saturn's atmosphere might be 1/20 of the radius.

He showed that the argument based on their low densities is as good an argument *for* the theory that the four planets are solid as against it. They must be composed of matter very different from the chief constituents of the Earth, and the fact that Saturn's density is less than that of ice, even when the atmosphere calculated is allowed for, shows that some still lighter substance is indicated. The low densities of satellites, such as Titan, with a density comparable to that of ice, also supports this idea of lighter substance, and the high albedo of Saturn's ring is suggestive of its being composed of a colourless finely divided substance, perhaps ice or some other non-metallic compound.

If the density of Saturn's solid part is similar to that of Titan (1·4) about half its volume would have to be gaseous, and the depth of the atmosphere would have to be about 1/5 of the radius. This might be the case, since there was a good deal of uncertainty in the theory of the general circulation of the atmosphere. Similar considerations applied to Jupiter. Dr. Jeffreys was of opinion that the evidence does 'indicate that such atmospheres must be cold and surround solid surfaces, and it suggests that the suspended clouds in them are probably composed of some material with much lower melting and boiling points than water'.

He admitted that this argument relating to temperature refers to the effective radiating surface, that is, the cloud surface, and that it might be argued, by analogy with the Earth, that the cloud surface may be cold without the solid surface being so, but the vertical variation of temperature in the Earth's atmosphere is due to the heating of the Earth by solar radiation absorbed by the solid surface. The cloud layers in the case of the great planets must be practically opaque, so that the solid surfaces cannot be heated directly; hence it seems probable 'that the atmospheres below the cloud layers are approximately isothermal'.

In this paper Jeffreys had given a new interpretation to the low mean densities of the great planets, and had shown them to have an entirely different constitution from the hot, gaseous state that had been generally assumed during the previous half century (see end of Chapter 14). Further than that, he had suggested in the course of his discussion several very fruitful lines of investigation which have since been followed up, and have resulted in important advances in the knowledge of the physical conditions on Saturn and the other great planets, for example: the likelihood of their atmospheres containing gases with very low freezing and boiling points, of their solid surfaces being covered with thick ice layers, and of their solid cores consisting largely of substances of very low density, even lower than that of ice.

Water-cell transmissions and planetary temperatures (Menzel)

In the same year, 1923, Donald H. Menzel wrote a very technical paper (*Ap. J.*, vol. 58, p. 65) on the ascertainment of planetary temperatures based on water-cell transmission measures made by Dr. Coblentz at Lick Observatory in 1914, and by Drs. Coblentz and Lampland at Flagstaff Observatory in 1921–22, and on the physico-mathematical theory worked out by Professor H. N. Russell. The principle of the method is based on the fact that, while reflected solar radiation of short wave-length from a planet is largely transmitted by a water cell, long wave-length radiation from a planet's surface is stopped completely. Menzel deduced from the observational data a surface temperature of $-110°C$ for Saturn and Jupiter (as against 50°C for Venus and $-16°C$ for Mars), which, though very low, he considered higher than could be maintained by solar radiation alone, and therefore confirmed the belief that Saturn and Jupiter are hot internally.

Internal constitution of Saturn (Jeffreys)

In his second paper presented in 1924 (*M.N.*, vol. 84, p. 534) Dr. Jeffreys applied the Radau–Darwin theory* of the figure of the Earth to determine from their ellipticities the moments of

* B. M. Peek (*The Planet Jupiter*, p. 216) has explained that if the ellipticity of a rotating spheroid is known, the Radau–Darwin theory enables its moment of inertia to be calculated. This can be compared in the case of a planet with the moments of inertia computed according to various assumptions regarding the distribution of shells of differing densities within the spheroid.

inertia of Jupiter and Saturn, and hence an indication of their internal constitution. From his calculations he confirmed Darwin's inference that Saturn and Jupiter must be more condensed towards the centre than the Earth is, and found in the case of Saturn that the surface density must be less (probably much less) than 0·31. He then considered the densities in the liquid and solid states of some of the lightest known substances, taking into account the temperature found for Saturn by Menzel, and found that the only ones less than the maximum surface density for Saturn are those of hydrogen and helium, whose boiling points are −253° and −269°C respectively. Hence it was impossible to reconcile the data for Saturn with a wholly solid constitution and a large fraction of the planet must be gaseous.

He pointed out that Menzel's values for the observed temperatures of Saturn and Jupiter supported his conclusions against the traditional view of those planets being very hot even near the surface. With regard to the point that Menzel's values for the temperature were higher than could be explained by solar radiation alone, he showed that the discrepancy could not be explained by original heat, and that the only remaining possibility was that the heat is new heat generated from radioactive sources within 300km. of the surface, and radiated soon after it is generated.

Jeffreys went on to show that, in view of the low temperature found by Menzel, it would be possible for Titan to be composed of rock and ice, with possibly some atmosphere; the lighter satellites of Saturn may be a white porous rock, or even snow, as suggested by Hepburn. Jeffreys concluded from the available evidence that Saturn (and Jupiter) may be composed of a rocky core, surrounded by a thick layer of ice and covered by an atmosphere of a depth in the case of Saturn of about 23 per cent of the radius of the planet. The gases in the atmospheres, he thought, probably consist mainly of hydrogen, nitrogen, oxygen, helium and perhaps methane; the clouds might consist of particles of some substance such as solid carbon dioxide.

Jeffreys faces the 'lions', 1926

In 1926 January Dr. Jeffreys opened a discussion at a meeting of the B.A.A. by outlining his theory of the 'Physical Condition of the four Outer Planets'; great interest was shown and a lively discussion ensued (*B.A.A.J.*, vol. 36, p. 98). He said that he felt 'rather like Daniel intruding into a den of lions' in speaking on such a subject before such an audience—since he confessed that he had never seen any of the planets through a telescope of more than 2 inches aperture. He had been working on the theory of the origin of the solar system from a stream of gaseous matter bursting out from the Sun, and had calculated that it would only have taken the Earth about 10,000 years to radiate away in the form of heat all the energy it possessed so that for a long time the Earth's temperature must have been sustained by solar radiation, not by its own original heat. This surprising result had led him to make similar calculations for the greater planets. He then went on to outline the results stated in his papers of 1923 and 1924. In the course of his statement he explained that the observational measurement of the planetary temperatures by the water cell method consisted in letting the light from the planet fall on a series of solutions, each absorbing known parts of the spectrum; the residual radiation fell on a thermopile, and from the amount lost on the way it was possible to reconstruct the spectrum.

Since the theory of Jeffreys that the major planets have cold atmospheres and solid interiors is now universally accepted, the opposing arguments in the ensuing discussion are of scarcely more than academic interest. Phillips, referring to Jupiter, showed analogies to the Sun, emphasised the great atmospheric disturbances observed on the planet, claimed that Menzel's results supported the argument of internal heat, and said that the suggestion by Jeffreys that the residual heat was produced by radioactive substances at or near the surface would conflict with the idea of a concentration of mass towards the centre of the planet and a surface layer of low density. Phillips also cited the theory that the great planets had ejected the short-period comets associated with them.

Dr. Crommelin suggested that the two views might be reconciled by supposing the interior of Jupiter to be hot and volcanic though the surface was very cold, so that great eruptions might give rise to the atmospheric disturbances and to the occasional ejection of comets. Hepburn expressed agreement with the theory of Jeffreys as regards the physical conditions of Saturn's ring and satellites, but did not feel able to admit that the appearance of Saturn itself was consistent with icy conditions.

Jeffreys, replying, considered that the atmospheric movements and changes observed on Saturn and Jupiter were to be expected if the temperatures are maintained principally by solar radiation; this would give rise to a general circulation with a system of moving cyclones similar in its main features to those on the Earth.

Photographs by light of different colours (Wright)

In 1927 July a paper by W. H. Wright (*P.A.S.P.*, vol. 39, p. 231) stated that in spite of the current low declination of Saturn ($-18\frac{1}{2}°$) placing it unfavourably for observation from Mount Hamilton, a few fairly good monochromatic photographs had been obtained during that summer (see Plate XII, fig. 2). Differences in Saturn's appearance by light of different colours are less striking than those found for Mars and Jupiter, the most interesting relating to the clear visibility of the planet through the crepe ring in the ultra-violet photographs, whereas in the red and infra-red images there is a broad, dark margin between the edge of the ring and the visible part of the globe. This indicates, very surprisingly, that the crepe ring is highly transparent to ultra-violet light and relatively opaque to infra-red. The darkness of the margin increases progressively as the colour used is changed through the range of colours from ultra-violet to infra-red.

In the tropical and temperate zones limb darkening is very pronounced with long waved colours, the limbs being bright in ultra-violet; this is also found with Mars and Jupiter. Limb darkening in longer waved colours is very pronounced in the vicinity of the (north) pole, and causes an appearance of obscuration on the widely open ring adjoining the pole. In the ultra-violet photographs limb-light is not seen near the pole.

Differences in the aspect of the belts and zones in the various colours are not marked, except that the ultra-violet photographs show a broad, dark belt approximately covering the equatorial zone. Wood, who saw this to the north of the equator on photographs taken in 1915 (see Chapter 29), suggested that it might, in part at least, be the shadow of a hypothetical extension towards the planet of the crepe ring, and by 1927 it had moved to the south of the equator, as would be expected if it were associated with the crepe ring. But the crepe ring's shadow is practically absent from the ultra-violet photographs where the band is shown, and strong in the red ones where the band is not found.

These observations do not confirm Dr. V. M. Slipher's opinion (*P.A.S.P.*, vol. 39, p. 149) that Cassini's division is much less evident in blue than in yellow light; the clearness of the division seems (in Wright's view) to depend on the conditions under which the plate was taken rather than on the colour employed.

Hydrogen abundance and constitution of the giant
planets (Menzel)

Discussing in 1930 the low densities of Saturn and the other great planets, Jeffreys's theory and the temperature results found by himself, D. H. Menzel (*P.A.S.P.*, vol. 42, p. 228) suggested that, as hydrogen is many times more abundant than all other atoms combined in the outer layers of the Sun and stars, the low densities of the major planets could easily be accounted for if they were largely formed of hydrogen, as solid hydrogen has a density only 0·07 that of water. The planets' much higher present densities could be explained by the escape of hydrogen during the first few thousand years after they were formed; the total amount lost would depend on the surface gravity and the time taken to cool to a certain critical temperature below which the loss would become negligible. The lower density of Saturn than that of Jupiter might be due to more rapid cooling counteracting greater escape through smaller gravitational pull. Menzel worked out that the height of the homogeneous atmosphere on Saturn need only be 50 km. if the atmosphere were chiefly hydrogen, but considered that there would be nothing against the existence of a highly compressed atmosphere of almost indefinite extent.

CHAPTER 34

WILL HAY'S WHITE SPOT (1933), AND OTHER OBSERVATIONS, 1932–35

Although the great white spot of 1933 was the outstanding visual feature of the early 1930s, certain members of the B.A.A. Saturn Section had already the year before recommenced observing the planet.

Satellites and globe spaces, 1932–34 (Barker)

Robert Barker, using his 12·6-inch aperture Calver reflector, made a series of observations 1932–34, chiefly of the variability of the satellites (*B.A.A.J.*, vol. 43, p. 56; 44, p. 74; 45, p. 41). Like Antoniadi a few years earlier, Barker found in 1932 that the western space between globe and rings looked smaller than the eastern, but in the discussion at a B.A.A. meeting (*ibid.*, vol. 43, p. 40) Instructor-Captain Ainslie, then Director of the Section, suggested that the appearance might be an optical effect due to the globe's shadow on the rings being on one side of the globe, and possibly also to a slight phase effect on the globe. He said that micrometric measurement of the spaces by G. F. Kellaway had shown them to be of the same size to within a few hundredths of a second of arc.

Barker's observations suggested variations of brightness for Rhea, Dione and Tethys, and in 1934 he watched the decreasing light of Iapetus as it moved eastward, its path appearing almost a straight line, since the Earth was passing through the plane of Iapetus's orbit during that opposition.

North equatorial spots, 1932 (Butterton)

Saturn was still, in 1932, more favourably placed for observers in the southern hemisphere, and M. S. Butterton, using the 9⅓-inch aperture refractor of the Dominion Observatory at Wellington, New Zealand, under almost perfect conditions, made numerous observations of black patches in the north equatorial belt. He deduced a rotation period of $10^h 20^m$ (*B.A.A.J.*, vol. 43, p. 405). This seems to have been the first spot outbreak of the period.

Discovery of the white spot, 1933 August 3 (Hay)

It might be thought that by the 1930s the development of astrophysics and the achievements of the observatories with their great telescopes had left no place in the study of Saturn for the amateur with his small instrument. Nevertheless, one of the most important discoveries of the decade was made with a 6-inch aperture Cooke refractor by a keen amateur observer of the B.A.A. Saturn Section, W. T. Hay, much better known to the public as 'Will Hay', the popular comedian of stage, screen and radio. His accounts of the observation (*M.N.*, vol. 94, p. 85; *B.A.A.J.*, vol. 43, p. 426) state that when observing Saturn at Norbury on 1933 August 3, $22^h 35^m$ U.T., he detected a large white patch on the equatorial zone. The patch or spot was oval in shape, and about one-fifth of the planet's diameter in length, both ends being well defined: (Plate XIII, Fig. 2 reproduces Hay's discovery drawing). Hay reported the find by telephone to Dr. W. H. Steavenson, who confirmed it a few minutes later. The spot was also observed at the same hour on the same evening by A. Weber at Berlin, but Will Hay was the first to announce the discovery.

Will Hay described how the spot, on the days after discovery, lengthened rapidly in the preceding direction, and from August 6 onward occupied the whole width of the zone in latitude, even at one time encroaching on the north equatorial belt. The following end remained moderately well defined for several weeks, but the preceding end became very indistinct a few days after discovery: this made the position of the spot's centre hard to determine for estimating times of transit over the central meridian. Hay found on August 9 that the spot took 51 minutes to cross the c.m., and that the part of the zone immediately following the spot was as dark as the north equatorial belt, and gave the impression that the dark belt extended right up to the crepe ring. He suggested that, 'following on some disturbance beneath the bright matter which forms the equatorial zone, there was a sudden rush or concentration of some of this bright matter towards a centre, uncovering, at each end of the concentration, the dark matter which possibly lies at a lower altitude'. He conjectured that the subsequent lengthening of the spot might be due to the subsidence of this concentrated bright matter to its normal level, and that this occurred at the preceding end first, perhaps followed by a general slide of the whole spot in the same direction. By September 13 he found that it had spread out so much that it could no longer be called a spot, but that it still seemed to be rather brighter than the rest of the equatorial zone.

Further observations of the white spot
(Steavenson, Waterfield, Phillips, Butterton, Lower)

A great deal of information about the striking changes in the physical aspect of the spot is provided by the observing notes of members of the B.A.A. Saturn Section, quoted in the Section's Interim Report (*B.A.A.J.*, vol. 44, p. 220) by B. M. Peek, at that time Director. He expressed the opinion that the beginning of the outburst was quite sudden; this is borne out by a statement of E. C. Slipher (mentioned in *B.A.A.J.*, vol 44, p. 401) that the surface of the planet had been closely examined at Lowell Observatory less than 48 hours before the spot's discovery, and that no indication of its presence had then been found. Peek also deduced, from subsequent observations, that at the time of discovery it was at about its maximum intensity. He mentioned too that subsequent but independent discoveries of the white spot were made on August 5 by J. E. Willis at Washington, and on August 7 at Wellington by I. L. Thomsen, a New Zealand member of the B.A.A.

Dr. W. H. Steavenson's impressions of the object in its early stages were in substantial agreement with those of W. T. Hay already recounted, but Steavenson gave some additional information, e.g. that on August 3 it was elliptical, of the same proportions as Jupiter's Red Spot, rather soft in outline though well defined, and not much brighter than the normal equatorial zone, but the zone was elsewhere much duskier than usual and faintly mottled or patchy following the spot. By August 6 the white patch's p.–f. diameter was about one-quarter that of Saturn, and three days later between one-quarter and one-third. On that date, August 9, it tapered with a southerly trend at the p. end, fading imperceptibly into the zone. On August 18 Steavenson found that the f. end was becoming less well defined and showing a tapering extension, this development being intensified by August 21.

On August 24 the Rev. T. E. R. Phillips considered that the spot had become a long white band, followed by a vague ligament crossing the zone, then another bright region tapering off in the following direction. Steavenson noted that the spot was stretching about halfway round the planet in longitude and tending to become mottled or break up into separate sections, while the rest of the zone seemed to be recovering its normal brightness. On the same date Dr. R. L. Waterfield noted that the north and south equatorial belts seemed to dip towards each other and to be joined by a fine ligament; following this neck in it the spot tapered rapidly like a funnel. On August 29 Waterfield found that an elliptic bright nucleus seemed to have developed in it preceding the f. end, this nucleus being about as large as the spot had originally been, and its centre possibly corresponding with the centre of the original object.

Phillips pointed out that it was only in the early days that satisfactory transit observations of the spot's centre were possible: it grew rapidly, apparently in the p. direction, and completely changed in character. Since mid-August the p. end had become too vague to fix, so he concentrated on observing the f. end, which, though less well defined than at first, could still be observed by September 12.

M. S. Butterton, observing in New Zealand, had the advantage of seeing the planet at a higher altitude in the sky, and his general comment, besides being a useful confirmation of the impressions of the observers in England, contains some additional information: 'It was soon seen that the activity on Saturn was not limited to a single spot, and . . . three other spots, besides bright streaks, have been observed. Observations . . . made . . . on numerous occasions . . . show that great changes were taking place in the equatorial zone . . . On August 18 and 19 the spot was well seen with the 9-inch refractor, and appeared as elliptical, bright, yellow, and large, nearly filling the whole of the equatorial zone. On August 30 it had changed slightly, and a smaller yellowish spot preceded the main one across the central meridian by nearly one hour and a half. The spot had changed on September 8, and finally, on September 29, it appeared changed out of recognition, and also developing a long bright streak on the following side, which extended for a considerable distance along the equatorial zone'.

Butterton's statement about the additional spots is in agreement with that of Dr. E. C. Slipher who pointed out that the original patch was only one of a series which developed in the zone along almost the entire perimeter (*B.A.A.J.*, vol. 44, p. 401). The Flagstaff observers, in fact, announced one of these, a white spot, on August 29, and it was subsequently observed, and its transits were recorded, by several members of the Saturn Section between that date and September 14.

The observations of Hay's white spot on seven dates from 1933 August 7 to 16 by Charles A. and Harold A. Lower, of San Diego, California (reported in *P.A.S.P.*, vol. 45, p. 243) gave some interesting particulars. Using mainly a 12-inch aperture Cassegrain reflector, they found the spot creamy white, oval and much brighter than the rest of Saturn. They estimated its rotation period at about $10^h 13^m$, and its size when first seen at about 12,000 miles long by 8,000 miles wide (north to south), thus making its shape rounder than that of Jupiter's Red Spot, to the proportions of which Steavenson had compared it. The Lowers noted its rapid increase in length—

378

to about 25,000 miles when last seen (August 16)—whereas they recorded that its width increased only slightly.

Photography of the white spot with colour filters (Wright)

Dr. W. H. Wright made a report (in *P.A.S.P.*, vol. 45, p. 236) on the photography of Saturn and Hay's white spot with colour filters carried out at the Lick Observatory from 1933 August 6 to September 14; he also mentioned that it was sketched by Dr. G. P. Kuiper on August 5. The telescope used was the 36-inch aperture Crossley reflector. The first photographs showed the spot more or less round in violet and ultra-violet light, but elongated in yellow. Two days later the length as shown in red, yellow and violet had increased, and an effect suggestive of a 'smearing' of the spot in the p. direction was noted; it merged into the zone at the p. end, while terminating abruptly at the f. end. This lengthening and combing out of the spot on the p. side persisted, leading to its disintegration and a building up of the zone at its expense. Wright compared the smear to 'a trail of drifting smoke in a steady wind'. By mid-September the spot had faded and seemed broken and curdled; in one photograph it looked like little more than the ragged f. end of a bright section of the zone.

Wright pointed out that Hall's spot of 1876 must have behaved similarly, because Hall wrote that towards the latter part of the observations it had become drawn out on the p. side into a bright belt; also that Barnard's spot of 1903 lengthened during the period of observation (see Chapters 15 and 24).

Wright considered that the behaviour of Hay's spot suggested an eruptive nature, and that the forward brushing of zone material seemed to imply an overrunning wind in the direction of the rotation. Hence there seemed to be something in Saturn's atmosphere moving even faster than spots such as this one from which the high equatorial speed was deduced. The spot's colour reaction suggested that it must be at least an appreciable depth below the atmospheric surface: this was deduced from the fact that it could be observed by ultra-violet only when comparatively near the central meridian—since ultra-violet light has low penetrative power, anything at a depth would, if viewed obliquely, be observed through a considerable thickness of atmosphere which ultra-violet would not penetrate. Yet the fact that the spot was visible by ultra-violet, while the zone was not and appeared dark, suggested that the visible part of the spot was above the general level of the zone material.

Wright also mentioned a point of interest regarding spots and their rotation periods in general, namely, that a comparison of ascertained rotation periods showed that a spot on the equator of Saturn gains on one in the north temperate region at approximately 21,000 miles per day; if speed decrement with latitude were uniform, the result would be nearly a mile of longitudinal drift for every mile of difference in latitude, but of course this might be much modified, as spots may have considerable proper motions.

Rotation periods of Hay's spot and the Flagstaff spot (Rowland, Peek)

A paper (in *M.N.*, vol. 94, p. 86) by the Rev. J. P. Rowland, S.J., gave a full account of the method followed and the precautions taken at Stonyhurst College Observatory to get accurate transits of Hay's spot by filar micrometer. From the Stonyhurst observations August 23 to September 16 he worked out a mean rotation period for the spot of 10^h 14^m12. He also gave a list of 30 observations by more than a dozen observers in various countries extending from 1933 August 3 to September 16. He stated that, owing to its rapid change in character during its early life, the spot made the determination of a consistent rotation period difficult: the period appeared to shorten from about 10^h 16^m in the first days of observation to about 10^h 13^m on September 10. The mean value for August 6–18 was 10^h 15^m4, and from August 9 to September 10 was 10^h 14^m25, with a sudden discontinuity about August 23.

He suggested that the rapid extension of the brightness of the zone and the acceleration of the spot might be due to a rise of the latter by about August 23 to a higher level in the planet's atmosphere. He also pointed out that a little confusion was caused by the subsequent appearance of a number of other spots in the equatorial zone: some of the transit times given by observers evidently related to subsidiary spots rather than the original one.

These difficulties and others were explained in Peek's very detailed discussion of the observations in the Section's Interim Report, which listed a very large number of observations by some two dozen different observers on about that number of dates in 1933 August and September, with diagram (see Figure 21). In order to try to make the observations fit he suggested three possible alternative solutions, and he seems to have reached a similar conclusion to that of Father Rowland for the motion of the spot as a whole prior to September 10. Even its following end which had been much observed and seemed at first sight to provide a reliable basis for a rotation period, became confused between August 21 and 24, the grey material which indicated its position dividing into two portions with

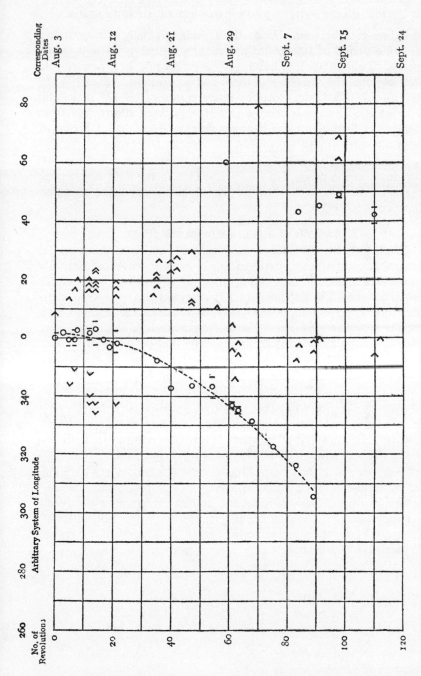

FIGURE 21. Will Hay's white spot 1933: B. M. Peek's chart showing the movement of the spot and its ends in longitude. The object on the right of the chart is the Flagstaff spot. (From *B.A.A.J.*, vol. 44, p. 231.)

a lighter area between. Hence the f. end, originally having a slow rotation period of 10^h 16^m or more, soon settled down to a steady period of about 10^h 15^m, but shortly after August 20 it divided into two objects. The following of these had a period exceeding 10^h 15^m but quickly faded; the preceding one a period of 10^h 13^m or less till August 30. There was then a gap in the observations, but from September 8 to 20 an object resembling the latter had a period of approximately 10^h 15^m.

The Flagstaff object followed the original spot by some 80° longitude at the time of its discovery; though the data were rather meagre to determine a rotation period, the mean value seems to have been 10^h 14^{m}4.

Latitude of north equatorial belt (Peek)

In 1934 B. M. Peek made measurements on eight dates of the saturnicentric latitudes of the north and south edges of Saturn's prominent northern belt, using a Cooke position micrometer mounted on his 12$\frac{1}{4}$-inch aperture reflector (*B.A.A.J.*, vol. 45, p. 121). The Rev. T. E. R. Phillips also made two micrometer readings with an 8-inch aperture Cooke refractor. The mean results, given below, were in satisfactory agreement considering that the edges of the belt are not sharply defined:

	Peek	Phillips
North edge	+30°7	+29°5
South edge	+16°4	+16°1

Peek pointed out that the situation of this belt on Saturn differs considerably from that of Jupiter's N.E.B., the zenocentric latitudes of the edges of which are +15°2 and +6°3 respectively.

Peek (*The Planet Jupiter*, p. 270, fig. 13a) has shown in a diagram the difference between zenocentric and zenographical latitude; for Saturn corresponding saturnicentric and saturnigraphical latitudes can be calculated, allowing for the somewhat greater oblateness of Saturn. (Crommelin's formulae for Saturnian latitudes are given in Appendix II of *this* book.)

CHAPTER 35

METHANE IN SATURN'S ATMOSPHERE, AND OTHER ASTROPHYSICAL STUDIES IN THE 1930s

---

Satellite orbits and perturbations (G. Struve)

In 1931 appeared a paper (*P.A.S.P.*, vol. 43, p. 377) by Georg Struve, Astronomer of the University Observatory of Berlin-Babelsberg, reviewing the progress made in the previous 50 years in knowledge of the motions of Saturn's satellites from the time when Hermann Struve began his fundamental researches, since continued at Königsberg, Berlin and Babelsberg. He stated that all this work, with American and (more recently) South African contributions had built up a large and very homogeneous collection of material for the investigation of the theory of the satellites of Saturn, but he pointed out that no final solution can ever be reached, further refinements always giving rise to more and more problems, and that in the theory of perturbations only successive approximations are possible.

Herschel's observations of the conjunctions of satellites with the ends of the major axis of the rings had been the first that could be used in part for determining the longitudes of satellites; other series of measures by Bessel, Bond, Lassell and Marth, Asaph Hall and others were mostly estimated measures of conjunctions or elongations and only to a small extent micrometer measures, and until the 1880s there was no real theory of these satellites. H. Struve made the improvement (1884–92) of connecting the satellites micrometrically with one another. His measures had a probable error of only $\pm 0''06$, compared with $\pm 0''84$ of Herschel a century before. It was very important to get a proper distribution of observations over different parts of the orbit and to secure as far as possible the elimination of systematic observational and instrumental errors.

The systematic series of measures made at Pulkovo produced the first highly accurate satellite orbits and made it possible for H. Struve to investigate the secular and periodic variations of their elements. This led to the discovery of laws of motion and of libration–perturbations existing in pairs of satellites. Such perturbations are generally of long period: a close commensurability in the mean motions of two bodies produces variations within narrow limits

383

repeating themselves precisely after definite time intervals. G. Struve stated that the close commensurability in the case of Saturn's satellites appears not simply in the mean motions of the satellites, but only when the secular motions of the nodes and lines of apsides are also taken into account.

He gave as examples H. Struve's laws of motion for Mimas and Tethys, and for Enceladus and Dione, as follows: (a) 'The conjunctions of Mimas and Tethys always take place at the point on the equatorial plane of Saturn which is midway between the ascending nodes of their orbits: their departure from this point may at most amount to but 49° and the libration is completed in 71 years.' (b) 'The conjunctions of Enceladus and Dione always coincide with the perisaturnium of Enceladus, or at least must oscillate about that point.'

He went on to say that the dynamic masses of the four inner satellites could be determined very accurately in terms of Saturn's mass from the inclinations and eccentricities of the orbits which can be derived from observations. The mass of Titan is known accurately from the libration–perturbations of the system Hyperion–Titan. The mass of Rhea can be derived from the theory of secular motions, but only with a very large uncertainty; there was a similar uncertainty about the mass of Iapetus, and for that of Hyperion only rough approximations were yet possible.

He gave a table of magnitude, mass, density and diameter which showed a rapid increase in density from the inner to the outer satellites (as shown in Chapter 40). He stated that the results of the table suggested that the innermost satellites may have been formed by a gradual condensation and concentration of the outermost ring particles. This is interesting as being the antithesis of Roche's theory of the formation of the rings by the break-up of a former fluid satellite that came too near Saturn and was consequently disrupted by tidal forces.

G. Struve concluded his paper by giving examples of eccentricities of satellite orbits which have been forced by libration–perturbations. The eccentricity of the orbit of Enceladus is not free but largely forced by the mass of Dione. The mass of Titan (more than sixty times that of Rhea) and Titan's highly eccentric orbit have caused a forced eccentricity for Rhea, whose perisaturnium nearly coincides with that of Titan, oscillating about it in a period of 38 years (as mentioned in Chapter 23). G. Struve's paper shows what complexities have been revealed in the motions of Saturn's satellites by the

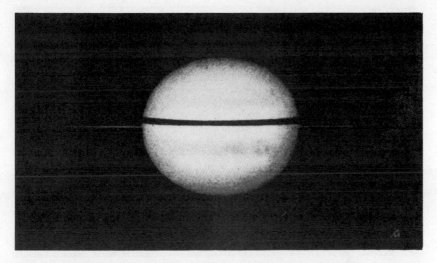

Fig. 1.—1936 July 2, soon after appulse of Earth to ring-plane. Note the greater apparent length of the p. ansa than the f. ansa which was remarked by several observers.

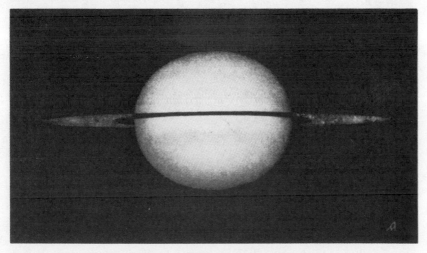

Fig. 2.—1936 December 19.

PLATE XV.—Drawings of the almost edgewise ring by Antoniadi, 33-inch refractor of Meudon Observatory (from *B.A.A.J.* vol. 47, frontispiece, figs. 1, 2).

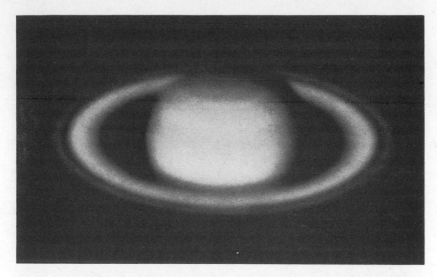

FIG. 1.—By H. Camichel, Pic-du-Midi Observatory, 1942 September 19.

FIG. 2.—By G. H. Herbig, 1943 December 14 (Lick Observatory 36-inch refractor)—a striking example of the 'peaked' shadow on the rings (from *B.A.A.J.* vol. 55, plate facing p. 125).

PLATE XVI.—Photographs of Saturn.

refined methods of observation and research carried out at the great observatories.

Discovery (1932) of methane and some ammonia in Saturn's atmosphere (V. M. Slipher, Wildt, Dunham)

Brief accounts of the wonderful discovery through the spectrum of some of the principal constituents of the atmosphere of Saturn (and of the other major planets) were given in 1933 and 1934 by Theodore Dunham, Junr., of Mount Wilson Observatory (*P.A.S.P.*, vol. 45, pp. 42, 202; vol. 46, p. 231; and summary of first paper in *B.A.A.J.*, vol. 43, p. 307). Dunham's first paper referred to the photography of the spectra of Saturn and Jupiter in the visible red region by Dr. V. M. Slipher in 1905 at the Lowell Observatory (see Chapter 25), the band $\lambda6450$–6507 showing evidence of structure and being much weaker in Saturn's spectrum; also to the discovery by Slipher in 1909 of strong bands near $\lambda\lambda7190$ and 7260 in the infra-red part of the spectra of those planets. Slipher however, had been unable to identify the bands.

The next, and crucial step, was taken by Dr. Rupert Wildt (now Professor Wildt of Yale University Observatory) who, in 1932, drew attention to the agreement between six of Slipher's lines in the red region with some of the lines that had been measured in the absorption spectrum of ammonia. Wildt also obtained objective prism spectra of Jupiter and Uranus, and suggested that the bands near $\lambda\lambda5428$, 6191, 7260 and 8780–9110 may be due principally to methane. (Veröff. der Universitäts–Sternwarte zu Göttingen, Heft. 22, 1932.)

An effort was therefore made at Mount Wilson Observatory in 1932 to secure a conclusive proof by obtaining spectra of Saturn and Jupiter on a scale sufficient to show the structure of the bands. The 100-inch aperture reflector was used, with plane-grating Littrow spectrographs at its Coudé focus, and the spectra photographed in the regions $\lambda\lambda6400$–6500 and $\lambda\lambda7800$–7950. For comparison an absorption tube 20 metres long was built, filled in turn with ammonia and methane at atmospheric pressure, light from a ribbon filament lamp being passed twice through the tube and focused on the slit of a spectrograph, the grating of which was the one used in the planetary exposures; hence an absorption spectrum of 40 metres of each gas was obtained. A striking similarity in the structure of the bands was evident, though, probably owing to differences in temperature, the relative intensities of lines in the planetary and laboratory spectra were different, suggesting an independent method of determining planetary temperatures. In the two regions 69 ammonia lines were identified in Jupiter's spectrum, but in that of Saturn the ammonia bands were much weaker. The identification of methane was established with certainty in the spectra of both Saturn and Jupiter by the agreement in position of eighteen lines in the laboratory spectrum with lines in the planetary band near λ8640. Bands of the two gases were found to overlap near λ7900.

Dr. Dunham commented that, in view of the high surface gravity, Saturn (and the other major planets) would be expected to have retained an excess of hydrogen, and that no known substance except hydrogen could exist at the high pressures below the surfaces of those planets with a density low enough to explain their mean densities. As the planets cooled, carbon, nitrogen and oxygen must have combined with part of the hydrogen. Of the hydrogen compounds only the most volatile, methane, would be expected to show in the atmospheres; ammonia, more easily condensed, is scarcely detectable except on Jupiter; all free oxygen must have combined with hydrogen and fallen below the clouds as ice. Therefore the atmospheres of the major planets turned out to be almost exactly as expected from theories of cosmogony and knowledge of physical chemistry.

In a lecture in 1934 December on the atmospheres of the planets (summarised in *B.A.A.J.*, vol. 45, p. 215) Professor H. N. Russell pointed out that the presence of neither hydrogen nor inert gases such as nitrogen and helium could be shown by planetary spectra; that methane, equivalent in Jupiter to one mile-atmosphere, was shown by the spectra to be more plentiful on Saturn, and more still on Uranus and Neptune; and that, as Saturn is at a lower temperature than Jupiter, nearly all the ammonia must be frozen out of Saturn's atmosphere, which would consist mainly of methane and hydrogen.

Plate XIV shows in the spectrum of Saturn some of the most intense bands found in 1905 (see Chapter 25) and 1909 by Dr. V. M. Slipher and identified as due to methane by Wildt: λ5428 in the green, λλ6145–6191 in the orange, and λ7260 in the red; λλ8780–9110 are in the infra-red outside this photograph.

Various other lines and bands in the Saturn spectrum should be discounted, as not being produced by Saturn's atmosphere: *A*, *B* in the red are due to oxygen and *a* to water vapour in the terrestrial atmosphere; *D*, *E* and *b* in the yellow and green are due to sodium, iron, calcium and magnesium in the solar atmosphere, while *C* in the orange–red and *F* in the blue are the well-known hydrogen α and β lines also produced by the solar atmosphere, and of course reflected by Saturn.

[Plate XVI, fig. 1 in Peek's *The Planet Jupiter* shows comparison spectra of Saturn, Jupiter and ammonia gas in the near infra-red, the ammonia lines, so clear in Jupiter's spectrum, are faint or lacking in that of Saturn.]

Supposed eccentricity of the rings (Stroobant)

In two papers (in *L'Astronomie*, 1934, pp. 57, 121) P. Stroobant, Director of the Belgian Royal Observatory, gave a detailed history of the observations relating to the suspected eccentricity of the position of Saturn's globe in relation to the rings, and a lengthy analysis of the micrometric measures made at various dates by leading astronomers. He found the evidence rather contradictory, and concluded that the micrometric measures carried out in the best conditions by observers (e.g. Barnard, H. Struve, Lowell) using powerful instruments show that the *mean* eccentricity of the various edges of the rings is extremely slight (a few hundredths of a second of arc). In his opinion the apparent eccentricity may arise from the asymmetry shown by the images of the planet, especially the shadow cast by the globe on the ring system. He also considered that the eccentricity does not show a short-period variation, but its changes could be attributed to the theory of the rotation of the apsides of the orbits of the ring particles, due to the perturbing force of the planet's equatorial bulge, when the perisaturnia of the particles are found in the same direction with reference to the centre of Saturn.

Rotation of the satellites (Antoniadi)

In a paper (*B.A.A.J.*, vol. 45, p. 195) E. M. Antoniadi pointed out that J. D. Cassini, in 1705, had suggested that it might ultimately be found that it was a property of satellites to have rotation periods approximately equal to their revolution periods round their principal planets, and that Laplace had expressed the belief, a century later, that equality of duration of rotation and revolution appeared to be a general law of the movement of satellites. More recently, in his book *The Earth* (1924, p. 234), Dr. Jeffreys had stated that the rate of change of velocity of rotation in a satellite is proportional to the fourth power of its mean motion, and that it would therefore be expected that all satellites with periods less than that of the Moon would turn the same faces permanently toward their primaries. Jeffreys had added that, although satellites with longer periods may not yet have reached that state, all satellites of known rotation period, including Iapetus, do actually keep the same faces to their primaries.

Having fully discussed the probable reasons for the cautious wording of the pronouncement of Jeffreys, Antoniadi, relying on the law of inverse sixth powers of the distance, deduced the following principle: 'The tides of Saturn on Iapetus having annulled the rotation of this satellite at 62 mean radii of the planet, all the other satellites which are inside this distance from their planets must . . . have durations of rotation equal to those of their revolution.'

Antoniadi also considered the cases of the minute and very distant satellites, such as Phoebe (in the case of the Saturnian system), and concluded that at their enormous distances the tidal drag is far too weak to annul the rotation.

Fading of rings under low illumination (Hargreaves, Peek)

A very interesting discussion took place at the 1936 December meeting of the B.A.A. (*B.A.A.J.*, vol. 47, p. 102) on a question raised by F. J. Hargreaves, a leading amateur observer of Jupiter and other planets, as to why the diminution of the angle between the line of sight and the ring-plane (during the approach to the edgewise position) should cause a very large decrease in the apparent surface brightness of the rings, in view of the fact that the rings are made up of discrete particles. Dr. W. H. Steavenson said that the difficulty was a very real one, because, while theory and the known transparency of the rings suggested a discrete structure, there was no doubt that the observed fading was of a kind that would be produced by a smooth, unbroken sheet of matter.

B. M. Peek offered a qualitative explanation. As the ring is partly opaque, it could be assumed that the Sun's illumination would penetrate to a certain effective depth, and that the penetrating power of the eye would reach all particles down to a similar depth. Under ordinary circumstances all the particles that can be seen are effectively illuminated, but when the elevation of the Sun above the ring-plane is small, and less than that of the Earth, the effective illumination will not penetrate the entire thickness of the ring, whereas, Peek suggested, effective vision will extend to a greater depth than the illumination, embracing numbers of particles not effectively illuminated, so that the surface brightness of the ring will appear reduced.

Shortly afterwards Hargreaves wrote a short paper (*B.A.A.J.*, vol. 47, p. 185) suggesting that the decrease in brightness of the rings as the Sun approaches the ring-plane must apparently be due to eclipses of some particles by others, and showing (with diagrams) that the configuration that best satisfied the requirements was that of a single layer of spherical bodies of appreciable diameter and of equal size.

Spectroscopic observations of Saturn's rotation (J. H. Moore)

In 1939 Dr. J. H. Moore (Director of Lick Observatory) gave an account (*P.A.S.P.*, vol. 51, p. 274) of his work in 1936 and 1937 when the ring was nearly edgewise, using a 3-prism Mills spectrograph attached to the 36-inch aperture Lick refractor, and obtaining spectrograms with the slit at the latitudes of the equator of Saturn and of about 27°, 42° and 57°. By measuring the slope of the lines he obtained a comparison between the rotation periods at those latitudes.

He calculated that the rotation period relative to that at the equator is 1·06 at latitude 27°, 1·08 at 42° and 1·11 at 57°, confirming that the period is longer at 36° than at 0°, but also suggesting an even longer period at still higher latitudes. He considered that the values he had obtained had a possible error of 1 or 2 per cent, the results at 57° being the least reliable as the spectrum there was only half the width of that at the equator, and the effects of poor seeing, imperfect guiding and unequal illumination of the slit consequently would become serious. His result for the actual rotation period at the equator from the spectrograms was $10^h 2^m \pm 4^m$, which is nearly 2 per cent shorter than that given by observations of spots.

Atmosphere and temperature (Dunham)

Of Theodore Dunham's excellent review entitled 'Knowledge of the Planets in 1938' (*P.A.S.P.*, vol. 51, p. 233) only a few items need be stated here, since most of the information relating to Saturn has already been given in this and previous chapters. He mentioned that in spite of all the recent advances in astronomy, it was still (in 1938) true that a good eye with a telescope of moderate size could detect finer detail on the planets than any photographic process. In relation to photography with colour filters, he said that E. C. Slipher noted a marked change in the colour of Saturn's equatorial region between 1927 and 1929: during that interval the surface became so red that blue photographs were dark compared with those made with yellow light. In 1930 the two kinds of photograph were of almost equal density.

In the course of his account of the use of vacuum thermocouples attached to telescopes of large aperture for the measurement of planetary temperatures, he pointed out that great difficulties were encountered in making such measurements of the visible layers of

Saturn and the other major planets, because at their very low temperatures the thermoelectric current to be measured is very small, and even more serious is the blocking of most of the radiation by the terrestrial atmosphere and the absorption by methane in the planetary atmospheres. Nevertheless, from data obtained by Coblentz and Lampland, Menzel had computed temperatures for Jupiter and Saturn of $-135°$ and $-150°C$. respectively, the figure for Jupiter agreeing closely with that deduced from spectroscopic determinations of the amount of gaseous ammonia. The figures $-150°$ for Saturn and $-185°$ or lower for Uranus agreed reasonably well with what would be expected from solar radiation falling on them, and there was no evidence that internal heat makes any important contribution.

Dealing with knowledge of planetary atmospheres obtained from spectra, Dunham pointed out that Saturn and the other major planets present quite a different problem from the terrestrial planets. With the latter, one has to search for bands of vanishing intensity, while with Saturn and its neighbours, it is a question of identifying bands of such tremendous absorbing power for red light that they completely alter the colour of the outer planets, causing the greenish tinge of Uranus and Neptune. After a certain number of the spectral lines and bands had been identified with ammonia and others with methane, there still remained many unidentified, but Slipher and Adel had recently shown that methane accounts quite satisfactorily for all the other bands so far observed.

Model of Saturn (Wildt)

Geophysical theories about the internal structure and core of the Earth, and the low mean densities that had been calculated for the major planets (extraordinarily low in the case of Saturn) led Dr. R. Wildt in 1938 (*Ap. J.*, vol. 87, p. 508) to propose models of the major planets which for some years found wide acceptance by astronomers. He assumed that the various components of the originally gaseous mixture separated under gravity in a way similar to that which was supposed to have formed an iron core in the Earth. He presumed each major planet to have a central core of metallic and rocky substances with an estimated mean density of 6 grams per cubic centimetre, a little greater than the Earth's mean density, surrounded by a thick layer of 'ice', consisting of frozen water, carbon dioxide, and other gases that condense easily. This layer would have a mean density of about 1·5 g./cm.^{3}. The outermost layer would consist of highly compressed permanent gases, the chief constituent being solid hydrogen; this layer's mean density was assumed to be 0·3 g./cm.^{3}. The attractiveness of this model is that from the known mean density and moment of inertia of each major planet the thickness of the various layers could be calculated. In the case of Saturn the dimensions according to this model would appear to be: diameter of solid core 28,000 miles; thickness of 'ice' layer 6,000 miles; depth of atmosphere 16,000 miles. Similar calculations for Jupiter, Uranus and Neptune appear to make the depth of atmosphere in their cases 8,000, 3,000 and 2,000 miles respectively.

Unfortunately, as Dr. W. H. Ramsey pointed out in 1951 (*M.N.*, vol. 111, p. 431) there were fatal flaws in Wildt's proposed model; the excessive amount of free hydrogen it would give Saturn compared with the other major planets, and the too low densities assumed for the core and outer layers. Wildt did, however, make the valuable suggestion, also made independently in the same year by D. S. Kothari (*Proc. Roy. Soc.*, *A*, vol. 165, p. 486) that hydrogen may be metallic in the interiors of the major planets, which was confirmed in 1946 by the detailed calculations of R. Kronig and others (*Physica*, vol. 12, p. 245). This suggestion and these calculations were the basis of Ramsey's new theory of the constitutions of the major planets (see Chapter 40).

CHAPTER 36

APPULSE OF THE EARTH TO THE RING-PLANE 1936 JUNE 28, AND OTHER OBSERVATIONS 1936–40

Prediction of appulse (Comrie and Peek)

In 1936 May a paper by L. J. Comrie and B. M. Peek (*B.A.A.J.*, vol. 46, p. 261) showed that, according to H. Struve's elements of the ring-plane which had been used in the *N.A.* prior to 1936, the Earth would have been expected to make passages of the ring-plane on 1936 June 23 and July 3. From 1936, however, the elements used by the *N.A.* were the revised ones by G. Struve, and the figures of the Earth's elevation above the ring-plane derived from these did not indicate any change of sign (from north to south): the elevation north of the plane declined to a minimum on June 28 and 29, and thereafter increased again. To determine whether or not there would be a change of sign for a few hours on June 28, Comrie and Peek had recomputed the values to a higher order of accuracy, and they thus found the minimum to be $+0°0001$ on June 28 at 18^h. This meant that the Earth would enter the ring-plane from the north, remain in it for a few hours, and then recede from it to the north without passing through.

Comrie and Peek pointed out that when the Earth's elevation above the ring-plane is less than $0°001$ there should be no sensible contribution to the light of the ring from the fact that the elevation is not exactly zero, and the ring should disappear entirely; also that the fact that the Earth would approach and leave the ring-plane from the side on which the Sun also lay, should provide observational data of exceptional interest.

Observations around date of appulse (Coleman and Knight, J. H. Moore, F. O'B. Ellison)

An observation which was noteworthy as showing what could be seen with a small telescope on the date of the predicted appulse was made at Rye on 1936 June 28 at 2.30 a.m. by A. Coleman and John Knight (*B.A.A.J.*, vol. 47, p. 30). A 5-inch aperture refractor with an excellent object-glass was used (presumably the same instrument with which Knight had observed the passage of a star behind ring A in 1917—see Chapter 30). Seeing was absolutely steady and definition perfect. The ring was visible throughout its length, 'as a fine, sharp-cut line' across the disk, and 'on each side in the form of a slender line of beads, resembling a dew-besprinkled spider's web seen in sunlight'. The powers used were 150 and 80. The visibility of the ring with such a small aperture and the peculiar beaded appearance were both unexpected by the observers.

Several observations made with a large telescope, the 36-inch aperture Lick refractor, powers 270 to 500, were reported by Dr. J. H. Moore (*P.A.S.P.*, vol. 48, p. 225). On June 28·5, 29·5, 30·5 and July 1·5 he found the rings easily visible as a fine line to east and west of Saturn and immediately north of the alignment of the shadow of the rings on the planet; this shadow crossed south of the centre, the Sun's elevation being 2°5 north of the ring-plane. He estimated the width of the fine bright line of the ansae at less than 0″1 on June 29 and 30; it seemed slightly wider on June 28 and July 1, but this was uncertain owing to poorer seeing.

A report by Dr. F. O'B. Ellison (*B.A.A.J.*, vol. 50, p. 213) is of particular interest because he was able to observe Saturn at 1936 June 28ᵈ·85 U.T., only 1¾ hours after the time of the Earth's minimum elevation above the ring-plane; he was at Colombo, Ceylon, and at the actual time of the Earth's nearest approach to the plane, Saturn was above the horizon only in countries farther east, such as the E. Indies, Japan, Australia and New Zealand. He found the ring visible in his 12-inch aperture reflector as a very thin thread of light; it was quite conspicuous, extending equally on both sides of the disk, 'The shadow of the ring appeared as a very black line S. of the ring where it crossed the ball.' He could see the six inner satellites, Enceladus preceding the ring and Mimas on the following side intersected by the ring, 'looking like a diamond on a thread of silver'. Bad weather prevented him from making further observations until the autumn.

Difference in apparent length of ansae, 1936 June–July
(Peek, Quénisset, Antoniadi)

In the 1936 report of the B.A.A. Saturn Section, the Rev. Theodore E. R. Phillips, then Director of the Section, called attention to 'an extraordinary phenomenon . . . extremely difficult to account for in the case of a ring rotating like that of Saturn', namely, . . . 'the superior visibility reported by various observers of the p. ansa as compared with the f. ansa near the time of the Earth's close appulse to the ring-plane'. (*B.A.A.J.*, vol. 46, p. 362.) Photographs taken by

F. Quénisset at Juvisy on June 21 and July 17 showed the ring more extended on the p. than on the f. side. Peek found the p. ansa the more conspicuous on June 23 and 25, and also when next observed on July 4; in fact he was doubtful whether the f. ansa was seen at all on June 25. On 1936 July 2, E. M. Antoniadi, who had found the ring invisible half an hour earlier in a 12½-inch aperture refractor, saw it quite easily with the Meudon 33-inch refractor, power 650, and described it in a report on his observations (*B.A.A.J.*, vol. 47, p. 252) as 'extremely thin, like a spider's web, uniform and without knots, and much shorter to E. than W.' (see Plate XV, Fig. 1). This difference in the apparent length of the ansae was noted by H. Camichel, observing first, and confirmed by Antoniadi and Grenat. Antoniadi estimated the ring as showing its whole length of 2·3 radii to the west (preceding ansa), but as having about one-fifth of its length missing on the outside to the east. Phillips stated that Antoniadi had used reversing apparatus to eliminate any possible illusion that might arise from a physiological cause.

Other experienced observers, including W. H. Steavenson, P. M. Ryves and H. Tompkins, also found during that period that the preceding ansa appeared the brighter or more visible of the two (*B.A.A.J.*, vol. 50, pp. 230–1). On the other hand, as Phillips pointed out, this curious phenomenon does not seem to have been noted at Lick; Ellison (at Colombo) found the ansae equal in length on June 28, but, as will appear from his later observations, he found the p. ansa the brighter at times in 1936 December and 1937 January.

Globe features in 1936 (Antoniadi)

The Saturn Section's report mentioned that the north equatorial belt was narrower, darker and more sharply defined than the south equatorial belt, which seemed diffuse and in much higher latitude. The equatorial zone showed changes in intensity, and the southern part became very much brighter than the part north of the equator: this was confirmed by Quénisset's photographs taken on July 17, showing the north part of the zone dark, and recalling the Lick results of 1927.

Antoniadi described Saturn as looking greyish in the southern hemisphere, but more dusky and orange in the northern on July 2; the orange colour of the equatorial zone was particularly intense north of the equator; the roseate brown north equatorial belt contained a highly complex dark brown spot with three dusky nuclei. In another colourful description of the aspect on December 19, he described the southern half of the disk as cream, the northern half purplish orange, the polar regions being the duskiest parts of the planet. He drew particular attention to the contrast between the two parts of the equatorial zone—south of the equator pale lemon or cream-yellow, but an intense purplish orange north of the equator.

Sun's passage through ring-plane, 1936 December 28
(Antoniadi, Ellison, Phillips, Steavenson)

A number of interesting observations were made around the date of the Sun's passage through the ring-plane. On December 19 Antoniadi, using the Meudon 33-inch aperture refractor, found the ring visible all along as a very narrow faint ellipse, containing the inner part C, which, viewed so obliquely, appeared much more luminous than usual. It was brightest at the parts corresponding to the outer border of ring B, but the Cassini division could only be glimpsed as a dark dot on the east ansa. 'The Ring touches the disk by two thread-like arms on either side, and its blackish projection, plus its most thin shadow, on the globe, is concave to N. Owing to a double effect of diffraction, the upper thread-like arms rise slightly above the S. edge of the dark projection of the ring.' (See Plate XV, Fig. 2.) On December 30, two days after the Sun's passage, Grenat and Antoniadi glimpsed the ring, finding its unilluminated surface very faint.

Ellison found the ring gradually becoming much fainter during the period preceding December 28. On that date it had been computed that it would be exactly edgewise to the Sun at 13h 45m U.T.; at the critical time Ellison was able to observe it at Colombo and found the ring quite visible, but there appeared to be more of it on the p. than on the f. side. The ansae appeared to be north of its projection, which showed as a black line along the equator of the globe. The ring appeared brighter adjoining the planet and was only seen as far out as about the outer edge of ring B. The parts of the disk north and south of the equatorial zone looked darker than the zone, and 'of a distinct steel-blue colour'. As Saturn would be in daylight farther west and below the horizon in the far east, it was probable that it could only have been observed, at the actual time of the Sun's passage, in Ceylon and India.

Phillips, on December 27, found the ring system quite easy to detect but very faint; he thought the increased faintness might have been due to the Sun being already partly south of the plane. Steavenson agreed with this, and had also found the ring quite an easy object on that date; he had even glimpsed the ansae on

December 29, when very little of the Sun remained to illuminate the near side (*B.A.A.J.*, vol. 47, pp. 102–4).

The unilluminated face of the rings, 1937 January–February
(Antoniadi, Ellison, Hargreaves, Congreve-Pridgeon)

There was a theoretical disappearance of the rings from 1936 December 28, when the Sun crossed the plane to the south side, until 1937 February 21, when the Earth also crossed to the south side of the ring-plane. Antoniadi on 1937 January 8 found the unilluminated face a very narrow and very faint ellipse, without luminous spots. The ring, therefore, never disappeared from sight in the great Meudon refractor during 1936–37. He noted that the globe appeared pale lemon or cream in the southern hemisphere, pink-orange all over the northern one; the polar regions were dusky; the north equatorial belt had a jagged south edge.

Ellison had no difficulty in seeing the ring on January 3, 4 and 24, but he found it a good deal brighter and more easily seen next to the globe than farther out; on January 16 it was still visible but very difficult; on that date and on January 3 he noted that the p. ansa was the brighter and more easily seen, but they appeared equal on the 24th. Later, on 1937 September 23, he noted a very unusual appearance of the globe, which seemed bluish all over except for the narrow bright northern half of the equatorial zone (this was the part of the zone that had been darker the year before).

F. J. Hargreaves, using a 14½-inch aperture reflector, observed the unilluminated side of the ring twice during the early part of 1937, and on both occasions the rings had been invisible except for two very faint nebulous spots, one on either side, at a distance of about 2/5 of the planet's equatorial diameter from the limbs; F. M. Holborn also saw them (*B.A.A.J.*, vol. 47, p. 145; 50, p. 231).

As stated in the Saturn Section's report for 1937 (*B.A.A.J.*, vol. 47, p. 352) the Earth's passage through the ring-plane on 1937 February 21 occurred at a time when the planet was too near conjunction with the Sun to be generally observable. Nevertheless several members continued observing until February, and R. Congreve-Pridgeon, using the 6-inch aperture Cooke refractor at Hampstead Observatory, managed to glimpse the ring as a fine silver line shortly after sunset on February 22.

Other observations in 1936 and after (Haas, Barker)

Walter H. Haas reported (*J.R.A.S. Canada*, vol. 33, p. 247) that he and other observers in America, and one in Germany, using telescopes of apertures from 6 to 10 inches, had on various occasions 1936 June–September and December 25 seen bright spots on the ansae of the rings, and in the autumn of 1936 and again in 1937 had been struck by the pale blue colour of the rings. In 1938 Haas noted repeatedly that ring A seemed brighter and whiter than ring B: this is a phenomenon that can occur when the rings are narrowly open and the Sun is at a lower elevation than the Earth above the ring-plane.

In a report on observations of satellite phenomena in 1936 (*B.A.A.J.*, vol. 47, p. 152) R. Barker stated that he had observed the shadow transits of Dione that had been predicted for August 22 and October 24, finding them extremely difficult. In a discussion (*B.A.A.J.*, vol. 47, p. 170) Dr. Steavenson, congratulating the observer, said that the observation of Dione's shadow must have been very rarely, if ever, made before, since the shadows of the more distant satellites are largely penumbral—about 25 per cent in the case of Dione.

Saturn 1937–40 (Phillips)

In the Section's report for 1937, the Rev. T. E. R. Phillips stated that the rings, presenting their sunlit surface, were conspicuous, but that the Earth's elevation above the ring-plane, which reached $5\frac{1}{2}°$ in 1937 mid-July, was temporarily diminishing to $2\frac{3}{4}°$ in mid-December, after which it would again increase. A few times in excellent seeing he had found ring C visible in the ansae and the Cassini division easy near the extremities of the system. During the earlier part of the apparition the rings in front of the globe appeared as a dusky band, closely double. 'The outer edge of this band seemed sensibly continuous with the outer edge of ring A as projected on the sky, and it was sharply bounded on the north by a very bright zone' (north part of equatorial zone). 'The inner component (of the dusky band) was taken to be ring C combined with its shadow.' He added that the contrast between the duskiness of the rings and the brightness of the equatorial zone had since been diminishing, and by mid-September the rings in front of the globe, apart from C, were invisible. He attributed this, not to any fading of the bright zone, but to a brightening of the rings to an equal luminosity with the contiguous surface of the planet, owing to the Sun's elevation above the ring-plane gradually increasing till, about September 20, it reached equality with the diminishing elevation of the Earth. Another phenomenon noted was an apparently disproportionate breadth of ring C projected on the globe; he thought that the dusky band there might possibly include a portion of the inner part of B as well as C.

Phillips's Section Report in 1939 (*B.A.A.J.*, vol. 49, p. 379) stated that during the latter part of the 1937–38 apparition the belts were extremely faint and the bright equatorial zone had also faded. In the autumn of 1939 the S. polar area was dusky, the south equatorial belt was feeble but broad (suspected to be double), with its N. edge sharply defined by a rather narrow dark line; the rings were well open, the crepe ring easily seen extending nearly half way to the limbs in the ansae; Encke's division was well seen as a very fine line on several occasions.

His Section Report in 1940 (*B.A.A.J.*, vol. 50, p. 362) called attention to the dark S. polar cap, described by H. M. Johnson as the darkest part of the globe, and a light spot at its centre detected by F. J. Hargreaves, and confirmed by Phillips himself. A light zone separated the S. polar area from the broad, faintly dusky south equatorial belt, which was bounded at its north edge by a narrow dark band. The equatorial zone, at first moderately bright, was recorded as white in October.

Bright projecting limb spot observed from Odessa, 1938 September (Andrenko)

Léonid Andrenko, Director of the Second Observatory of Odessa, U.S.S.R., reported (*L'Astronomie*, 1938, p. 473) that when observing Saturn on 1938 September 26, using the $6\frac{1}{2}$-inch aperture refractor of the Observatory, he was struck by the presence on the following limb, at 20° above the exterior edge of the ring, of a small and brilliant white projection, which seemed separated from the globe by a faint grey border. Three other observers gave independent confirmation. On the next day these and other observers saw the spot again, but on the preceding limb, to which it had been carried by the planet's rotation. It continued to be seen on one or other limb on each of the next four days, but on October 1, when last visible, had considerably diminished in size.

B. M. Peek (*The Planet Jupiter*, p. 163) recounts a number of instances of the observation of spots of this kind on Jupiter, seen in turn on both limbs and appearing to project beyond the limb by irradiation; such phenomena are rare on Saturn.

CHAPTER 37

SATURN'S RING SUBDIVISIONS, BELTS AND SPOTS, AND TITAN'S MARKINGS, IN THE 1940s

---

Belts, spots and subdivisions observed by American amateurs, 1939–46 (Haas)

During the war years, although Saturn was well placed in the sky, the observation of the planet rather naturally languished in Europe, and the B.A.A. Saturn Section virtually ceased to operate. In America, however, regular observations were carried out during those years by a small group of amateurs under the leadership of Walter H. Haas, who collected and reported their results (in *J.R.A.S. Canada*, 1944, and *Popular Astronomy*, vol. 55, no. 9), and who, in 1947, formed them into the Association of Lunar and Planetary Observers, with observing sections after the B.A.A. model. Most of their observations were made with smallish apertures ranging from 4 to 12 inches, though Haas and others occasionally used the 18-inch aperture refractor of Flower Observatory.

Though in 1942–43 they found the double south equatorial belt fainter than usual and difficult to separate, it was at other times conspicuous, the northern component generally appearing the darker. From measurements of drawings, Haas calculated that the saturnicentric latitudes of the components were −17° and −28°. Small rather difficult humps were detected on the edges of this belt, and from transits of two of these on the north edge of the north component obtained during the autumn of 1942, Haas calculated a rotation period of 10ʰ 17ᵐ 44ˢ (*B.A.A. Saturn Section report*, 1942–43). They found that a belt at the edge of the south polar shading became dark and conspicuous in 1943 and 1944, subsequently fading; from measures on drawings Haas calculated the saturnicentric latitude of this south polar belt to be −73°. During 1943–46 delicate narrow belts, sometimes rather dark, were detected, one near the equator and the other in the south temperate region.

Apart from Cassini's division and Encke's (which they often saw, near the middle of ring A), from 1942 onwards these observers frequently saw a rather wide, dusky subdivision, which they called 'the third division', on ring B near the inner edge. During the same period they also glimpsed at times a delicate shading, which they called 'the fourth division', at the inner edge of the bright outer third of ring B. From 1943 onwards they also claimed to detect, with their larger telescopes, a black gap between rings B and C. Previous chapters have shown that the existence of this feature has been much disputed: Dawes, and later Lowell, claimed to have seen it, whereas Lassell, and afterwards Barnard, denied that there was any boundary line between the two rings.

The American group also devoted some attention to a white spot on the rings adjoining the globe's shadow and a similar spot adjacent to the limb on the unshadowed arm of the rings, both of which Haas considered were merely contrast-effects. H. M. Johnson wrote an interesting paper quoting 1889 observations and those of himself and other American amateurs in the early 1940s (*B.A.A.J.*, vol. 51, p. 309). Such phenomena have, of course, long been recognised as being purely optical, and the one adjacent to the shadow on the rings is known as 'Terby's white spot' (see Chapter 17).

Photography of Saturn at Lick and Pic-du-Midi Observatories, 1939–43 (Jeffers, Herbig, Camichel)

In 1945 Dr. J. H. Moore, Director of Lick Observatory, sent to the B.A.A. two very fine photographs of Saturn, taken with the Lick 36-inch aperture refractor, the first on 1939 October 21 by H. M. Jeffers, the second on 1943 December 14 by G. H. Herbig. These were reproduced on a plate and commented on by M. A. Ainslie, then Director of the Saturn Section (*B.A.A.J.*, vol. 55, p. 125). He pointed out that, on the earlier photograph, showing the rings partly open, the southern hemisphere (except the bright equatorial zone) seemed dusky and featureless, and there was no indication of Encke's division, though it had been an easy object visually in 1939 for apertures of more than 8 inches; though both photographs showed a contrast between the brighter inner part and duskier outer part of ring A, neither showed contrast between the outer and inner parts of ring B, or any sign of the crepe ring in the ansae. Ainslie drew attention to the striking changes of tone shown on the globe in the 1943 photograph (see Plate XVI, Fig. 2), in which the south polar region was remarkably well-defined and dark, there was a lighter south temperate zone, and the south equatorial belt looked prominent and double. One striking feature of the 1943 photograph, with the rings very wide open, was the indistinctness of a small part of the Cassini division in front of the globe, due to light filtering through from the globe (as Dr. Moore suggested). A small part of ring A in

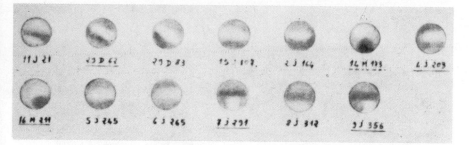

FIG. 1.—Drawings of Titan by B. Lyot and colleagues (Pic-du-Midi 24-inch refractor × 1000 and × 1250).

FIG. 2.—Lyot's diagram of Saturn's belts and markings on the Southern face of the rings, observed 1943 with Pic-du-Midi 24-inch refractor. The crepe ring is shown translucent.

PLATE XVII

FIG. 1.—1948 March 24.

FIG. 2.—1949 April 16.

PLATE XVIII.—H. Camichel's photographs of Saturn (Pic-du-Midi 24-inch).

front of the globe appeared to be similarly illuminated, and hence translucent, but Ainslie pointed out that, owing to the very wide opening of the ring, this part of ring A would be below (in the photograph) the north limb and not in front of it.

Perhaps one of the most interesting things about the 1943 photograph is the extraordinary shape of the shadow of the south polar region on ring B and the Cassini division, very similar to some of Coolidge's drawings at Harvard Observatory in 1854–57 (mentioned in Chapter 10).

Ainslie considered these photographs quite the best of Saturn he had ever seen, and remarked on their brilliance and 'sparkle' not often seen in the best planetary photographs.

Bernard Lyot reported (*Ap. J.*, vol. 101, p. 258) that in 1941 Henri Camichel at the Pic-du-Midi Observatory, using an object-glass of 0·38 metre aperture (about 15 inch) and contrasty, fine-grain panchromatic plates, and by the superposition of some ten negatives, had obtained very fine composite pictures of Saturn, e.g. on 1941 October 21. These brought out: (i) the transparency of ring A with the globe showing through it; (ii) a bright line on the edge of ring B's shadow on the globe, probably due to light passing through between rings B and C; (iii) two minor divisions in ring B and one in A.

Revival of the B.A.A. Saturn Section, 1946

In the autumn of 1946, M. A. Ainslie, who had made such notable observations of the planet and had been so prominent in Saturn affairs for over thirty years, resigned the directorship of the Section owing to age and ill-health and the author of this book was appointed as his successor. Increased interest in astronomy and the great post-war influx of new members of the Association seemed to offer a favourable opportunity to revive the active observation of Saturn. A revised observing programme (see Appendix II) was drawn up, including certain new features, such as the eye estimation of relative intensities of belts and zones, with a view to encouraging frequent routine inspection of Saturn, and an appeal was made, both for new observers and for experienced observers of other planets, to embark on the regular observation of Saturn. There was a very good response from observers, both from those in England and Wales and from others abroad.

The results of the many observations of spots, intensities and belt latitudes will be given in the following sections of this chapter; other special points noted in the appearance of the planet in the later 1940s may be summarised as follows: In 1946–47 the crepe ring across the globe seemed unusually intense, and the south temperate zone patchy; there was a small light area on the south limb. The most surprising feature at the next apparition was the abnormal thinness and faintness of the crepe ring across the globe during the autumn of 1947, a reduction too great to be wholly due to the positions of Earth and Sun relative to the ring-plane. This was observed many times by W. H. Haas, E. J. Reese and other Americans from 1947 October, and confirmed in December, when observing became possible in England, by E. F. Coney and F. H. Thornton; in the new year the ring became more normal in width. During both apparitions the north polar region looked less dusky than the area of the south pole (*B.A.A.J.*, vol. 57, p. S5; 58, p. 240; 59, p. 209).

Visual estimates of intensity

A very large number of these observations, mainly with telescopes of about 6-inch aperture but some larger, were made by many observers of the B.A.A. Saturn Section during the apparitions of 1946–47 and 1947–48, using a scale from 1 (brightest) to 10 (darkest). The main results (from *B.A.A.J.*, vol. 59, p. 208) are summarised below. A comparison between the two apparitions is unsure, as few observers made estimates in both: the letters D, L show cases where a majority of observers of both apparitions found a feature darker or lighter in the later one.

Similar observations by the Section, though fewer, for the apparitions of 1949, 1951 and 1952 were analysed by A. P. Lenham and averaged by month and apparition (*B.A.A.J.*, vol. 63, p. 140). Apparition means unfortunately gloss over variations occurring during an apparition, and perhaps the making of intensity observations has proved its usefulness most in helping the observer to note small changes in the intensity of a feature, e.g. the equatorial zone, from month to month in the course of an apparition. Good examples of this are given in M. B. B. Heath's Section reports of the 1950s (see Chapter 39).

South equatorial belt spots and rotation periods (Fox, Haas, Reese)

In 1947 March, while carrying out a routine observation of Saturn with a $6\frac{1}{2}$-inch aperture reflector, W. E. Fox, of Newark, who is now Director of the B.A.A. Jupiter Section, noticed several small dark spots on the northern edge of Saturn's south equatorial belt. He promptly reported this by telephone, so members of the B.A.A.

were informed, by letters and circular, of dates and times when the objects should be looked for. Although some experienced observers with larger telescopes were unable to confirm the spots, at least a dozen others (some of them with 12-inch aperture reflectors) in England, Czechoslovakia and America recorded transits, and W. H. Haas and E. J. Reese in the last-named country, using 6-inch aperture reflectors, had, in fact, been recording transits of those and other spots on the southern edge of the belt for some time previously. As there is no standard system of longitudes for Saturn, an *ad hoc* system had, as usual, to be devised, based on an assumed rotation

Mean Intensities of Features on or across Globe (South to North)

	1946–47			1947–48			
	No. of Observers	*No. of Estimates*	*Mean of all observations*	*No. of Observers*	*No. of Estimates*	*Mean of all observations*	
S. Polar Region	16	238	4·1	9	170	4·4	D
S. Polar Band	13	188	4·7	8	115	5·0	
S. Temp. Zone	15	224	4·0	9	164	3·6	
S. Equatorial Belt	16	233	4·75	9	163	4·7	
S.E.B. (South)			4·7			4·6	
Intermed. zone	6	98	4·2	4	90	4·4	
S.E.B. (North)			4·8			4·8	
Limbs of Globe	7	80	3·6	3	68	3·2	L
Equatorial Zone	15	226	1·9	8	144	1·6	L
C_m (Ring C projection)	15	214	7·9	9	170	7·6	L
B_m (B across Globe)	6	83	2·1	2	83	3·1	D
A_m (A across Globe)	6	59	3·1	6	110	3·9	D
Ring shadow on Globe	1	11	7·4	3	26	6·8	
N. Polar Region	12	161	3·4	9	161	3·8	

Mean Intensities of Features in Ansae (outer to inner)

A1	16	231	3·8	9	170	4·0	
Encke's Division	3	6	5·4	2	8	5·6	
A2	16	231	3·8	5	126	4·1	
Cassini's Division	12	142	8·8	5	98	7·6	
B1	16	222	1·1	9	170	1·3	
B2	11	194	2·0	7	135	2·6	D
B3	6	57	3·5	3	68	3·7	
C	14	158	8·2	8	137	8·3	
Globe shadow on Rings	14	186	9·9	6	109	9·5	

period of 10^h 14^m, in order to plot the observations and determine rotation periods for the spots. The result was a normal period of 10^h 14^m for the objects on the southern edge, but a somewhat longer rotation period of 10^h $15^m.9$ for those (nearer the equator) on the northern edge of the belt (*B.A.A.J.*, vol. 57, pp. 192 and S6).

Haas and Reese continued to record transits of irregularities on the south equatorial belt in the next apparition, 1947–48, the former finding that a dark column on the south edge accelerated in 1948 March–April from a period of 10^h 14^m to one of 10^h 9^m (*B.A.A.J.*, vol. 58, p. 240).

During 1949 November–December, Haas, Reese and other American observers recorded many transits of the preceding end of a dark area on the north temperate belt, finding that in 76 rotations (nearly five weeks) the rotation period of the dark patch decreased from 10^h 14^m to 9^h 55^m (*B.A.A.J.*, vol. 60, p. 217). This may have been due to a rapid growth or spreading of the patch in the preceding direction, rather reminiscent of the behaviour of Will Hay's spot in 1933.

Latitude of south equatorial belt, 1946–47 (Heath)

M. B. B. Heath wrote an interesting paper (*B.A.A.J.*, vol. 62, p. 202) on the determinations of the saturnicentric latitudes of the components of the south equatorial belt in 1946–47 by four members of the Saturn Section: himself, Haas, Reese and the author of this book. At that apparition the belt was often seen triple—with two dark components separated by a somewhat lighter zone—but good definition was required to separate them. The four observers made in all 169 determinations, by measurements of sketches made from careful eye estimations. Their mean results were:

South component	S. edge $-25°6$	N. edge $-21°4$
North	S. edge $-15°9$	N. edge $- 9°4$

This would make the mean total width of the belt $16°2$, of the south component $4°2$, of the intermediate zone $5°5$, and of the north component $6°5$. The curious thing was that, though three of the four observers were in fairly good agreement for each of these results, it was a different combination of three in each case.

As a check, Heath measured and worked out the results from the few drawings each, not made specially for this purpose, sent in by fourteen other observers. The mean values derived from these drawings were:

South component	S. edge $-25°4$	N. edge $-21°2$
North component	S. edge $-14°1$	N. edge $- 8°9$

It will be seen from this that there was a satisfactory agreement between the mean results obtained from the two sets of observers.

Saturn in 1947 (Focas)

Most of the evidence for the markings on, and activity shown by, Saturn in the 1940s, as already described in this chapter, has been based on co-operative observation by amateur astronomers, equipped usually with telescopes of small, or at most medium, aperture, and the discriminating reader will doubtless wish to know what (if any) verification is forthcoming from larger instruments. The rest of this chapter will, it is hoped, provide the answer.

J. E. Focas, assistant at the National Observatory of Athens, gave an interesting account (in *L'Astronomie*, 1948, p. 1) of his observations in 1947 January and February using the Doridis equatorial refractor of 16-inch aperture, with powers not exceeding 400. The south polar cap was lightish, surrounded by a wide dusky band; there were two dusky belts in the south temperate region, the more northerly at latitude about −50°. There were two large elliptical areas, whitening at intervals, in contact with the south component of the south equatorial belt. He found this belt double, the south component very irregular and broken with projections on the edge, the north component less broken, but with very dark lumps ('nodosités') on its north edge, and projections connected with filaments crossing the equatorial zone. Between these filaments were patches on the equatorial zone whiter than the brightest part of ring B, and he observed one of these white spots, very well-defined, crossing the central meridian on February 7, at 7ʰ G.M.A.T. He saw Encke's division sharply defined several times, but on other occasions in good seeing it was completely invisible.

Pic-du-Midi photography of planet and white spots (Camichel)

The Pic-du-Midi Observatory, situated in the Pyrenees at an altitude of over 9,000 feet, and therefore above the dust layer of the district, often enjoys almost ideal seeing conditions for the photography of planets and the observation of fine planetary detail. H. Camichel obtained photographs in 1941 which, in addition to the features already mentioned earlier in this chapter, showed some half dozen belts and zones in the southern hemisphere of Saturn, and on 1941 September 16 and 22 showed a diffuse white spot at about latitude −40° (*Trans. I.A.U.*, vol. 7, p. 164).

In 1943 a 24-inch aperture object glass was installed in the great refractor, tested and perfected. With this equipment Camichel photographed a fine white spot by the northern edge of the south equatorial belt on 1946 February 11 and 14, and calculated a rotation period for it of $10^h 21^m4$. Its latitude was about $-12° 30'$, it was small and brilliant on the first date, but by the second had become less bright and obliquely elongated; on that date it was also observed at Meudon Observatory by A. Danjon and B. Lyot (*L'Astronomie*, 1946, p. 161; 1953, p. 12; *Trans. I.A.U.*, vol. 10, p. 250).

Camichel's photograph of Saturn of 1949 April 16 (Plate XVIII, Fig. 2) shows the globe through ring A, and a narrow dark belt in the northern hemisphere parallel to one in the southern; that of 1950 April 7 (Plate XX, Fig. 1) the rings nearly closed and their brightness strikingly diminished, the Sun lighting them up very obliquely from an altitude of only 2° above their plane, while the Earth was at an altitude of 4°, the equatorial zone appearing brilliant. After the passages of the Earth and Sun through the ring-plane in 1950 September, the next photograph, 1951 April 15 (Plate XX, Fig. 2) shows an oblique view of the northern face of the rings and their shadow across the disk, together with a very brilliant equatorial zone and unsymmetrical belts. Following his description of these composite photographs, Lyot's notes pointed out (*L'Astronomie*, 1953, p. 13) that while they show large objects such as belts at least as well as these could be seen by visual observation, they could not display the finest details, such as certain of the ring divisions, because of the slight oscillation, often more than $0''3$ of arc, produced on the telescope by the wind in the course of an exposure of about 20 seconds, an effect by which the observer's eye would not be disturbed.

Pic-du-Midi observations of ring subdivisions (Lyot)

Bernard Lyot, inventor of the coronagraph and celebrated for his solar and planetary observations, made in 1943 precise measurements of the rings and their divisions, at first with a filar micrometer, and then with a new double image micrometer (*Trans. I.A.U.*, vol. 8, p. 213), using the 24-inch aperture refractor of the Pic-du-Midi Observatory with a power of 900. Unhappily he did not live to publish the full results, but his assistant, Audouin Dollfus, presented (in *L'Astronomie*, 1953, p. 3) Lyot's draft report on the planetary photographs and observations at the Pic-du-Midi and his remarkable diagram of Saturn's belts and ring markings (see Plate XVII, Fig. 2). Lyot's description of the aspect of the southern face of the rings was as follows:

Ring A. At the outer edge a very brilliant strip $0''4$ wide, then a narrow black line, next another narrow light zone; inside that, covering most of the middle and inner part of the ring, a wide shaded region, in which three minima of light could be distinguished, with a rather narrow light zone between it and Cassini's division.

Ring B. Separated into two almost equal parts by a rather definite division, the exterior part containing a rather wide lightly shaded diffuse band. Interior to the central division there is the lightest zone of the ring. Near the interior edge of B there is a double division, the two quite definite components being a quarter of a second of arc apart. A division a little wider than half that of Cassini separates ring B from ring C.

It is rather surprising that apparently Lyot found the part of ring B just interior to the middle the brightest, whereas the outer edge adjoining Cassini's division is usually considered the brightest part of the whole of the rings, but this impression is doubtless due partly to contrast with the darkness of the division. For the most part Lyot's diagram and notes seem to represent the ring subdivisions as diffuse deeper shadings, or 'minima of light' rather than sharp 'pencil lines', and modern opinion seems to look upon them as surface ripples or temporary displacements among the crowding ring particles rather than true gaps like Cassini's division. It will be remembered that Dawes, as long ago as 1855, had come to doubt whether Encke's division was a true gap (see Chapter 11). The large aperture and perfect definition at the Pic-du-Midi probably showed more nearly than most other observations the true aspect of these markings, as they did for the 'canals' of Mars, and it is as well to remember Antoniadi's conclusion in 1930 (*L'Astronomie*, vol. 44, p. 164) that any narrow subdivisions seen on the rings with a small telescope should, if they were objective, appear much wider and darker and more easily visible in a large instrument. He also, however, expressed the opinion (contrary to that of Lowell and Lyot) that his own observations and those of Lassell and Barnard had proved that there is no division between rings B and C.

Lyot's ring subdivisions compared with Kirkwood's gaps

The author of this book wrote a paper in 1953 (*B.A.A.J.*, vol. 64, p. 26) attempting to show a correlation between Lyot's diagram of the ring markings, the ring positions where subdivisions might be expected according to Kirkwood's theory of perturbations by Mimas and Enceladus, A.L.P.O. observations of Encke's division and of the '3rd, 4th and 5th divisions' in the 1940s, and Ainslie's observation of brightenings of the star behind ring A in 1917 (see Chapter 30). All these observations and those of Lowell and E. C. Slipher

1913–15 (see Chapter 29) were of the rings' southern face. Lowell and Slipher had observed, on ring A, Encke's division in two components and another subdivision further out; on ring B, other subdivisions exterior to and interior to the three whose positions they identified and which are marked with an asterisk on the chart shown below, which gives the approximate positions of Lyot's and some of the other observed markings, and of Kirkwood's theoretical gaps:

Lyot's Divisions		Kirkwood's Gaps		Other probable correlations	
	Position in ring (0)		Position in ring (0)		Position in ring (0)
RING A (outer edge)	(0)				
		$\frac{3}{7}$ E	10	Ainslie's star's 2nd brightening	5
Outer division	20	—		Ainslie's star's 1st brightening	18
Inner division:					
outer edge	36			Encke's Div. Encke's	
minima ⎫	40	$\frac{3}{5}$ M	39	⎱ (2 compo- Div.-White	40
of ⎬ {	62	$\frac{2}{5}$ E	57	⎰ nents)	
light ⎭	80			—Tombaugh	
inner edge	80				
(inner edge of ring)	(100)				
CASSINI'S DIVISION		$\frac{1}{2}$ M, $\frac{1}{3}$ E			
RING B (outer edge)	(0)		(0)		
Outer division:					
outer edge	11				
darkest part	17	$\frac{5}{11}$ M	24	4th division—Cragg	
—	—	$\frac{4}{9}$ M	31		
inner edge	39				
Middle division:	42	$\frac{3}{7}$ M*	41		
—	—	$\frac{2}{5}$ M*	57		
Inner division:					
outer edge	70				
darkest ⎫ {	80	$\frac{3}{8}$ M*	76	3rd division ⎱ 3rd division	
parts ⎭ {	89	$\frac{1}{4}$ E	89	—Cragg ⎰ —Johnson	
inner edge	96				
(inner edge of ring)	(100)		(100)		
DIVISION BETWEEN RINGS B AND C		$\frac{1}{3}$ M		5th division	

Notes. M = Mimas; E = Enceladus.
All the observations of faint divisions relate to the *ansae* of the rings.

The A.L.P.O. observations mentioned earlier in this chapter were verified in 1948 and 1949, in some cases with the use of large telescopes. L. T. Johnson in 1949 saw the 3rd division dusky, not black, and estimated its width as equivalent to 20 per cent of that of ring B, which seems in fair agreement with Lyot's opinion. T. Cragg, using in 1948 the 200-inch aperture Hale reflector, found Encke's division to be 'a band of definite width but not a gap', but the 3rd and 4th divisions narrow. C. W. Tombaugh (discoverer of Pluto) stated that Encke's division is approximately in the middle of ring A, and that he saw it double on three occasions of exceptionally good seeing, using the 24-inch aperture Lowell refractor, and estimated the separation of the components at the equivalent of about 14 per cent of the width of ring A (*Strolling Astronomer*, vols. 3 (4), p. 11; 2 (8), p. 1; 3 (3), p. 11).

Markings on Titan (Lyot, Bruch, Camichel, Dollfus and Gentili)

Of the outstanding work carried out by Lyot and his colleagues at the Pic-du-Midi in the photography and observation of Saturn in the 1940s, perhaps the most remarkable of all is their achievement in observing and making drawings of the markings on the tiny disk of the satellite Titan (Plate XVII, Fig. 1). Observations and drawings were made independently by five observers and afterwards compared, each of the views shown on the plate representing what had been detected by several observers. The sketches are arranged in order of increasing longitude of the satellite in its orbit, 0° representing superior conjunction with Saturn. Lyot stated (*L'Astronomie*, 1953, p. 14) that Titan shows, at opposition, a disk of apparent diameter 0″8 of arc, with very dusky limbs, and with the 24-inch aperture refractor, power 1,000, and especially with power 1,250, often shows markings on its disk. Some of the drawings made portray Titan with a wide light belt across the centre, either oblique or parallel to the equator, and a dusky north polar area. Others show a dark equatorial belt and a lightish north pole. No certain conclusion could be deduced from them as to the satellite's rotation, and Lyot pointed out that the markings might well be variable, since Titan is known to have an atmosphere of methane, discovered by Kuiper.

CHAPTER 38

ATMOSPHERES OF TITAN AND SATURN, AND STUDIES OF THE RINGS

Discovery of Titan's atmosphere, 1943–44 (Kuiper)

Dr. Gerard P. Kuiper, of the McDonald and Yerkes Observatories, during the winter of 1943–44 photographed the spectra of the ten largest satellites of the solar system, using a one-prism spectrograph attached to the 82-inch aperture McDonald reflector. Photographs of the spectra were obtained for the infra-red as well as the photographic regions. When Titan's spectrum was found to contain methane absorption bands, a number of plates with higher dispersion were taken. Kuiper's paper (*Ap. J.*, vol. 100, p. 378) is illustrated with plates showing Titan's spectrum compared with those of Saturn and the Ring and other planets and satellites, with low- and medium-dispersion on panchromatic film and infra-red plates, and with panchromatic film and medium-dispersion in the photographic as well as the infra-red regions. Kuiper found the methane absorption at λ6190 in Titan's spectra to be striking, and in marked contrast to the spectra of Rhea and the satellites of Jupiter. The results on Tethys and Dione were also definitely negative, though a trace of this absorption was suspected on Neptune's large satellite, Triton.

The infra-red spectra of low dispersion showed the λ7260 band of methane clearly present on Titan, but not on the satellites of Jupiter or on the ring of Saturn.

For medium dispersion a large Cassegrain spectrograph with two quartz prisms and a curved plateholder was used. Density measures of the spectra with the aid of a photometer showed that the methane band λ6190/A is slightly shallower on Titan than on Saturn. The presence of the ammonia band at λ6400/A was suspected on Titan's spectrum. The rings of Saturn showed the true solar spectrum with no evidence of atmosphere.

Kuiper's conclusion from the whole of the photographs was that there appeared to be a close resemblance between the spectrum of Titan and that of Saturn, but as the methane bands on Titan are definitely weaker, he considered that the total thickness of Titan's atmosphere is comparable to, but somewhat less than, that of the observable layers of Saturn and Jupiter, for which 0·5 mile-atmospheres of methane gas have been estimated by Slipher and Adel. He also found the colour contrast between Titan and Saturn shown by the spectra to be striking, agreeing with telescopic observation, which shows Titan to be orange, like Mars. He suggested that it seems likely that the colour of Titan, like that of Mars, may be due to the action of the atmosphere on the surface.

In the latter part of his paper Kuiper worked out the velocity of escape for gases on the various planets and chief satellites, and showed that if V is the velocity of escape in km./sec. and r the distance from the Sun, the value of $(V \times r^{\frac{1}{4}})$ is a parameter for judging the atmosphere stability; it takes into account the temperature which varies roughly as $r^{-\frac{1}{2}}$. The following shows the order of decreasing atmosphere stability among the smaller planets and larger satellites, with the value of the parameter in brackets: Earth (11·3), Venus (9·5), Triton (6·3 or 6·6), Mars (5·7), Titan (5·3), Ganymede (4·3 or 4·5), Callisto (3·4 or 3·8), Io (3·5), Europa (3·1), Mercury (2·9), Moon (2·4), Rhea (1·1 or 1·3), Dione (0·9 or 1·0), Tethys (0·6 or 0·8). Kuiper suggested that the boundary line between atmosphere and no atmosphere might lie between Titan and Ganymede.

He also pointed out that Titan's atmosphere would be endangered if its temperature of 100° to 125°K were doubled, and a still greater increase would cause rapid dissipation of its methane. Hence if Titan has gone through a period of high surface temperature, it follows that the atmosphere was formed after that period. He also considered that the similarity of Titan's atmosphere to Saturn's makes it highly probable that Titan was formed within the Saturnian system, and shows definitely that the satellite was not captured from an elliptical orbit extending to the interior regions of the solar system.

The importance of Kuiper's discovery is obvious: Titan's is the first case of any satellite in the solar system being proved to possess an atmosphere.

Possible disintegration of the rings through precession (Gregory)

In 1945 C. C. L. Gregory suggested that precession caused by the Sun's gravitational field was likely to bring about the disintegration of the rings, unless there be cohesion among the particles, or a replacement (from a tenuous outer atmosphere) of particles lost to space, or some compensating gravitational effect, such as might be found in the oblate spheriodal shape of Saturn. He calculated that the period of precession for particles at the inner edge of the ring (whose revolution period about Saturn is 0·228 days) would be

2·1 million years, and for particles at the outer edge (revolution period about 0·61 days) the period of precession would be approximately 0·8 million years (*B.A.A.J.*, vol. 55, pp. 136, 142).

Abundance of hydrogen in Saturn (Scholte)

In view of the low mean density of Saturn, and since the principal gas detected in its atmosphere, methane, is a hydrogen compound, and as hydrogen is undetectable by the spectrum in planetary atmospheres, it had already been assumed that free hydrogen must be a large constituent of Saturn's atmosphere.

In 1946 J. G. Scholte, in a mathematical paper (*M.N.*, vol. 107, p. 237) considered the effect of the increasing pressure in the interior of planets on the energy of the atoms and the density of the material, and calculated the maximum radius for a planet composed of hydrogen. He found that for any other element this maximum radius would be smaller than the radius of Saturn, whence he concluded that the two largest planets, Jupiter and Saturn, consist of hydrogen.

Roche's theory of the origin of the rings (Jeffreys)

In 1947 Dr. H. Jeffreys pointed out (*M.N.*, vol. 107, p. 260) that Roche's theory (see Chapter 13) of the break-up by tidal action of a small satellite if it approached within a critical distance of a planet, assumed the satellite to be liquid, and that all extensions of Roche's argument since given by others had dealt only with liquid or gaseous satellites. Jeffreys therefore considered mathematically the case of a solid satellite, and found that it would not be broken up, even if close to the surface of a planet, unless its diameter was less than a critical value of the order of 200 km. He considered as a special case the approach of a sphere of ice to Saturn. He deduced that if an ice satellite ever revolved about Saturn at the mean distance of the rings, and was broken up, its diameter was over 200 km., but that the fragments would not be broken up further once they had been reduced to this value. Hence the ring could not have been formed by the disruption of a solid satellite, since it appears equally bright all round, and its maximum possible thickness has been shown to be only 15 km. (see Chapter 28).

Effects of collisions on the rings (Jeffreys)

Another interesting paper by Dr. Jeffreys (*M.N.*, vol. 107, p. 262) considered the effects of collisions among the ring particles. He first of all set out the outstanding observational facts concerning the rings: (1) that the velocity at any point has been shown by spectroscopic evidence to be nearly that of a particle in a circular orbit at the same distance; (2) that the reflecting power is high, that of ring B exceeding, and of ring A approaching, the reflecting power of the planet; (3) that the rings are nevertheless not quite opaque; (4) that the reflecting power, even allowing for the difference in distance, is appreciably lower when Saturn is at quadrature than at opposition; (5) that the ring is extremely thin.

Jeffreys pointed out that the evidence for Clerk Maxwell's theory that a meteoric constitution of the ring would be the only stable one, was not quite decisive, since Maxwell had supposed a liquid or gaseous ring to be in uniform rotation like a rigid body. The hypothesis of a rigid-body rotation would suggest a very high viscosity for a fluid ring, which could be arbitrarily thin if this assumption were abandoned. Jeffreys then showed that the observational data would not support the supposition of a gaseous or liquid ring: a very thin gaseous ring would not give sufficient reflection; a liquid one would show images of the globe and of stars which could not have escaped observation. He therefore concluded that the meteoric theory is the only tenable one, and, moreover, H. Seeliger had shown that it explains observational fact no. (4) mentioned above.

Jeffreys considered in detail the very complicated problem of collisions among the particles, and deduced that if the rings were several particles thick, the damping effect of collisions would reduce the rings in less than a year to a state where the particles were piled one on another. Dissipation by friction and impact would continue, eventually extending the ring outwards and inwards in its own plane, until its thickness nowhere exceeded that of one particle, and the particles were just sufficiently spaced for collisions to be avoided. This, he felt, agreed with Seeliger's conclusion that a distance between particles decidedly more than their diameters is needed to explain the reduction of albedo near quadrature, but it could not be many times their diameters, in view of the high albedo when the rings are opened to the fullest extent. If, however, it were possible to observe the rings full face, a much smaller fraction of the area would appear occupied than as they are actually seen from the Earth.

Jeffreys calculated that the time needed to attain the state of particles being spaced out and the ring being one particle thick would be of the order of a million years for particles of 1 cm. diameter and less for larger ones. He considered that if there was a division in the ring when the particles were still in contact, the tendency would always be to fill it up; hence it is likely that the divisions have been formed since collisions became rare. He made some comments on Goldsbrough's theory of instability at the inner edge of ring B and the outer edge of ring A. This would suggest a tendency of particles near the inner edge of B to join the crepe ring; if this were accepted another crepe ring would be expected outside ring A. Though this had not been definitely confirmed some observers had suggested it (see Chapter 28).

Studies of the rings, satellites and atmosphere (Kuiper)

Infra-red spectra of the ring obtained by Dr. G. P. Kuiper, showing similarities to those of the polar caps of Mars, suggested the presence in the rings of ice crystals; such crystals could persist at a temperature of 70°K, their vapour pressure would be negligible, and so would their evaporation. From measurements of the diameters of the five inner satellites, it would appear that their albedos are very high, about 0·8, and that their densities are close to that of water.

Ultra-violet spectra showed Kuiper that the quantity of sulphur dioxide present in Saturn's atmosphere is very small, less than 10^{-8} atmosphere. He analysed the vertical distribution of temperature and pressure in Saturn's atmosphere, finding the temperature of the stratosphere to be 65°K, the depth of the clouds 60 km., and their temperature 153°K.; and that methane could attain saturation in the neighbourhood of the tropopause.*

He suggested that if it were assumed that Saturn's atmosphere originally extended beyond Titan, the satellite could possibly have retained a small portion of this envelope. As the velocity of escape depends partly upon the temperature, varying as the square root of the absolute temperature at the top of the atmosphere, it is possible that the outer fringe of the atmosphere round Titan was cold enough to prevent its escape from the satellite. In addition, the five inner satellites, all well within the orbit of Titan, and also the ring, were probably formed inside the envelope when it was cooling.—(*Trans. I.A.U.*, vol. 8, p. 213; and *The Atmospheres of the Earth and Planets*, edited G. P. Kuiper, 1949, chap. XII).

Atmospheric absorption (Hess)

Seymour L. Hess, of the Lowell Observatory, studied particularly the variations in the strength of absorption of methane at λ6190 as shown by the spectra for different regions of Saturn and Jupiter. He concluded that the cloud surface on Jupiter rises towards the

* *Tropopause* is the height at which the normal decrease in atmospheric temperature ceases; above this, in the stratosphere, temperatures change relatively little for a considerable distance.

limbs, but is more nearly level on Saturn. The approximately known temperatures on Saturn and Jupiter are such that methane would remain uncondensed at the visible surface, while ammonia is almost certain to be in equilibrium between the solid and gaseous states. Hence the amount of methane found above any point would be a measure of that point's elevation. Hess's paper (in *Ap. J.*, vol. 118, p. 151) was based, in the case of Saturn, on spectra obtained in 1950 April and May, with a 3-prism spectrograph mounted on the Lowell 24-inch aperture refractor. He obtained figures for methane absorption on Saturn at latitudes 5° and 55° S. and 45° N. He found that the absorption clearly increases with latitude, and that this was partly, but not completely, accounted for by variation of path length with latitude. After allowing for this, he concluded that the cloud surface should be from 5 to 8 km. higher at 55° S. compared with 5° S., with a similar difference for 45° N., and that the observed increase in absorption towards the poles on Saturn suggests a more uniform atmospheric pressure along the cloud surface than for Jupiter. He pointed out that a variation in methane absorption due to a difference in height of the reflecting surface of as little as 10 km. is easily detectable.

CHAPTER 39

VISUAL OBSERVATIONS IN THE 1950s

The nearly edgewise rings (Steavenson)

In 1950 the single passage of the Earth through the ring-plane occurred on September 14, Saturn's conjunction with the Sun on September 16 and the Sun's passage on September 21: the interesting phenomena of the passages were therefore quite unobservable. In the spring of 1950, however, the elevation of the Sun above the ring-plane was for some weeks considerably less than that of the Earth, causing the apparent surface illumination of the rings to be unusually faint. This is well shown in a photograph taken by H. Camichel at the Pic-du-Midi Observatory with the 24-inch aperture refractor on 1950 April 7 (Plate XX, Fig. 1), and it attracted the attention of Dr. W. H. Steavenson early in May during a casual observation of the planet with his 30-inch aperture reflector (*B.A.A.J.*, vol. 60, p. 185). Steavenson therefore made a series of observations, using the 25-inch aperture Newall refractor of Cambridge Observatory.

On 1950 May 13 he found that the relative brightness of the various rings had undergone a change; the brightest part of the system was now the inner three-fifths of ring A which showed on each side as a small bright crescent just outside the Cassini division, the division appearing so pale as to be scarcely perceptible. Ring B and the outer parts of ring A were of about equal intensity, and he suggested that the relative faintness of ring B might be partly due to its particles casting shadows on one another. He thought that fore-shortening would, on the other hand, tend to give an increased brightness to ring C and the Cassini division by bunching the sparsely distributed particles together. He inferred from the paleness of the division that, as had been suspected before (notably by Barnard), it is evidently far from being a vacant space. He noted that the crepe ring appeared brighter in its outer parts, presumably through greater density of particles there: as a rule when the ring system is widely open this effect is, he stated, offset by contrast with the sky and ring B.

Steavenson said that the superior brightness of ring A was confirmed on May 22 by Professor R. O. Redman, Dr. E. H. Linfoot and M. W. Ovenden, but by May 28 the difference was much less

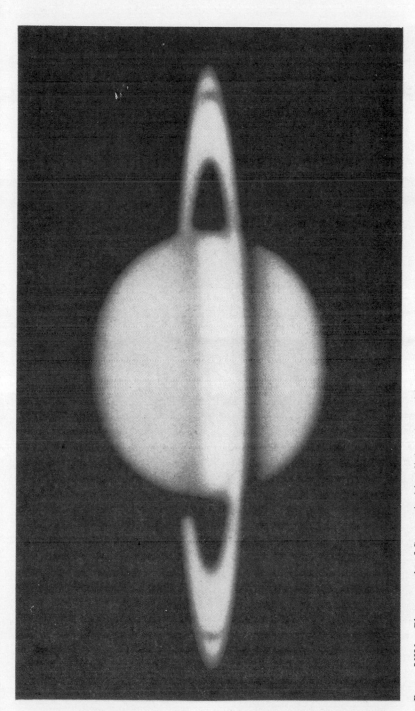

PLATE XIX.—Photograph of Saturn in blue light taken with the 200-inch Hale Telescope at Mount Palomar Observatory (1948 mid-Dec.). The approximate date can be deduced from the proportions of the ring ellipse which narrowed greatly by 1948 autumn but widened somewhat in 1949 spring, afterwards narrowing still further (cf. Camichel's photographs in Plate XVIII).

FIG. 1.—1950 April 7.

FIG. 2.—1951 April 15.

PLATE XX.—Photographs of Saturn a few months before and after the Earth's passage through the ring-plane, taken by H. Camichel at the Pic-du-Midi Observatory. Note the change in tilt from South to North.

marked. During the discussion of the observations Dr. R. d'E. Atkinson pointed out a difficulty regarding the explanation: the close correlation of the naked-eye magnitude with the degree of openness of the rings means that the surface brightness of at least a good fraction of the rings is nearly constant, as might be expected if the particles were so numerous that they covered the apparent area nearly completely, whatever the angle of opening; hence it was difficult to see why closing the ring up more should decrease the surface brightness, or why this should affect ring B differently from ring A. Dr. Steavenson, while admitting that he was not entirely satisfied with the explanation, stated that he felt no doubt as to the reality of the effect.

In 1951, after the passages of Earth and Sun through the ring-plane, Steavenson made another remarkable observation, again using the 25-inch aperture refractor: on June 6 he succeeded in glimpsing the sky as a 'slender grey line' between the globe and the ansae of the almost edgewise ring. The extreme difficulty of this observation may be judged from the estimated width ($0''2$) of the strip of sky (*B.A.A.J.*, vol. 61, p. 212).

These observations by Dr. Steavenson compensate somewhat for the planet's having been unobservable around the dates of the passages through the ring-plane, and also strikingly illustrate the necessity of large aperture for the more delicate and critical examination of the phenomena of Saturn.

Other 1950–51 observations

J. R. Bazin and other observers in the B.A.A. Saturn Section drew attention to the faintness of the belts, especially after opposition, and the brightness of the equatorial zone, while some of them noted condensations now and then on the south temperate and north equatorial belts (*B.A.A.J.*, vol. 61, p. 211).

The Saturn Section of a new German society of amateur astronomers, the Bund der Sternfreunde, made numerous observations 1951 February to June with telescopes of 10 inches aperture or less, finding the north and south equatorial belts fairly prominent and sometimes doubled, and containing knots and darker sections; they also detected a south temperate and a fragmentary north temperate belt. They agreed with the brightness of the equatorial zone until March, but found it losing brilliance and width thereafter (*B.A.A.J.*, vol. 62, p. 115).

On 1950 May 6, 21^h 45^m U.T., G. Ruggieri, observing at Venezia-Mestre with a refractor of aperture only 12 cm., had succeeded in detecting the shadow of Titan in transit over Saturn's disk, as shown in his drawing reproduced on Plate XXI, Fig. 1. Owing to the small aperture and the considerable apparent separation between the planet and the satellite he thought at first that the notch he saw on the disk might be due to a defect in the eyepiece, but he soon verified its objectivity. The shadow, at first elongated, became more circular as it approached the central meridian. The position of the shadow as shown on the drawing agrees well with the prediction of the transit in *B.A.A. Hdbk*. 1950, which gave the times of beginning and end of the shadow transit as May 6^d 21^h 17^m and May 7^d 0^h 43^m U.T. respectively, a prediction of which the observer was unaware.

In the summer of 1951 the author of this book, being called upon to take charge of the B.A.A. Jupiter Section, had to resign the directorship of the Saturn Section, being succeeded by one of its leading observers, M. B. B. Heath.

Saturn 1951–54 (Heath)

One of the characteristics noted by the Saturn Section in 1951–52 was the temporary extreme brilliancy and whiteness of the equatorial zone, and its subsequent fading. In 1951 November D. J. Fulcher made the intensity of the zone $\frac{1}{2}$ point brighter than the outer edge of ring B, and thenceforward, especially in 1952 March and April, intensity estimates and reports from Patrick Moore and numerous other observers emphasised its 'startling whiteness'. Heath found its intensity $1\frac{1}{2}$ points brighter than the outer part of ring B on April 11. Thereafter his and A. P. Lenham's intensity estimates showed a gradual decline to the end of June. Other features of 1951–52 were: the duskiness of the globe north and south of the bright zone, with few belts showing; the northern shadings usually darker than the southern; the darkness of the north equatorial belt, especially along its southern edge, considered not to be wholly due to contrast with the adjoining zone; faint mottlings in this belt and a slight irregularity along its southern edge (*B.A.A.J.*, vol. 62, p. 254; 63, p. 34).

In 1952–53 although the outer part of ring B was the brightest portion of the ring system, Heath found that its intensity had declined half a point from normal by 1952 December and a further half point in 1953 February, gradually recovering by early April. This was confirmed by the independent intensity estimates of Dr. A. J. Way (New South Wales). The crepe ring, though not an easy object, was seen by several observers; I. R. H. Brickett (South Africa), probably in better seeing conditions, estimated it a point less dark than did observers in England. The equatorial zone, though never attaining the unusual temporary brilliance of 1952, was for some time brighter than the brightest part of ring B, as shown by the estimates of Dr. Way and W. E. Fox in 1953 April and May, while other estimates appear to show an appreciable fading afterwards. By far the most prominent marking was the north equatorial belt, often with a very decidedly darker southern border; several observers drew dark spots within the belt, though Dr. Way could detect none. The evidence seems more conclusive that this belt was occasionally double. On the whole the north polar regions seem to have been darker than the south limb. It is doubtful whether any other belts than the north equatorial one were visible, though the generally uniform shading of the northern hemisphere beyond that belt sometimes allowed a suspicion of a very slightly lighter north tropical zone (*B.A.A.J.*, vol. 64, p. 23).

Heath's Section report for 1953–54 shows that a number of observers detected, though with difficulty, Encke's division in the ansae of ring A. The crepe ring was well seen by many, and Brickett's numerous estimates of intensity seem to show it to have been of the same darkness as in 1952–53. Several minor divisions in ring B and one between rings B and C were detected by the American observers, D. P. Avigliano and T. A. Cragg, using respectively the 24-inch aperture Lowell refractor and the 60-inch aperture Mount Wilson reflector. Cragg appears to have estimated the positions of the three subdivisions he detected on ring B at about one-third, half and two-thirds of the distance from the inner to the outer edge of the ring; these would be similar to the positions of those seen (also on the northern face of B) by Antoniadi in 1896 (see Chapter 21 and Fig. 12).

The tilt of Saturn's axis in 1953–54 rendered the south polar region invisible; the globe from the outer edge of ring A to the south limb was seen more or less shaded by most observers, though a few occasionally glimpsed a south temperate belt. From 1953 October until mid-April following, the ring's shadow fell on the north side and was involved with the trace of the crepe ring across the globe; for the rest of the apparition the shadow fell on the south side of the ring; the shadow's extreme thinness at the period of transference made it unobservable during 1954 April. The equatorial zone was certainly less bright than the brightest part of ring B (B_1) in the early part of the apparition, but much brighter in December and January, from then onwards slowly fading to a minimum below B_1, and finally rising to equality. A very thin faint equatorial band in the midst of the zone was sometimes seen. The north equatorial belt, again prominent with a dark southern edge, was sometimes seen double. Most observers distinguished a north tropical zone in the midst of the northern shading; some saw a north tropical belt; nearly all saw a more or less dark north polar cap (*B.A.A.J.*, vol. 65, p. 156).

Kuiper rejects minor 'divisions' in the rings

On a nearly perfect night in 1954 Dr. G. P. Kuiper examined Saturn's rings, using the 200-inch aperture Hale reflector at Mount Palomar with a power of 1,170. He was inspecting the rings for the reality of the numerous divisions that had been reported. His conclusion was that *only one division exists*, the Cassini division, whose width is one-fifth that of ring A. The other 'divisions' are, he considers, either minor intensity ripples, with some 10–15 per cent amplitude, or are non-existent. The Encke 'division' is a ripple at the position where ring A changes its intensity abruptly. There are three ripples in ring B, and there is *no* gap between ring B and the crepe ring. In his opinion there is an incredible difference between the large-scale luminous image of the great telescope, without disturbing chromatic effects, and with an effective resolving power of 0″05 or better, and that seen in a refractor of 20 or 40 inches aperture (*Trans. I.A.U.*, vol. 9, p. 255).

The importance of this observation and pronouncement by one of the world's leading planetary observers, using the world's largest telescope, is self-evident, and clearly refutes any idea that may exist of there being any gap through the rings other than Cassini's division. It confirms the presence of a ripple near the middle of ring A and three in ring B, however, and as past observations seem to show such markings to be variable in number and position, does not entirely exclude the possibility of others being observable at other times. Also, Kuiper's observation was of the markings on the northern face of the rings, while those of Lowell and Lyot were of markings on the southern face.

It would seem that Kuiper, having established the difference in character between Cassini's division and the minor markings, wishes to discourage the use of the term 'division' for the latter since it might be taken to imply a gap through the ring. Evidently in the early nineteenth century, markings such as Encke's division were so termed because they were really believed to be gaps, as foreshadowed by Laplace's theory, but by 1854 Dawes had begun to doubt this interpretation (see Chapter 11). Since then it is probable that most observers of these markings have employed the term 'division' in a purely technical sense as a convenient description, established by long usage, for a narrow, dusky, concentric marking on a ring, without intending to imply anything about its real character. There is a certain analogy here to the use of the equally unfortunate, but equally convenient, term 'canal' to describe a particular type of marking on Mars.

Saturn in 1955 (Heath, Ruggieri)

The B.A.A. Section Report (*B.A.A.J.*, vol. 66, p. 312) on apparition owed much to a detailed account from Guido Ruggieri of his observations in 1955 May–July, using the 20-inch aperture refractor of Merate Observatory (drawing on Plate XXI, Fig. 2) and his own 10-inch aperture reflector at Venezia-Mestre; owing to Saturn's low declination most other reports came from the southern hemisphere: R. W. Boggis (Western Australia, 12-inch reflector); J. H. Botham and I. R. H. Brickett (Johannesburg, 9- and 6-inch apertures).

On ring A, Ruggieri was able to trace Encke's division as a narrow dusky line round most of the ring; he estimated its position at from 33 to 50 per cent of the distance from the outer to the inner edge of the ring, agreeing well with the positions of the two outer markings measured by Lyot on the southern face of that ring (see Chapter 37 and Plate XVII, Fig. 2). He also noted, especially in steady seeing, some 'pale, clear radial streaks' on both ansae of that ring from the Cassini division to the outer edge. As appearances of this kind had formerly been drawn by Trouvelot, Antoniadi (see Figure 12), Rudaux and Maggini, and especially at the Jarry Desloges Observatory in 1909 and 1911, Ruggieri made some research on the subject, and concluded that this is undoubtedly an optical illusion, disappearing when the rings are too closed or too open; similar appearances can be noted when observing at a certain obliquity an illuminated disk with very delicate concentric streaks.

Nearly all the observers agreed that the inner part of ring B was the darker; they did not, however, detect any sign of a division or ripple on this ring, even at the line of demarcation between the brighter and darker parts. Ruggieri reported that a careful examination of the junction of rings B and C at the ansae showed no division between them, only a sharp difference in brightness. He found the crepe ring merely grey, not brownish, and he and the South African observers verified the transparency of ring C across the globe. In good conditions the globe's shadow on the rings was always seen very sharply, perfectly black, and without any irregularities, as was Cassini's division.

A thin sliver of the south limb was visible throughout the apparition, the limb not disappearing completely behind the rings until 1955 November 19, but a portion of it was blacked out by the shadow of the rings on the globe from May 1, and was shown as a dark line adjoining the outer edge of ring A on observers' drawings. Most estimates made the equatorial zone slightly less bright than the outer part of ring B, at least from April onwards. Ruggieri found the zone mottled with very tenuous diffuse shadings, and detected in it an equatorial band as a faint broken line.

The most prominent globe feature, the north equatorial belt, had irregularities on its southern edge noted by several observers; always undulating and notched, that edge once appeared to Ruggieri, in very steady seeing, as a delicate fringe; within the belt also he found a structure of condensations and irregular shadings. He considered the intensity of the north tropical and temperate zones variable. Several observers occasionally noted a north temperate belt as a diffuse streak with uncertain edges. Though the north polar region seems to have been devoid of detail, intensity estimates and drawings agreed in making it slightly darker than the rest of the northern shadings.

Saturn in 1956 and 1957 (Heath)

Protracted breaks in the continuity of observing Saturn had occurred in the later 1860s and 1920s (see Chapters 14 and 33), in each case mainly because of the planet's southern declination. An equally bleak prospect faced the B.A.A. Saturn Section in the 1950s: observing would, as the Director pointed out (*B.A.A.J.*, vol. 62, p. 173), become steadily more restricted until, from 1956 to 1961, the planet would be observable from 52° N. latitude only during summer twilight at opposition. Happily the fear of stagnation has not so far been realised, largely because of the valuable contributions received by the Section, not only from observers in Britain and others in the southern United States but also from several in South Africa, Australia and South America.

The Section report for 1956 (*B.A.A.J.*, vol. 67, p. 239) shows that the south limb was now completely hidden behind the widely open rings. Little difference was found in brightness between the inner and outer parts of ring A, and perhaps less than usual between those of ring B. Encke's division was seen in the ansae by some; A. P.

Lenham, working at American observatories and using large apertures, never saw it as a hard black division but as a 'grey shade' at the junction of A_1 and A_2 (the outer and inner parts of the ring). Observers, including Lenham with the 82-inch aperture McDonald reflector, all found the crepe ring very faint during the apparition; R. W. Boggis, using 10- and 12-inch apertures at Perth Observatory, Western Australia, found ring C consisting of two distinct sections, the inner nearly as dark as the sky, the outer rather brighter than usual. V. A. Firsoff, observing through a blue filter, found the space between globe and ring C intensely dark, which would seem to indicate the almost complete absence of any finely divided matter there. I. R. H. Brickett, using the 9-inch aperture refractor at the Union Observatory, Johannesburg, was able to observe the black shadow of globe on rings on the preceding side up to and including the day of opposition (1956 May 20); two days later he found no shadow visible, but with high magnification the part of ring A north of the globe appeared considerably darkened. After May 8 the shadow of the rings fell south of them and so was invisible, but up to that date the shadow was involved with the crepe ring where it crosses the globe, though that part of the crepe ring always appeared dark whether the shadow fell there or not.

The equatorial zone is shown by visual and intensity observations to have deepened in tone considerably in 1956 May, it having been only slightly less bright than the outer part of ring B in March and early April; this dullness of the zone was noted by observers in S. Africa and Australia as well as by P. A. Moore in England, Boggis describing it on May 22 as 'very deep yellow'. During June and July it slowly lightened, losing the deep yellow tint, and by August was of about the same intensity as in April. The north equatorial belt, fairly dark throughout the apparition, was occasionally seen double by some observers with a slightly lighter zone between the components; its south edge was sharp, but the north edge sometimes seemed to fade gradually into the slightly lighter temperate zone, which was seen by Lenham, using the 82-inch aperture reflector, to be of a beautiful pink colour. Faint mottlings in the interior of this belt and slight undulations on the edges seem to have been glimpsed from time to time. A north temperate belt, probably not permanently visible, was occasionally seen by most observers; Firsoff found both these northern belts emphasized by using a green monochromatic filter. Most observers noticed a rather dark north polar cap, and two also noted a still darker north polar band.

The Section report for 1957 (*B.A.A.J.*, vol. 68, p. 57) states that rings A and B and Cassini's division presented no unusual features at this apparition, during which the northern tilt of Saturn was great enough for the outer part of ring A to be visible the whole way round. Lenham, using large apertures at the Yerkes and McDonald Observatories, could see no sign of any division except Cassini's. Ring C seems to have been very faint in the ansae even to observers in the southern hemisphere, but across the globe it was invariably estimated darker than the north equatorial belt; according to Botham's estimates it was darkest in April, when the shadow of the rings on the globe was involved with it, and the impression of its darkness may have been strengthened by the fact that the northern edge of the south equatorial belt was noted closely contiguous to the outer edge of C^m (the crepe ring across the globe) in May and June by Boggis and Botham—hence part of that belt lay directly behind C^m. Heath has pointed out in this report that to be so visible the belt must have extended considerably towards the equator in recent years, as the saturnicentric latitude of its northern edge could not have been greater than $-5°6$ at the time of these observations; in the latitude of this belt the Sun would have been eclipsed by the rings for several years.

In 1957 the equatorial zone was unanimously considered definitely duller than the outer edge of ring B; Lenham noted in it a faint, diffuse, rather broad equatorial band, once very faintly seen also by Boggis. Using an orange filter, Boggis and C. R. Edwards, at Perth, Western Australia, drew some faint detail of streaks interspersed with light areas on the north equatorial belt, showing it double, but Lenham, both with and without a blue-minus filter, observed no spots or other irregularities on it. He described the north temperate zone as 'still pink'. A north temperate belt seems to have been occasionally glimpsed by some observers. All intensity observations and drawings, and the photographs taken by H. E. Dall ($15\frac{1}{2}$-inch aperture reflector) and Botham, show the north polar region very dark. (Plate XXII, Fig. 1 shows a photo of Saturn in 1957 taken by Dr. H. Camichel.)

Saturnicentric latitudes of belts, 1957. The following mean figures were obtained from measurements of photographs by Dall and Botham and of drawings by Boggis and Edwards:

	Dall	Botham	Boggis	Edwards
North Eq. Belt south edge	$+12°2$	$+13°3$	$+15°2$	$+13°6$
South edge N. Polar shading	$+59°1$	$+60°7$	$+65°5$	$+62°1$

The latter set of figures shows very fair agreement in view of the foreshortening at high latitudes.

424

SATURN IN 1956 AND 1957 (HEATH)

Occultation of star BD −20° 4568. The occultation of this star by the rings on 1957 April 28 was reported on by J. E. Westfall, using the 20-inch aperture refractor at Chabot Observatory, California. According to his report the star, orange-red before occultation, assumed a peach colour when glimpsed through the rings, and during the 3½ hours while passing behind the ring system showed many rapid rises and falls in brightness. Dr. M. Bobrov, of the U.S.S.R. Academy of Sciences, computed from Westfall's observations an optical thickness of rings A and B of the order of 1·0 to 1·5, which he considered possibly too high, since Ainslie's observation of 1917 gave an optical thickness of ring A of 0·6 to 0·7. He thought it might be accounted for by the comparatively low altitude of Saturn when observed by Westfall (*B.A.A.J.*, vol. 67, p. 277; 68, pp. 58, 287).

Saturn in 1958 (Heath)

At the period of the opposition of 1958 June 13 the southern declination of Saturn was nearly 22°, so the report of the B.A.A., Saturn Section (*B.A.A.J.*, vol. 70, p. 29) was based mainly on the work of numerous observers in the United States, Brazil, S. Africa and W. Australia; a set of observations came also from the Auckland Astronomical Society, New Zealand. In 1958 (late October) the rings attained their maximum opening, nearly 27°, and Botham, at Johannesburg, noted on October 2 that the Cassini division dipped just below the north limb. The crepe ring was usually very faint, and often invisible even in good seeing, during the 1958 apparition, but more prominent with a green filter. A. P. Lenham, using the 40-inch aperture refractor of Yerkes Observatory, found the crepe ring uniform in intensity with a smooth inner boundary, and no division between it and ring B (*B.A.A.J.*, vol. 69, p. 165). He also saw once (July 24) a weak intensity ripple in ring A, probably representing Encke's division, and on August 27 two intensity ripples in ring B. Botham, using the 26½-inch aperture refractor of the Union Observatory, Johannesburg, on October 2, made a special examination of the structure of the rings, noting Encke's division at two-fifths of the way from the outer to the inner edge of ring A, which seemed to be of equal brightness on each side of the division, and two distinct steps in brightness, not seen clearly as breaks, in ring B. C. R. Edwards, using the 12½-inch aperture reflector of Perth Observatory, W. Australia, on October 13, also recorded these two steps in brightness.

R. R. de Freitas Mourão, using the 8-inch aperture refractor at the National Observatory, Rio de Janeiro, in 1958, only had a glimpse of Encke's division and a comparison of his intensity estimates, made in 1957 and 1958, showed both parts of ring A darker in 1958 with more intense contrast between them, and the part of Cassini's division crossing the globe darker than it had been in 1957 (*Publ. Obs. Nac.*, No. 6).

The equatorial band was frequently invisible, and, when seen, often very faint even in large apertures (e.g. 26½- and 40-inch). The north equatorial belt often appeared narrow, the southern component only being visible, and even that barely continuous. Brickett (Johannesburg) found it in June more like a series of irregular dark markings than a continuous belt, but in the autumn of 1958 he saw it broad and featureless but with well-defined edges. On July 23 D. P. Cruickshank, using the Yerkes 40-inch aperture refractor, saw the north equatorial belt completely double, and the Section report shows that partial duplicity, as well as faint condensations in the belt, were noted and drawn by numerous observers in several countries; some drew irregularities along the southern edge. Lenham (40-inch) described the southern component of this belt as rather narrow, and the northern as thin, very weak and seen only in good conditions.

The north temperate belt, fairly broad and diffuse according to Lenham, may not have been a permanent feature extending all round the planet; it was seen faintly or with difficulty by some observers in W. Australia, S. Africa and Brazil. Lenham seems to have been the only one who also detected in good seeing a very weak and narrow north north temperate belt as well. Intensity estimates and drawings showed that the north polar region was usually but not always dark in 1958, and Lenham found the north polar cap rather small.

Observations with filters. R. W. Boggis (W. Australia, 12-inch aperture reflector) made visual estimates of comparative intensities of the equatorial zone, north equatorial belt, north polar region and the rings with six different colour filters. He found no marked difference with Dufay B or W47b, but W61 made the globe features and ring C brighter; K2 made the belts darker, but the zone and rings A and C brighter; W15g made the belts darker but ring C brighter; while W29 darkened everything except the equatorial zone.

The increasing use of filters evident from observers' reports calls for a word of warning. The application of filters ought to be limited to large or at least moderate apertures; the danger with small ones is that the filters reduce the brightness of the image so much that any detail is only just above the threshold of vision, and so the way is opened to error and self-deception, against which observers of Saturn have to be always on their guard in any case.

Saturnicentric latitudes of belts. The measurement of drawings and a photograph by eight observers in four countries gave values ranging from about 8° to 12° N. for the saturnicentric latitude of the north equatorial belt, the general mean being about 10° N. Lenham, using the Yerkes 40-inch aperture refractor, made micrometric measures, and also measures on a plate taken at the principal focus, with the following results: E.B. 5°, N.E.B$_s$ edge 11°, N.E.B$_n$ edge 26°, N.T.B. 44°, N.P.C. edge (low weight) 76° saturnicentric north latitude.

Subdivisions (light minima) on northern face of rings, 1957 and 1958 (Dollfus)

Dr. Audouin Dollfus has kindly given the author by letter, in advance of publication, some particulars of his recent observations of the details on the northern face of the rings. In the summer of 1957 he was able to make a careful study of the minima of light on the northern face of Saturn's rings, using the 82-inch aperture McDonald reflector, stopped down to about 25 inches to suit the seeing conditions. He found almost exactly the same results as in Lyot's drawing of the southern face (see Plate XVII, Fig. 2). There were no gaps except Cassini's division, but several minima of light, or narrow shadings of low contrast, in perfectly good agreement with Lyot's observations. Dollfus saw the same features again in

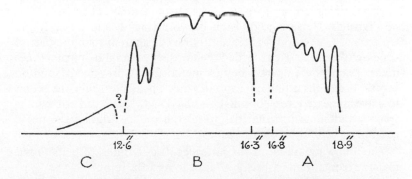

FIGURE 22. A. Dollfus: photometric diagram of the northern face of Saturn's rings based on his observations at the McDonald and Pic-du-Midi Observatories, 1957 and 1958. The scale shows apparent distances of ring edges from Saturn's centre, the planet being at 10 A.U. from the observer. (Redrawn after his diagram).

excellent seeing in the spring of 1958, using the 24-inch aperture refractor at the Pic-du-Midi Observatory at full aperture. He also made micrometric measurements of rings A and B with his bi-refringent micrometer, and considers that the accuracy of the determinations with this instrument are probably far superior to the classical ones obtained with filar micrometers. He considers that his observations and size determinations of the minima of light seem to fit well with the improved theory derived from Kirkwood's explanation of the ring markings as due to a resonance effect produced by the perturbations of the satellites on the ring particles, and that this would explain with a good degree of accuracy the most prominent of the details observed.

Dollfus's diagram (Figure 22) shows a deep trough (very dusky marking) in the outer part of ring A, three fainter ones around the middle of that ring, two very slight ones near the middle of ring B, and a pair of very dusky ones on ring B near its inner edge.

Saturn in 1959 (Mourão)

Early information of Saturn in 1959 has been kindly sent to the author by letter from Brazil. Professor R. R. de F. Mourão has supplied the main particulars of his observations made with an 18-inch aperture Cooke refractor at the Observatório Nacional, Rio de Janeiro. Encke's division seemed to him merely 'the separation of two parts of different brightness', but his colleague Dr. L. M. Barreto noted it as 'a weak thread at the limit of visibility'. Cassini's division was clearly outlined and perfectly dark. Ring B, as in his previous observations, showed three parts of different intensity but no divisions. He saw no division between rings B and C; he found the part of C next to B lighter and more transparent than the part of C nearer the globe. He recorded the north equatorial belt (sometimes double), a north temperate belt (sometimes very prominent), and on several occasions a thin equatorial band. The equatorial zone seems to have appeared very bright compared with the northern hemis-sphere, and some irregularities were seen on the edge of the north equatorial belt adjoining the zone.

The following are means of saturnicentric latitudes of belts found from micrometric measures by Mourão and Barreto:

| N.E.B. (S. edge) | $+12°3$ | N.T.B. (mean of 2 measures) | $+43°5$ |
| N.E.B. (centre) | $+15°8$ | N.P.R., S. edge (mean of 3 measures) | $+64°8$ |

They also made a series of intensity estimates.

Saturn was in Sagittarius at about $22\frac{1}{2}°$ S. Declination.

These Brazilian observations with some from A. W. Heath and

D. V. Wraige (England), W. H. Haas and O. C. Ranck (U.S.A.) and C. R. Edwards (Australia), and two good drawings by Edwards, made up the B.A.A. Saturn Section's report for 1959 (*B.A.A.J.*, vol. 71, p. 105), which also noted the equatorial zone as fainter than the outer edge of ring B, and the north polar region as variable with the cap small.

Chapter 40

HYDROGEN CONTENT, MASS, AND OTHER RECENT THEORETICAL AND ASTROPHYSICAL STUDIES OF THE PLANET AND SATELLITES

Saturn a hydrogen planet (Ramsey)

Dr. W. H. Ramsey's study of the internal constitutions of the major planets is contained in three papers (*M.N.*, vol. 111, p. 427; vol. 112, p. 234; and *Occ. N.*, vol. 3, p. 87), of which the first was produced in 1951 May and the second, written in collaboration with B. Miles, in the following January. The following résumé of some of the main points relating to Saturn can, as with other fine papers summarised in this book, do but scant justice to the original.

With regard to the models of the major planets proposed by Wildt in 1938 (see Chapter 35), Ramsey pointed out that Wildt's theory would have given Saturn, intermediate in mass, relatively about three times as much free hydrogen as the other major planets, which would imply that the planets were originally of different chemical compositions. An even more serious flaw in the model was that no allowance was made for transitions to metallic phases that occur at the high pressures in the planets' interiors. Hence the density of the core would be far greater than Wildt assumed, and the densities used for the other layers are also too low. The heavy elements therefore cannot be so abundant in the major planets as the model suggests.

Furthermore, scientific opinion had changed since 1938 on two matters taken for granted when Wildt proposed his theory: the internal constitution of the Earth and the feasibility of large-scale chemical separations under gravity. Seismic observations indicate a discontinuous change in structure at about halfway to the centre of the Earth, and in the 1930s this had been interpreted as indicating a separation under gravity into an iron core and a surrounding mantle of less dense material—a mixture of magnesium and iron orthosilicates. In the newer theories the deep interior of the Earth is assumed to be chemically homogeneous; the core is attributed to the enormous pressures at great depths partially destroying the molecular structure of the material, so that it becomes dense and metallic. Moreover, since 1938, it has also been shown that geological time is too short to allow the formation of an iron core in the Earth, and the time required for chemical separation under gravity would be far longer for planets such as Saturn owing to their greater size. Hence the major planets may be chemically homogeneous. Ramsey pointed out, however, that whether they are so at present depends on their mode of formation: each major planet would possess a small central core of terrestrial composition if they were formed by an aggregation process, whereas they could be chemically homogeneous if they were formed from material ejected from a star. In his first paper Ramsey assumed them to be chemically homogeneous; in his second he re-examined his theory on the assumption that the heavier elements are in part concentrated in a central core and in part distributed uniformly.

A comparison of mean densities of the major planets shows that Saturn, though intermediate in mass, is only about half as dense as the others. He drew attention to two opposing factors regarding density: the greater compression in the larger planets tends to make mean density increase with mass, but their probable retention (or capture) of a higher percentage of lighter materials tends to have the opposite effect. He calculated and considered the ratio (e/m), where e is the ellipticity and m the ratio of the centrifugal and gravitational forces on the planet's equator. The ratio (e/m) is smaller for planets whose mass is more concentrated towards the centre: 0·5 where the planet's mass is concentrated in a small central core, 1·25 for one of uniform density. For the Earth which has a dense central core the figure is 0·975, for the major planets a range from 0·65 (Saturn) to 0·78. Hence all of these planets, and Saturn especially, have strong central concentration of mass.

From an argument by Jeffreys he derived an upper limit to the surface density (Saturn 0·39 g./cm.³) and a lower limit to the central density (Saturn 1·7 g./cm.³) for each of the major planets and the Earth; as hydrogen and helium are the only solids which have densities as small as 0·39 g./cm.³, he inferred that the outer layers of Saturn must consist mainly of these elements.

As all gases in the atmospheres of the major planets are highly compressed under their own weight, and the density of a perfect gas would exceed that of its solid at a depth of less than one per cent of the radius below the visible surface, the thin layer of gases in a truly gaseous state would only contribute negligibly to the mass and volume of the planet. Hence Ramsey felt that, for the purpose of his calculations, he could assume that the planets are solid throughout.

In view of the suggestion of Wildt and Kothari that hydrogen may be metallic in the interiors of the major planets, Ramsey drew up a table, based on his own calculations and those of others, showing the pressure-density relationship for solid hydrogen at absolute zero temperature. He found that the critical pressure for the transition of hydrogen to a metallic phase is $0\cdot8 \times 10^{12}$ dynes/sq. cm., which is equivalent to 700,000 atmospheres, and about eight times the highest pressure attained in a laboratory (by 1954). The calculations showed that at this pressure the density jumped from $0\cdot35$ g./cm.^{3} (molecular phase) to $0\cdot77$ g./cm.^{3} (metallic phase). But the central pressure in Saturn is reckoned to be about seven times (in Jupiter about forty times) as great as the critical pressure for the production of metallic hydrogen, i.e. there is a pressure of about 5 million atmospheres at the centre of Saturn.

After a discussion of central temperatures he reached the conclusion that the major planets cannot have central temperatures greater than of the order of 10,000°C., either as a result of radioactivity or as a relic of original heat. This, he considered, would be too low a temperature appreciably to modify the pressure-density relationship.

He found that the computed moments of inertia are close to, but somewhat larger than the empirical values, suggesting a non-uniform distribution of the heavier elements in the planets and that such elements are to some extent centrally condensed. He therefore tested his calculations on this assumption in his second paper, but found that it had little effect on the estimated hydrogen contents, the final result being a hydrogen content for Saturn of from 62 to 69 per cent by mass (for Jupiter 76 to 84 per cent). This showed that these two planets are of roughly solar composition; their ellipticities suggested a central concentration of heavy elements of the order of ten times the Earth's mass, though the possibility of chemical homogeneity was not completely excluded. The difference in the hydrogen content of Jupiter and Saturn he considered due mainly to the fact that the central core accounts for a larger fraction of Saturn's mass. He considered that Saturn and Jupiter are the only mainly hydrogen planets: Uranus and Neptune, about ten times as dense as hydrogen planets of the same mass, are presumably composed mainly of water, methane and ammonia.

The fact that scientists in different countries are often found to be working on the same lines at about the same epoch is instanced by the paper by Professors V. G. Fesenkov and Mme A. G. Masevitch (*Ast. J. U.S.S.R.*, vol. 29, no. 5, 1951), who examined the

Fig. 1.—1950 May 6, with 12-cm. refractor, showing Titan's shadow on the planet, and Titan near left side of picture. See text p. 418.

Fig. 2.—1955 May–July, with 20-inch refractor of Milano-Merate Observatory and 10-inch reflector at Venezia-Mestre.

Plate XXI.—Saturn drawn by G. Ruggieri; crepe ring well shown.

Fig. 1.—Photograph of Saturn by H. Camichel (Pic-du-Midi Observatory) 1957 August 28.

Fig. 2.—Photograph of Saturn by J. H. Botham (9-inch refractor, Union Observatory, Johannesburg) 1960 April, showing N.N. temperate white spot spread along zone.

Fig. 3.—Saturn drawn by G. Ruggieri 1960 September 10, 10-inch reflector.

PLATE XXII

internal structure of Jupiter and Saturn on the basis of available data on the equation of state for solids at high pressure, including the transition to the metallic phase, and suggested a model with an outer layer of molecular hydrogen, an intermediate one of atomic metallic hydrogen, and an inner core of hydrogen with heavier atoms. They found the total hydrogen content to be about 80 per cent by weight, in close agreement with Ramsey's results for Jupiter (*Trans. I.A.U.*, vol. 8, p. 215).

Ramsey's conclusions have also been supported by the subsequent mathematical investigations into hydrogen planets of Sir Harold Jeffreys (mentioned later in this chapter) and of Wendell C. DeMarcus, who calculated for Saturn a hydrogen composition exceeding 63 per cent by weight and a number of hydrogen atoms more than seven times the number of helium atoms (complete paper in *Ast. J.*, vol. 63, p. 2, 1958).

Structure of the rings (Bobrov)

Dr. M. S. Bobrov, whose interest in the observation of a star occulted by the rings in 1957 was mentioned in the last chapter, studied the transmission properties of the rings and the probable size and shape of the ring particles in a series of papers published in 1940 and 1951–56 (*Russian Ast. J.*, vol. 17, no. 6, p. 1; 29, no. 3, p. 334; 31, no. 1, p. 41; 33, no. 2, p. 161; 33, no. 6, p. 904; and *Proc. Acad. Sci. U.S.S.R.*, vol. 77, p. 581, 1951). Bobrov's 1952 paper has been summarised (in *Ast. News Letter*, no. 67, 6) and there are references to it in two papers by A. F. Cook and F. A. Franklin (mentioned later in this Chapter). From these sources it seems that Bobrov determined a lower limit for the optical thickness of ring B of 0·83 from an occultation on 1939 November 24 in which the star (BD +6° 259) could not be seen through the rings, and from other occultation observations a value of 0·7 for the average optical thickness of ring B. He then found that the proportion of sunlight reflected backward was too large to be accounted for by the diffraction phenomena, and exceeded the amount theoretically reflected from a smooth spherical surface. He therefore felt warranted in concluding that the particles of ring B are fairly large and have pitted surfaces, which would account for a phase-coefficient that is similar to that of the Moon, Mercury and some minor planets. As the albedo of the particles is large, he inferred that the unevenness of the surfaces must be somewhat more pronounced than for the Moon, and that their predominant size must be large enough to cast geometrical shadows by the elevations of their surfaces.

Other studies of Saturn in U.S.S.R.

Brief reference has been made (in *I.A.U. Trans.*, vol. 8, p. 215; vol. 9, p. 251) to a number of other fairly recent papers, of which details are unfortunately not readily available, written by Soviet astronomers about Saturn and other planets:

V. V. Sharonov (*Pulkovo Obs. Circ.*, nos. 26, 27, 1939) carried out an absolute photometry of Saturn during the 1937 opposition, using the Pulkovo 30-inch aperture refractor.

B. D. Furdylo (*Kharkov Obs. Publ.*, no. 7, 1941) suspected a reddening of Saturn's rings when the Sun and Earth approached the ring-plane.

N. P. Barabashev and A. T. Tchekirda (*Publ. Kharkov Obs.*, 2, 9, 1952) on photometric data of Saturn and the ring observed in 1951.

M. V. Bannova (*Ann. Len. Univ.*, nos. 153, 155, 1952) concerning isophotes of Saturn.

N. P. Barabashev (*Kharkov Univ.*, 1952)—Investigation of the physical conditions on the Moon and the planets.

V. V. Sobolev (*Progress in Astronomy*, 6, pp. 25–80, 1954)—Theory of transfer of light in planetary atmospheres.

V. V. Sharonov (*Progress in Ast.*, 6, p. 181, 1954)—a survey of planetary and satellite photometries.

A paper by V. V. Radzievsky on the braking action of radiation in the solar system and the age of Saturn's rings (*Russian Ast. J.*, vol. 29, no. 3, p. 306, 1952) has been briefly summarised (in *Ast. News Letter* no. 67, 3). In section 4, Radzievsky gives the age of Saturn's rings as $(0 \cdot 7 \times 10^9 \ a)$ years, where a, according to Bobrov, is the radius of the smallest particles, about 1 cm. Allowing for the mutual screening of the particles, his estimate of the age of the rings is something between $(0 \cdot 7 \times 10^9)$ and $(2 \cdot 1 \times 10^9)$ years.

More recently (*Trans. I.A.U.*, vol. 10, p. 250) V. V. Sharonov, continuing visual determinations of planetary colours with a colour-wedge photometer, measured Saturn in 1956 (*Ast. Circ. U.S.S.R.*, no. 174, 1956), and V. N. Lebedinetz published (*Publ. Kharkov Obs.*, 4, 1957) results of photographic photometry of Saturn with colour filters in 1951–54: the brightness of the brightest part of ring B during opposition was found to be 0·43 of that of an ideal white surface in visual light.

Search for other satellites (Gehrels)

T. Gehrels at Boyden Observatory made a telescopic search for possible hitherto undiscovered satellites of Saturn, but found none down to magnitude 19·5 and within a radius of $1\frac{1}{2}°$ of the planet (*Trans. I.A.U.*, vol. 10, p. 296).

Satellite masses and densities (Kuiper)

Dr. Kuiper (in *Atmospheres*, 1952, p. 307) gave the following figures for masses and densities of the inner satellites:

	Mass (Earth=1)	Mean Radius (Earth=1)	Mean Density
Mimas	0·000006	0·04	0·5
Enceladus	0·000014	0·05	0·7
Tethys	0·000109	0·08	1·2
Dionè	0·000176	0·07	2·8
Rhea	0·00038	0·102	2·0
Titan	0·0235	0·377	2·42

Kuiper pointed out that, whereas with Jupiter's four chief moons density decreases with distance from the planet, with Saturn's inner satellites the reverse is found.

Saturn's mass (Hertz)

In Chapters 16 and 29, various determinations of the value of Saturn's reciprocal mass by different astronomers have been given; in 1953 Dr. Hans G. Hertz worked out a figure based on observations of perturbations of Jupiter covering the period 1884–1948 (*Ast. J.*, vol. 58, p. 42). The value Hertz took as final is $3497·64 \pm 0·27$; this is close to H. Struve's determination (1888) of 3498, and 1/3497·64 is significantly higher than Bessel's value of 1/3501·6 for the mass of the Saturn system.

Satellite diameters (Kuiper, Camichel)

An important step towards the high-precision measurement of satellite diameters was the invention in the 1940s by B. Lyot and H. Camichel of the disk meter. This can be used for bodies showing very small apparent disks, the method being to measure the artificial disk produced in the focal plane of a large telescope: this artificial disk can be varied in brightness, size and colour until it appears exactly to correspond with the image of the satellite.

Using a disk meter on the 24-inch aperture refractor at the Pic-du-Midi Observatory, H. Camichel obtained the following results for satellites of Saturn: Dione 0″16; Rhea 0″22, Titan 0″76, all at 9 A.U. He considered his figures for Dione and Rhea as only estimates, but for Titan the mean of 30 settings gave an error of less than one per cent (*Ann. Astrophys.*, vol. 16, p. 41, 1953; *Ap. J.*, vol. 101, p. 259).

Kuiper used a disk meter on the 200-inch aperture Palomar

reflector on 1954 July 5 with the following results: Enceladus 0″08; Tethys 0″12; Dione 0″12; Rhea 0″24; Titan 0″67; Iapetus 0″195, all at 9·43 astronomical units (*Trans. I.A.U.*, vol. 9, p. 250).

The extreme difficulty, if not impossibility, of determining the diameters of Saturn's satellites more closely than to the nearest hundred miles is illustrated by comparing the results in miles from these measures, in columns (*a*) and (*c*) below, with those derived from Kuiper's table of mean radii (Kuiper; *Atmospheres*, p. 306):

	(*a*) From Kuiper's measured diameters miles	(*b*) From Kuiper's table of mean radii miles	(*c*) From Camichel's measures miles
Mimas		320	
Enceladus	340	400	
Tethys	510	630	
Dione	510	550	650
Rhea	1020	810	890
Titan	2850	2950	3080
Iapetus	830		

Satellite masses (Jeffreys)

A long mathematical discussion of the very difficult subject of the masses of Saturn's satellites by Sir Harold Jeffreys (in *M.N.*, vol. 113, p. 81) resulted in the following values expressed as fractions of the mass of Saturn:

Mimas $(6·69 \pm 0·20)10^{-8}$; Dione $(1·825 \pm 0·061)10^{-6}$;
Enceladus $(1·27 \pm 0·53)10^{-7}$; Rhea $(0·4 \pm 3·8)10^{-6}$;
Tethys $(1·141 \pm 0·030)10^{-6}$; Titan $(2·412 \pm 0·018)10^{-4}$.

The margin of error for some of them, especially Rhea, is very large, showing the indeterminacy of the mass. The formula for Rhea merely indicates the limits within which the theory would be satisfied on purely mathematical grounds: negative solutions should therefore be rejected. Jeffreys said that the result for Rhea might be interpreted as a probability of about 2/3 that the mass is less than 1/270,000 of Saturn's mass, one of about 1/3 that it exceeds this value, and of 1/20 that it exceeds 1/135,000; on the whole he considered that a value near 1/270,000 would be plausible. He has since suggested (in *M.N.*, vol. 114, p. 434) what is, in his opinion, probably a better value, namely: $(3·2 \pm 3·8)10^{-6}$.

Jeffreys also gave an estimate for the mass of Iapetus, derived from the inclination of the orbit to Saturn's equator: $2·46 (1 \pm 0·78)10^{-6}$ of Saturn's mass. The uncertainty is large, and he considered the result to be only an indication that the mass is comparable with those of Tethys and Dione.

Saturn's mass and mass-distribution (Jeffreys)

In two mathematical papers (*M.N.*, vol. 113, p. 97; 114, p. 433) Sir Harold Jeffreys in 1953–54 suggested a value for the reciprocal mass of Saturn of $3494 \cdot 8 \pm 2 \cdot 0$, allowing for uncertainty in the calibration of the micrometer screw, and considered the various alternative models for Saturn proposed by Miles and Ramsey. His conclusion favoured their model S_2, in which it was assumed that there were 18 times as many hydrogen as helium atoms, and in which 17 per cent of the mass of the planet would be concentrated in the central core. The absolute masses of the central bodies of Saturn and Jupiter would seem to be nearly equal (*Trans I.A.U.*, vol. 10, p. 254).

Orbit of Phoebe (Zadunaisky)

Since 1905 the elements and perturbations computed by F. E. Ross, based on observations 1899–1904, have been used for Phoebe, but observations have sometimes shown deviations from the ephemeris by as much as a minute of arc. In 1954 P. E. Zadunaisky rediscussed the orbit of Phoebe, using observations made between 1907 and 1942, and checking the new orbit with earlier and later observations (*Ast. J.*, vol. 59, no. 1, 1954). The calculations included solar perturbations, important in the case of an outer satellite, and the results showed a considerable improvement over Ross's orbit for Phoebe.

Formation of the planets (Kuiper)

The nebular theory of Kant and Laplace was outlined (in Chapter 7) because it tried to account for the existence of Saturn's ring system. Subsequent theories of the general origin of the solar system hardly come within the scope of this book and are readily available in others. Early in this century Laplace's theory was abandoned by most cosmogonists, and a succession of 'catastrophic' theories attributed the origin of the planets to material torn from the Sun or exploded from a companion star, but to practically all of these hypotheses objections were found. In 1955 Dr. G. P. Kuiper in three lectures (*R.A.S. Canada J.*, vol. 50, pp. 57, 105, 158) reviewed the history of and analysed many recent contributions to the subject showing how the two objections to the Laplacian theory (based on density and the distribution of angular momentum) though once thought insuperable, were in process of disappearing. Kuiper propounded a neo-Laplacian theory, according to which a nebula formed by the contracting Sun broke up into proto-planets, each developing at its centre a planetary body, and farther out minor satellites. Then the brightening of the Sun cleared out interplanetary space and caused the gradual evaporation of the proto-planet envelopes. This led to somewhat increased orbital eccentricities, resulting in the more distant satellites being dropped into inter-planetary space, some being eventually recaptured as irregular satellites. During the later stages each of the outer proto-planets collected quantities of material near its equatorial plane, which slowly spiralled inward and led to the formation of regular satellites close to the planets. With regard to Saturn's ring system, Kuiper considered that the gaseous disk surrounding the Sun would flatten until it reached a critical density, and then break up; in the case of Saturn's rings this density was never attained, not even upon con-densation of the ring, so that the flattening process has here been carried to the extreme. If it is supposed that proto-Saturn was less concentrated than proto-Jupiter, this would account for the greater mass loss of Saturn, and the somewhat greater extension of Saturn's system of regular satellites.

Recent chemical and spectrographic analysis of the metal contents of meteorites has indicated a reliable figure for the age of the solar system, namely 4·5 thousand million years (*Trans. I.A.U.*, vol. 9, p. 261).

Suspected radio emission from Saturn
(Smith and Douglas, Drake and Ewen)

Radio emissions were identified as coming from Jupiter in 1955 and at about the same period Venus was suspected of being a radio source. A planetary radio installation, operating at 21·1 mc., at Yale University Observatory, was used for two months in the early summer of 1957 by H. J. Smith and J. N. Douglas to follow Saturn, and thirteen events were noted which satisfied several criteria for Saturnian origin. These first results were, however, considered inconclusive pending more data and better discrimination against terrestrial atmospherics and galactic background (*Ast. J.*, vol. 62, p. 247).

From 1956 microwave observations of the planets have been made in America at the Naval Research Laboratory and at Harvard, and emissions have been recorded which probably have no direct relation to those mentioned above, since the wave-lengths of 3 to 4 cm. used in this case border on the infra-red region of the spectrum, so that the microwave emissions are presumably of thermal origin. Micro-wave emissions from Mars and strong ones from Jupiter and Venus have been recorded, and in 1957 F. D. Drake and H. I. Ewen, using a wave-length of 3·75 cm. at Harvard, received feeble ones from Saturn. Temperature measures for Jupiter and Saturn at this wave-length showed the temperature of Saturn to be the lower. The total emission from Saturn and its rings was 4·3 times weaker than that from Jupiter.

In calculating Saturn's temperature they assumed the rings to be transparent at 3·75 cm., which would mean that the ring particles are much smaller, with diameters of the order of a few millimetres or less. With improving apparatus microwave reception may reveal much about the lower atmosphere and surface of Saturn and other planets (J. D. Kraus: *Radio Obs. Report*, no. 8, Ohio Univ., 1958, p. 76; *Ast. J.*, vol. 64, p. 43).

Photometric measures of Saturn and ring (Camichel)

At the Pic-du-Midi Observatory H. Camichel had photographed Saturn at nearly all oppositions from 1942 to 1957 on panchromatic plates through a yellow filter, using 15-inch and later 24-inch aperture refractors. He thus had photographs extending over half a revolution of the planet, and showing the ring up to its maximum opening on both faces. He made measures of photographic density with the microphotometer at Toulouse Observatory and in 1958 published the results (*Ann. d'Astroph.*, vol. 21, p. 231).

He took the brilliance of the equatorial zone as the standard at unity, not because its brightness is constant but because it is least affected by difference of tilt and was the only region common to all the photographs. He found that the ratio of brightness (ring B: ring A) decreased 1942–49 from 1·75 to 1·3, but increased 1953–57 from 1·44 to 1·72, so that when the rings were only slightly open the difference in their brightness was least. He considered that the variation was due to varying brightness of ring B, that of A staying almost constant. He found no systematic variation from opposition to opposition for the belts, and no difference greater than the errors of measurement between belts of the two hemispheres. He found, however, a decrease in the brilliance of ring B when going from the extremity of an ansa towards that of the minor axis, and that this decrease was much greater when the opening was small.

He also found certain variations between the quadrants of ring A, and a possible similar but feebler difference for B, but was inclined to doubt these and certain other phenomena as being too near the margin of error due to diffraction and errors of measurement. Owing to the translucency of ring A, part of it in front of a polar region was appreciably brighter than A in the ansae.

In order to study ring C he had taken some long exposure photographs, and the crepe ring was visible on most of these; only by Barnard had the crepe ring previously been photographed. Camichel found that ring C is not uniform: it becomes steadily feebler from B to its edge, part of this effect being due to light diffused by ring B. He considered that his measures showed within an error of about 1/20 the relation in brilliance between rings A and B and the parts of the globe.

Optical properties of the rings and size of particles
(Cook and Franklin)

In 1958 A. F. Cook II and F. A. Franklin, both of Harvard College Observatory, published (*Smithsonian Contrib. to Astroph.*, vol. 2, no. 13) the first of a projected series of papers on the rings, this paper dealing with transmission properties. Using the 1920 observation in South Africa of a star occulted by ring B (see Chapter 31) and more especially the eclipse of Iapetus by the shadow of the rings in 1889 observed by Barnard (see Chapter 18), Cook and Franklin deduced from a long mathematical discussion the following results: a new value ($13''22 \pm 0''06$) for the inner radius of ring B at the mean distance of Saturn; an optical thickness of ring B of 0·58; a lower limit for the optical thickness of B of 0·45; an optical thickness of the inner edge of ring C of 0·0 and of its outer edge of 0·185. They also deduced the presence of a cloud of neutral gas, evidently associated with the crepe ring, and were able to show the shape of the pressure curve of the gas cloud as a function of the distance from the planet's centre.

Franklin and Cook (*Ast. J.*, vol. 63, p. 398) discussed two quite different explanations which have been offered to account for the phase curve of Saturn's rings. The first, originally put forward by von Seeliger (in 1884, 1887, 1895) suggested the rings to be composed of large particles, a metre or so in diameter; on this assumption the characteristic brightening of the rings at opposition would be a shadowing phenomenon. Bobrov (1954) modified this by removing Seeliger's assumption that the particles cast cylindrical shadows. The alternative explanation by Schönberg (1933) was that the phase curve might result from scattering by extremely small particles, so that diffraction and interference effects would be important. Schönberg deduced a value of $3·6\mu$ for the average particle size.

Bobrov (1952) concluded that the observed ring brightness is too great to be understood on the basis of Schönberg's hypothesis. Bobrov adopted for the maximum brightness of ring B the value

0·72, measured photographically by Scharonov (1939) as the brightness of the equatorial zone. After a mathematical discussion, Franklin and Cook appear to be unconvinced by Bobrov's argument, and they pointed out that Camichel's photometry of photographs taken at the Pic-du-Midi Observatory showed the average brightness of ring B to be 0·76 of that of the centre of the disk. While admitting that Bobrov's suggestion of scattering by large particles with pitted surfaces might meet the difficulty, Franklin and Cook concluded that the question of the average size of the ring particles is still quite an open one, and no final decision between the two alternative theories can be reached without photometry of a high quality over an extended period.

Probable shape of ring particles (Dollfus)

A remarkable discovery was made at the Pic-du-Midi Observatory in the spring of 1958 by Dr. Audouin Dollfus, and communicated by him to the author shortly afterwards. In his letter Dollfus stated that he had been fascinated by the results of polarimetric measures he had made on ring B. It appeared that the polarisation had two components, one being a direct normal component due to the diffusion of solar light. The other was a multiple component with plane of polarisation parallel or perpendicular to the radius of the ring, turning 180° from the upper part to the lower part of the minor axis. Dollfus wrote that this secondary component seems to show that the particles forming the ring are elliptical or cigar-shaped, with their major axis parallel to their orbit. He added that statistical study of the redistribution of energy during collisions between these particles shows that this orientation is a stable state.

CHAPTER 41

OBSERVATIONS IN 1960: OCCULTATION OF A STAR; BOTHAM'S WHITE SPOT

Saturn, far south in Sagittarius, provided striking phenomena in 1960, and fortunately observers in low and southern latitudes were available to follow them. The courtesy of Professor Mourão and Dr. Dollfus in supplying details of their observations promptly by letter, and that of Mr. Heath in putting at the author's disposal the B.A.A. Saturn Section observers' reports have enabled this chapter to be written.

Passage of star BD –21° 5359 behind ring A (Mourão)

The occultation by Saturn and its rings of the ninth magnitude star BD –21° 5359 (situated roughly one-third of the way from 50 Sgr to 43 Sgr) had been predicted for 1960 April 30—May 1, and the circumstances, computed in the Nautical Almanac Office, had been published in the *B.A.A. Hdbk.* for 1960 (p. 25). As Saturn was then near a station, with apparent motion only 1″ of arc per hour, no accurate prediction was possible and the times given were only approximate; immersion at the outer edge of ring A, position angle 258°, was expected at about 4ʰ U.T. on 1960 April 30; after passing behind rings and globe, the star was expected to emerge finally 38 hours later from the outer edge of the rings at position angle 75°. The extremely slow apparent motion of Saturn and the fact that the star was nearly 2 magnitudes fainter than those observed passing behind the rings in 1917 and 1920 (see Chapters 30, 31) made this observation very difficult, especially in determining the exact time of the immersion.

So far as is known the only observation of this phenomenon was made by Professor R. R. de F. Mourão and his colleague P. Mourilhe, at the National Observatory of Rio de Janeiro, Brazil, using the 18-inch aperture Cooke refractor with power 650. They were able to observe only the immersion and the first hour of the star's passage behind ring A, but they deserve great credit in view of the afore-mentioned difficulties. On April 30 at 3ʰ 10ᵐ U.T. when they found the star at the expected place, the atmosphere had become clear, transparent and stable around Saturn, and the principal details on the planet, including Encke's division on ring A, were distinct; further help was afforded by the contrast between the very blue colour of the star and the silver whiteness of the ring. At $3^h\ 30^m$ the separation between the star and the outer edge of ring A was very distinct, estimated at $0''8$. Their micrometric measurement made the position angle 267°.

By $4^h\ 45^m$ the separation between star and ring edge was no longer distinct. At $4^h\ 47^m$ it was impossible to separate them and a slight reduction of the star's brightness (by less than half a magnitude) was noted. At $4^h\ 55^m5$ half the diffraction disk of the star was visible, and at $4^h\ 59^m5$ the star appeared as a projection on the ring. By $5^h\ 13^m\ 13^s$ immersion seemed to be complete, the loss of brightness being half a magnitude. About $5^h\ 20^m$ the star seemed to 'beat', i.e. to move inside, then outside the ring, a phenomenon repeated several times.

At $5^h\ 32^m$ the star was conspicuously visible through the outer part of ring A but dimmed by one magnitude.

At $5^h\ 35^m3$ the star completely disappeared for 12 seconds, then reappearing $1\frac{1}{2}$ magnitudes fainter than normal. It disappeared and reappeared several times. By $5^h\ 45^m$ it appeared much weaker as a diffuse blue spot in ring A.

Mourão drew attention to the loss of magnitude of the star a few moments before it became completely occulted, to the greater translucency of ring A near its edge than further inside, and to the sudden disappearances and reappearances of the star during its passage behind the ring, suggesting non-uniformity in the density of the ring particles and possibly non-uniformity in their size.

A later report from Mourão states that Dr. Bobrov considers that Mourão's observation leads to a value of about 0·7 for the optical thickness of ring A.

Botham's white spot (Botham, Dollfus, Cragg, Whitaker)

Saturn's other manifestation in 1960 was quite unexpected. J. H. Botham, a South African amateur of the B.A.A. Saturn Section who though a regular observer had never before seen any striking markings on the planet, happened to observe Saturn in the early hours of 1960 March 31 at Johannesburg, and was astonished to see a large intensely white oval spot in the N.N. temperate zone, approaching the central meridian. He timed transits of the slightly diffuse p. edge, the centre, and the sharply-defined f. edge, finding that the whole spot took 70 minutes to cross the meridian. He estimated its intensity as brighter than the equatorial zone but less bright than the brightest part of ring B. He again took transits on April 4–5, when the p. edge of the spot seemed to fade into the p. area, and the brightest part was between the centre and the f. edge. On April 8–9 in poor conditions he again took transits, that of the p. edge being doubtful as it seemed to spread along the zone in the p. direction. He deduced a rotation period from his observations of 10^h 39^m. By April 12 the spot had much lengthened on the p. side into a long band and become more pointed at the f. end; the transit of the whole spot took more than 1^h 20^m. Absence followed by bad weather prevented Botham from making further observations until April 28 and May 2; the spot then lacked sharp outline and transits only of the brightest part and the f. edge were possible. The whole zone then looked much lighter on either side of the spot, which appeared not much brighter than the rest of the zone; a large area of lighter matter seemed to stream away from the spot in the p. direction. For his observations Botham used 6-inch Cooke and 9-inch Grubb aperture refractors with power 200. Spreading along the zone had also occurred with the white spot of 1933 (see Chapter 34). (Plate XXII, Fig. 2 shows Botham's photograph of Saturn with 1960 spot.)

On April 27 Dr. A. Dollfus, unaware of Botham's observations, independently discovered the brilliant spot on the edge of the north polar region of Saturn, using the 24-inch aperture refractor at the Pic-du-Midi Observatory. After a further observation on May 1 he announced the discovery by I.A.U. telegram, giving the latitude as 60°N. and the provisional rotation period as 10^h 40^m5. He later described its original appearance as a brilliant well-defined condensation, probably at the end of its evolution; on May 1 it was slightly more diffuse. On May 5 when it should have recrossed the meridian, he saw only a bright regular zone with no knots. The following day there were two smaller spots; as it was unlikely that the rotation period would have so soon lengthened by $6\frac{1}{2}$ minutes, he assumed that the original spot had disappeared and been replaced by two new ones. It seemed evident to him that these white spots evolve by spreading, finally to form a brilliant very white and regular zone all round the globe at their latitude; he remarked on the sharp contrast between the bright zone and the adjoining dusky polar region.

Probably for the first time in the history of spots on Saturn, polarimetric measures of the spot, the bright zone and the polar region were taken (by Dollfus). He found that the spot had a proportion of polarised light of only 2 or 3 thousandths under the angle of vision 5°5, much feebler than the rest of the disk and similar to that of the brilliant zones on Jupiter. On the other hand, the polar regions of Saturn are very strongly polarised owing to simple and multiple diffusions in the great thickness of atmosphere traversed. The bright zone therefore represented an elevated region in the atmosphere of Saturn.

T. A. Cragg, observing in California mainly with a 12-inch aperture reflector, obtained transits of a spot on five dates from May 6 to May 26, deducing a rotation period of $10^h 39^m8$; on the last occasion he described the spot as 'weak'.

E. A. Whitaker, using the 36-inch aperture reflector of McDonald Observatory with power 420, tried to follow up Dollfus's observations as often as possible during May, but had only four good nights for observing Saturn. On the first of these, May 5, he found at about latitude 60°N. simply a light featureless zone of almost constant width (as Dollfus did on the same date). But Whitaker also noticed the similar duskiness of the north polar region and the contiguous part of ring A, making the northern limb of the globe hard to see, and helping to produce the illusion of the 'square-shouldered' aspect of the globe. That illusion was enhanced by the brightness of the N.N. temperate zone, which looked like a white ring of cloud high above the rest of the planet. He considered this zone the next brightest after the equatorial zone. On May 10 he saw little change except a slight broadening of the bright zone, but he obtained the transit of a vague spot, which seemed to fit approximately with those obtained by Botham and Dollfus. On May 14 he could find, in excellent seeing, no trace of the white spot. Three days later he estimated the transit time of a vague enlargement of the bright zone, but as this feature preceded the predicted time by 110 minutes it was presumably another vaguer spot.

The disturbed condition of the zone is evident from the report of J. E. Focas, who, using the 25-inch aperture refractor of the Pentele Station of Athens Observatory, observed white and whitening areas on it separated by dark streaks, and obtained transits of some of these features on April 26 and May 3, 11 and 13.

C. R. Edwards (at Perth, W. Australia) recorded on June 4 a dull white spot in latitude about 60°N. that appeared to be about to divide into two, the p. one being slightly the smaller; the north and south edges of the spots seemed connected by several dusky linear condensations.

D. P. Cruickshank, using the 40-inch aperture refractor of Yerkes Observatory, found the light zone bordering the north polar region to be somewhat wider by mid-June than it had appeared to Whitaker in May, and to be getting duskier. On July 12 in near-perfect seeing with a high power, Cruickshank saw a number of loops in the zone, an appearance he confirmed two days later.

Cragg, in observations after May, found the zone very mottled: many ill-defined bright spots not quite definite enough for good transits. Towards the end of July the zone was becoming very 'knotty', and on August 5 Cragg and others at Los Angeles, using a 9-inch aperture refractor observed a fine bright spot, which they believed to be the original one, and suspected another larger one nearly 2^h preceding it; both spots looked quite impressive.

An interesting point about Botham's white spot is that it was found in a much higher Saturnian latitude than any previous ones of which transits had been obtained, yet though at least 20° further north than Barnard's and the other northern spots of 1903 (see Chapter 24), it had a very similar rotation period. Previous chapters have shown that large white spots have appeared in the northern hemisphere of Saturn in 1876, 1903 and 1933, in each case within a very few years before the return of the ring to the edgewise phase. Is the white spot of 1960 the precursor of the edgewise phase of 1966, or will another large spot appear before that phase is reached?*

It will be clear from these concluding chapters that Saturn from time to time provides very dramatic surprises for its observers, that the planet and satellites still have many unsolved problems, and that new techniques are developing which promise further revelations in the future.

* The B.A.A. Saturn Section's report for 1960 suggests that the activity in the N.N. temperate zone which produced Botham's spot may have started by 1959 July. After the disappearance of Botham's spot another, similar in brightness and size, reached its peak on 1960 August 25 and had a rotation period of $10^h 38 \cdot 7^m$. Of lesser spots in that zone two were bright enough and lasted long enough to give a period of $10^h 38 \cdot 5^m$. Much variation was observed in 1960 in the width and darkness of the adjoining north temperate belt.

Appendix 1

Oppositions of Saturn, 1956 to 2000

By M. B. B. Heath, F.R.A.S.

(from B.A.A.J. vol. 66, p. 167, 1956)

Shortened and approximate methods were used in the predictions but only when the computed time of the opposition falls near 0^h U.T. may there be a difference of a day from the date calculated from more rigorous methods.

The stellar magnitude of the planet at an opposition varies appreciably, being affected not only by the distances from the Sun and from the Earth but also by the apparent opening of the rings.

It is given by Müller's formula as

$$-8 \cdot 68 + 5 \log r\Delta \pm 0 \cdot 044 (U' + \omega - U) \pm 2 \cdot 60 \sin B + 1 \cdot 25 \sin^2 B$$

where r is the radius vector of the planet and Δ its distance from the Earth, both expressed in A.U.; and U' is the heliocentric longitude of Saturn measured in the plane of the rings from its ascending node on the ecliptic, ω the angular distance in the plane of the rings from its ascending node on the Earth's mean equator to its ascending node on the ecliptic, U the geocentric longitude of Saturn measured in the plane of the rings from its ascending node on the Earth's mean equator, and B the saturnicentric latitude of the Earth with respect to the plane of the rings and reckoned as positive towards the north, all expressed in degrees. The signs of the third and fourth terms must be taken so as to make these terms positive and negative respectively.

Date	Mag.	Date	Mag.	Date	Mag.
1956 May 20	+ 0·2	1971 Nov. 25	− 0·2	1986 May 27	+ 0·2
1957 June 1	+ 0·2	1972 Dec. 8	− 0·2	1987 June 9	+ 0·2
1958 June 13	+ 0·2	1973 Dec. 22	− 0·3	1988 June 20	+ 0·2
1959 June 26	+ 0·2	1974 —		1989 July 2	+ 0·2
1960 July 7	+ 0·2	1975 Jan. 6	− 0·2	1990 July 14	+ 0·3
1961 July 19	+ 0·3	1976 Jan. 20	− 0·1	1991 July 26	+ 0·3
1962 July 31	+ 0·4	1977 Feb. 2	0·0	1992 Aug. 7	+ 0·4
1963 Aug. 13	+ 0·5	1978 Feb. 16	+ 0·3	1993 Aug. 19	+ 0·5
1964 Aug. 24	+ 0·6	1979 Mar. 1	+ 0·5	1994 Sept. 1	+ 0·7
1965 Sept. 6	+ 0·8	1980 Mar. 13	+ 0·8	1995 Sept. 14	+ 0·8
1966 Sept. 19	+ 0·8	1981 Mar 27	+ 0·7	1996 Sept. 26	+ 0·7
1967 Oct. 2	+ 0·6	1982 Apr. 8	+ 0·5	1997 Oct. 10	+ 0·4
1968 Oct. 15	+ 0·3	1983 Apr. 21	+ 0·4	1998 Oct. 23	+ 0·2
1969 Oct. 28	+ 0·1	1984 May 3	+ 0·3	1999 Nov. 6	0·0
1970 Nov. 11	− 0·1	1985 May 15	+ 0·2	2000 Nov. 19	− 0·1

APPENDIX II

Observing Saturn

So much information has been given in this book about past observations and the methods employed that a complete statement on the visual observing of Saturn would be superfluous: for example, observing occultations of stars is fully dealt with in Chapters 30, 31 and 41; phenomena of the edgewise ring in Chapters 10, 14, 26, 32 and 36; spot transits in Chapters 15, 24, 34 and 41. With regard to transits of globe markings it should be remembered that outbreaks of spots on Saturn are too infrequent for transit observations to form the basis of visual work as they do in the case of Jupiter, and that there are no standard systems of longitude for Saturn such as there are for Jupiter. Bearing these points in mind, and also the fact that phenomena of and connected with the rings are unique to Saturn, much of the advice given by Mr. B. M. Peek (in *The Planet Jupiter*, chaps. 4, 5, 6) on visual observations of Jupiter applies also to Saturn. The full observing programme of the B.A.A. Saturn Section is contained in the pamphlet *The B.A.A.—its Nature, Aims and Methods*, and copies of the programme are obtainable by members from Mr. M. B. B. Heath, at present Director of the Section. In connection with it his valuable Saturn Section Notes (*B.A.A.J.*, vol. 65, p. 32) should be studied. To round off the subject a few extracts from the Programme and Notes are appended with brief comments.

Estimates of relative Intensity, of rings, belts, zones, (and the various parts of all these) and shadows, on a scale from 'zero' (brightest) to '10' (darkest), using the halves and quarters if desired (e.g. $2\frac{1}{4}$, $3\frac{1}{2}$).

Two standards are to be regarded as fixed: '1' for the brightness of the outer part of Ring B in the ansae adjoining Cassini's Division, anything brighter being recorded as $\frac{3}{4}$, $\frac{1}{2}$, $\frac{1}{4}$, or even zero if very exceptionally bright; '10' for the blackness of a very dark sky or deep black shadow.

The quickest and easiest method of recording intensities is to write the numbers down on outline sketches of the globe and rings which can be quickly drawn with a stencil. A good plan is to make one series of estimates taking the features in order from brightest to darkest, and another series taking them in the reverse order, adopting means where the two results differ.

The Notes point out that in numerical estimates of intensities even experienced observers show a certain amount of personal equation, some rating features too dark and others too light, but this can be allowed for provided they are consistent; unsteady air usually reduces the intensity of all dark markings.

Intensity estimates are not only useful in themselves, but they provide an incentive for the frequent and regular inspection of the planet. There can be little doubt that many spot outbreaks and other unexpected changes in, e.g. the intensity, position, width and visibility of belts and zones have been missed in the past even when Saturn has been well placed for observation, owing to neglect to observe the planet and the facile but false assumption that Saturn displays no activity.

Drawings. Heath has pointed out that good drawings of the complete planet are usually preferable to partial ones, and often give a better idea of what is seen than long descriptions. They should show everything that the observer is quite sure of, but no more. Partial drawings on a larger scale can be made to show fine detail of any unusual feature. The correct placing and shaping of the shadows on rings and globe is a useful criterion of the quality of the observation. The shadow of the globe may fall on the rings either to the east or west of the globe, usually becoming invisible for a time at the change, and the shadow of the rings may fall either inside the inner edge of ring B where it crosses the globe (being then involved in the crepe ring) or outside the outer edge of ring A where that crosses the globe, or occasionally be entirely hidden by the rings.

It is important to draw the globe the correct elliptical shape and the rings at the correct tilt—see Heath's paper on 'Drawing the Planet Saturn' (*B.A.A.J.*, vol. 63, p. 342).

Brightness of satellites. The programme suggests comparing the satellites with one another by estimating steps of one-tenth of a magnitude, assuming the magnitude of Titan to be constant at 8·3. Then from an estimate such as T (5), I (10), R (10), Tys (5) D, the magnitudes of Iapetus, Rhea, Tethys and Dione could be deduced as: 8·8, 9·8, 10·8, 11·3 respectively. Observers who find this step estimation too difficult, can still give useful information by recording simply the *order* of brightness as between Iapetus, Rhea, Tethys and Dione.

In reporting the visibility of satellites a small sketch showing their positions relative to the planet must be made, and the date and exact time of the observation given, as those nearest the planet have an appreciable hourly motion.

Saturnicentric latitudes of belt edges. Measurement by micrometer is best, but quite useful results can be obtained by measuring reasonably accurate drawings with a finely graded scale, and calculating the saturnicentric latitudes of the edges of the belts. A large number of such figures should give averages (for the Section) that will be useful, in comparison with those of other apparitions, to indicate any marked changes in the positions of the belts.

The measurements required are: the polar radius (r) and the distance (y) of each belt edge from the centre of the disk, taken along the C.M., measures of (y) northward from the centre being plus, and southward minus. Members are then asked to work out the saturnicentric latitudes from the measurements as follows, using the Crommelin formulae:

Get B, the saturnicentric declination of the Earth, from the *Astronomical Ephemeris.*

From $\tan B' = 1 \cdot 12 \tan B$, find angle B';

then from $\sin (b' - B') = y/r$, find $(b' - B')$, hence b';

finally, from $\tan C = \dfrac{\tan b'}{1 \cdot 12}$, find C, the required saturnicentric latitude.

It has been decided that the Section ought to use *saturnicentric* rather than saturnigraphic latitudes (L). The latter can readily be converted to the former by the formula:

$$\tan C = \frac{\tan L}{(1 \cdot 12)^2}, \text{ or logtan } C = \text{logtan } L - 2 \log (1 \cdot 12).$$

Latitude returns are wanted for: edge of ring C across globe; centre of E.Z.; N. and S. edges of S.E.B. (and of each component); N. and S. edges of S.P. Band; N. edge of S.P. area (if differing from S. edge of S.P. Band); and similar measures for belts of the N. hemisphere when observable.

Finally, although many important discoveries have been made in the past with small telescopes, and useful observations and an occasional discovery are still being made with them, it cannot be too strongly emphasised that for the vast majority of observations of Saturn, at least moderate apertures are essential, and large apertures are best of all.

A CATALOGUE OF SELECTED DOVER BOOKS
IN ALL FIELDS OF INTEREST

A CATALOGUE OF SELECTED DOVER
BOOKS IN ALL FIELDS OF INTEREST

CONDITIONED REFLEXES, Ivan P. Pavlov. Full translation of most complete statement of Pavlov's work; cerebral damage, conditioned reflex, experiments with dogs, sleep, similar topics of great importance. 430pp. 5⅜ x 8½. 60614-7 Pa. $4.50

NOTES ON NURSING: WHAT IT IS, AND WHAT IT IS NOT, Florence Nightingale. Outspoken writings by founder of modern nursing. When first published (1860) it played an important role in much needed revolution in nursing. Still stimulating. 140pp. 5⅜ x 8½. 22340-X Pa. $2.50

HARTER'S PICTURE ARCHIVE FOR COLLAGE AND ILLUSTRATION, Jim Harter. Over 300 authentic, rare 19th-century engravings selected by noted collagist for artists, designers, decoupeurs, etc. Machines, people, animals, etc., printed one side of page. 25 scene plates for backgrounds. 6 collages by Harter, Satty, Singer, Evans. Introduction. 192pp. 8⅞ x 11¾. 23659-5 Pa. $4.50

MANUAL OF TRADITIONAL WOOD CARVING, edited by Paul N. Hasluck. Possibly the best book in English on the craft of wood carving. Practical instructions, along with 1,146 working drawings and photographic illustrations. Formerly titled *Cassell's Wood Carving*. 576pp. 6½ x 9¼. 23489-4 Pa. $7.95

THE PRINCIPLES AND PRACTICE OF HAND OR SIMPLE TURNING, John Jacob Holtzapffel. Full coverage of basic lathe techniques—history and development, special apparatus, softwood turning, hardwood turning, metal turning. Many projects—billiard ball, works formed within a sphere, egg cups, ash trays, vases, jardiniers, others—included. 1881 edition. 800 illustrations. 592pp. 6⅛ x 9¼. 23365-0 Clothbd. $15.00

THE JOY OF HANDWEAVING, Osma Tod. Only book you need for hand weaving. Fundamentals, threads, weaves, plus numerous projects for small board-loom, two-harness, tapestry, laid-in, four-harness weaving and more. Over 160 illustrations. 2nd revised edition. 352pp. 6½ x 9¼. 23458-4 Pa. $5.00

THE BOOK OF WOOD CARVING, Charles Marshall Sayers. Still finest book for beginning student in wood sculpture. Noted teacher, craftsman discusses fundamentals, technique; gives 34 designs, over 34 projects for panels, bookends, mirrors, etc. "Absolutely first-rate"—E. J. Tangerman. 33 photos. 118pp. 7¾ x 10⅝. 23654-4 Pa. $3.00

YUCATAN BEFORE AND AFTER THE CONQUEST, Diego de Landa. First English translation of basic book in Maya studies, the only significant account of Yucatan written in the early post-Conquest era. Translated by distinguished Maya scholar William Gates. Appendices, introduction, 4 maps and over 120 illustrations added by translator. 162pp. 5⅜ x 8½.
23622-6 Pa. $3.00

THE MALAY ARCHIPELAGO, Alfred R. Wallace. Spirited travel account by one of founders of modern biology. Touches on zoology, botany, ethnography, geography, and geology. 62 illustrations, maps. 515pp. 5⅜ x 8½.
20187-2 Pa. $6.95

THE DISCOVERY OF THE TOMB OF TUTANKHAMEN, Howard Carter, A. C. Mace. Accompany Carter in the thrill of discovery, as ruined passage suddenly reveals unique, untouched, fabulously rich tomb. Fascinating account, with 106 illustrations. New introduction by J. M. White. Total of 382pp. 5⅜ x 8½. (Available in U.S. only) 23500-9 Pa. $4.00

THE WORLD'S GREATEST SPEECHES, edited by Lewis Copeland and Lawrence W. Lamm. Vast collection of 278 speeches from Greeks up to present. Powerful and effective models; unique look at history. Revised to 1970. Indices. 842pp. 5⅜ x 8½. 20468-5 Pa. $6.95

THE 100 GREATEST ADVERTISEMENTS, Julian Watkins. The priceless ingredient; His master's voice; 99 44/100% pure; over 100 others. How they were written, their impact, etc. Remarkable record. 130 illustrations. 233pp. 7⅞ x 10 3/5. 20540-1 Pa. $5.00

CRUICKSHANK PRINTS FOR HAND COLORING, George Cruickshank. 18 illustrations, one side of a page, on fine-quality paper suitable for watercolors. Caricatures of people in society (c. 1820) full of trenchant wit. Very large format. 32pp. 11 x 16. 23684-6 Pa. $4.50

THIRTY-TWO COLOR POSTCARDS OF TWENTIETH-CENTURY AMERICAN ART, Whitney Museum of American Art. Reproduced in full color in postcard form are 31 art works and one shot of the museum. Calder, Hopper, Rauschenberg, others. Detachable. 16pp. 8¼ x 11.
23629-3 Pa. $2.50

MUSIC OF THE SPHERES: THE MATERIAL UNIVERSE FROM ATOM TO QUASAR SIMPLY EXPLAINED, Guy Murchie. Planets, stars, geology, atoms, radiation, relativity, quantum theory, light, antimatter, similar topics. 319 figures. 664pp. 5⅜ x 8½.
21809-0, 21810-4 Pa., Two-vol. set $10.00

EINSTEIN'S THEORY OF RELATIVITY, Max Born. Finest semi-technical account; covers Einstein, Lorentz, Minkowski, and others, with much detail, much explanation of ideas and math not readily available elsewhere on this level. For student, non-specialist. 376pp. 5⅜ x 8½.
60769-0 Pa. $4.00

MUSHROOMS, EDIBLE AND OTHERWISE, Miron E. Hard. Profusely illustrated, very useful guide to over 500 species of mushrooms growing in the Midwest and East. Nomenclature updated to 1976. 505 illustrations. 628pp. 6½ x 9¼. 23309-X Pa. $7.95

AN ILLUSTRATED FLORA OF THE NORTHERN UNITED STATES AND CANADA, Nathaniel L. Britton, Addison Brown. Encyclopedic work covers 4666 species, ferns on up. Everything. Full botanical information, illustration for each. This earlier edition is preferred by many to more recent revisions. 1913 edition. Over 4000 illustrations, total of 2087pp. 6⅛ x 9¼. 22642-5, 22643-3, 22644-1 Pa., Three-vol. set $24.00

MANUAL OF THE GRASSES OF THE UNITED STATES, A. S. Hitchcock, U.S. Dept. of Agriculture. The basic study of American grasses, both indigenous and escapes, cultivated and wild. Over 1400 species. Full descriptions, information. Over 1100 maps, illustrations. Total of 1051pp. 5⅜ x 8½. 22717-0, 22718-9 Pa., Two-vol. set $12.00

THE CACTACEAE,, Nathaniel L. Britton, John N. Rose. Exhaustive, definitive. Every cactus in the world. Full botanical descriptions. Thorough statement of nomenclatures, habitat, detailed finding keys. The one book needed by every cactus enthusiast. Over 1275 illustrations. Total of 1080pp. 8 x 10¼. 21191-6, 21192-4 Clothbd., Two-vol. set $35.00

AMERICAN MEDICINAL PLANTS, Charles F. Millspaugh. Full descriptions, 180 plants covered: history; physical description; methods of preparation with all chemical constituents extracted; all claimed curative or adverse effects. 180, full-page plates. Classification table. 804pp. 6½ x 9¼. 23034-1 Pa. $10.00

A MODERN HERBAL, Margaret Grieve. Much the fullest, most exact, most useful compilation of herbal material. Gigantic alphabetical encyclopedia, from aconite to zedoary, gives botanical information, medical properties, folklore, economic uses, and much else. Indispensable to serious reader. 161 illustrations. 888pp. 6½ x 9¼. (Available in U.S. only) 22798-7, 22799-5 Pa., Two-vol. set $11.00

THE HERBAL or GENERAL HISTORY OF PLANTS, John Gerard. The 1633 edition revised and enlarged by Thomas Johnson. Containing almost 2850 plant descriptions and 2705 superb illustrations, Gerard's *Herbal* is a monumental work, the book all modern English herbals are derived from, the one herbal every serious enthusiast should have in its entirety. Original editions are worth perhaps $750. 1678pp. 8½ x 12¼. 23147-X Clothbd. $50.00

MANUAL OF THE TREES OF NORTH AMERICA, Charles S. Sargent. The basic survey of every native tree and tree-like shrub, 717 species in all. Extremely full descriptions, information on habitat, growth, locales, economics, etc. Necessary to every serious tree lover. Over 100 finding keys. 783 illustrations. Total of 986pp. 5⅜ x 8½. 20277-1, 20278-X Pa., Two-vol. set $10.00

AMERICAN BIRD ENGRAVINGS, Alexander Wilson et al. All 76 plates. from Wilson's *American Ornithology* (1808-14), most important ornithological work before Audubon, plus 27 plates from the supplement (1825-33) by Charles Bonaparte. Over 250 birds portrayed. 8 plates also reproduced in full color. 111pp. 9⅜ x 12½. 23195-X Pa. $6.00

CRUICKSHANK'S PHOTOGRAPHS OF BIRDS OF AMERICA, Allan D. Cruickshank. Great ornithologist, photographer presents 177 closeups, groupings, panoramas, flightings, etc., of about 150 different birds. Expanded *Wings in the Wilderness*. Introduction by Helen G. Cruickshank. 191pp. 8¼ x 11. 23497-5 Pa. $6.00

AMERICAN WILDLIFE AND PLANTS, A. C. Martin, et al. Describes food habits of more than 1000 species of mammals, birds, fish. Special treatment of important food plants. Over 300 illustrations. 500pp. 5⅜ x 8½. 20793-5 Pa. $4.95

THE PEOPLE CALLED SHAKERS, Edward D. Andrews. Lifetime of research, definitive study of Shakers: origins, beliefs, practices, dances, social organization, furniture and crafts, impact on 19th-century USA, present heritage. Indispensable to student of American history, collector. 33 illustrations. 351pp. 5⅜ x 8½. 21081-2 Pa. $4.00

OLD NEW YORK IN EARLY PHOTOGRAPHS, Mary Black. New York City as it was in 1853-1901, through 196 wonderful photographs from N.-Y. Historical Society. Great Blizzard, Lincoln's funeral procession, great buildings. 228pp. 9 x 12. 22907-6 Pa. $7.95

MR. LINCOLN'S CAMERA MAN: MATHEW BRADY, Roy Meredith. Over 300 Brady photos reproduced directly from original negatives, photos. Jackson, Webster, Grant, Lee, Carnegie, Barnum; Lincoln; Battle Smoke, Death of Rebel Sniper, Atlanta Just After Capture. Lively commentary. 368pp. 8⅜ x 11¼. 23021-X Pa. $6.95

TRAVELS OF WILLIAM BARTRAM, William Bartram. From 1773-8, Bartram explored Northern Florida, Georgia, Carolinas, and reported on wild life, plants, Indians, early settlers. Basic account for period, entertaining reading. Edited by Mark Van Doren. 13 illustrations. 141pp. 5⅜ x 8½. 20013-2 Pa. $4.50

THE GENTLEMAN AND CABINET MAKER'S DIRECTOR, Thomas Chippendale. Full reprint, 1762 style book, most influential of all time; chairs, tables, sofas, mirrors, cabinets, etc. 200 plates, plus 24 photographs of surviving pieces. 249pp. 9⅞ x 12¾. 21601-2 Pa. $6.50

AMERICAN CARRIAGES, SLEIGHS, SULKIES AND CARTS, edited by Don H. Berkebile. 168 Victorian illustrations from catalogues, trade journals, fully captioned. Useful for artists. Author is Assoc. Curator, Div. of Transportation of Smithsonian Institution. 168pp. 8½ x 9½. 23328-6 Pa. $5.00

CATALOGUE OF DOVER BOOKS

THE COMPLETE BOOK OF DOLL MAKING AND COLLECTING, Catherine Christopher. Instructions, patterns for dozens of dolls, from rag doll on up to elaborate, historically accurate figures. Mould faces, sew clothing, make doll houses, etc. Also collecting information. Many illustrations. 288pp. 6 x 9.   22066-4 Pa. $4.00

THE DAGUERREOTYPE IN AMERICA, Beaumont Newhall. Wonderful portraits, 1850's townscapes, landscapes; full text plus 104 photographs. The basic book. Enlarged 1976 edition. 272pp. 8¼ x 11¼.
23322-7 Pa. $6.00

CRAFTSMAN HOMES, Gustav Stickley. 296 architectural drawings, floor plans, and photographs illustrate 40 different kinds of "Mission-style" homes from *The Craftsman* (1901-16), voice of American style of simplicity and organic harmony. Thorough coverage of Craftsman idea in text and picture, now collector's item. 224pp. 8⅛ x 11.   23791-5 Pa. $6.00

PEWTER-WORKING: INSTRUCTIONS AND PROJECTS, Burl N. Osborn. & Gordon O. Wilber. Introduction to pewter-working for amateur craftsman. History and characteristics of pewter; tools, materials, step-by-step instructions. Photos, line drawings, diagrams. Total of 160pp. 7⅞ x 10¾.   23786-9 Pa. $3.50

THE GREAT CHICAGO FIRE, edited by David Lowe. 10 dramatic, eyewitness accounts of the 1871 disaster, including one of the aftermath and rebuilding, plus 70 contemporary photographs and illustrations of the ruins—courthouse, Palmer House, Great Central Depot, etc. Introduction by David Lowe. 87pp. 8¼ x 11.   23771-0 Pa. $4.00

SILHOUETTES: A PICTORIAL ARCHIVE OF VARIED ILLUSTRATIONS, edited by Carol Belanger Grafton. Over 600 silhouettes from the 18th to 20th centuries include profiles and full figures of men and women, children, birds and animals, groups and scenes, nature, ships, an alphabet. Dozens of uses for commercial artists and craftspeople. 144pp. 8⅜ x 11¼.
23781-8 Pa. $4.00

ANIMALS: 1,419 COPYRIGHT-FREE ILLUSTRATIONS OF MAMMALS, BIRDS, FISH, INSECTS, ETC., edited by Jim Harter. Clear wood engravings present, in extremely lifelike poses, over 1,000 species of animals. One of the most extensive copyright-free pictorial sourcebooks of its kind. Captions. Index. 284pp. 9 x 12.   23766-4 Pa. $7.50

INDIAN DESIGNS FROM ANCIENT ECUADOR, Frederick W. Shaffer. 282 original designs by pre-Columbian Indians of Ecuador (500-1500 A.D.). Designs include people, mammals, birds, reptiles, fish, plants, heads, geometric designs. Use as is or alter for advertising, textiles, leathercraft, etc. Introduction. 95pp. 8¾ x 11¼.   23764-8 Pa. $3.50

SZIGETI ON THE VIOLIN, Joseph Szigeti. Genial, loosely structured tour by premier violinist, featuring a pleasant mixture of reminiscenes, insights into great music and musicians, innumerable tips for practicing violinists. 385 musical passages. 256pp. 5⅝ x 8¼.   23763-X Pa. $3.50

TONE POEMS, SERIES II: TILL EULENSPIEGELS LUSTIGE STREICHE, ALSO SPRACH ZARATHUSTRA, AND EIN HELDEN-LEBEN, Richard Strauss. Three important orchestral works, including very popular *Till Eulenspiegel's Marry Pranks,* reproduced in full score from original editions. Study score. 315pp. 9⅜ x 12¼. (Available in U.S. only)
23755-9 Pa. $7.50

TONE POEMS, SERIES I: DON JUAN, TOD UND VERKLARUNG AND DON QUIXOTE, Richard Strauss. Three of the most often performed and recorded works in entire orchestral repertoire, reproduced in full score from original editions. Study score. 286pp. 9⅜ x 12¼. (Available in U.S. only)
23754-0 Pa. $7.50

11 LATE STRING QUARTETS, Franz Joseph Haydn. The form which Haydn defined and "brought to perfection." (*Grove's*). 11 string quartets in complete score, his last and his best. The first in a projected series of the complete Haydn string quartets. Reliable modern Eulenberg edition, otherwise difficult to obtain. 320pp. 8⅜ x 11¼. (Available in U.S. only)
23753-2 Pa. $6.95

FOURTH, FIFTH AND SIXTH SYMPHONIES IN FULL SCORE, Peter Ilyitch Tchaikovsky. Complete orchestral scores of Symphony No. 4 in F Minor, Op. 36; Symphony No. 5 in E Minor, Op. 64; Symphony No. 6 in B Minor, "Pathetique," Op. 74. Bretikopf & Hartel eds. Study score. 480pp. 9⅜ x 12¼.
23861-X Pa. $10.95

THE MARRIAGE OF FIGARO: COMPLETE SCORE, Wolfgang A. Mozart. Finest comic opera ever written. Full score, not to be confused with piano renderings. Peters edition. Study score. 448pp. 9⅜ x 12¼. (Available in U.S. only)
23751-6 Pa. $11.95

"IMAGE" ON THE ART AND EVOLUTION OF THE FILM, edited by Marshall Deutelbaum. Pioneering book brings together for first time 38 groundbreaking articles on early silent films from *Image* and 263 illustrations newly shot from rare prints in the collection of the International Museum of Photography. A landmark work. Index. 256pp. 8¼ x 11.
23777-X Pa. $8.95

AROUND-THE-WORLD COOKY BOOK, Lois Lintner Sumption and Marguerite Lintner Ashbrook. 373 cooky and frosting recipes from 28 countries (America, Austria, China, Russia, Italy, etc.) include Viennese kisses, rice wafers, London strips, lady fingers, hony, sugar spice, maple cookies, etc. Clear instructions. All tested. 38 drawings. 182pp. 5⅜ x 8.
23802-4 Pa. $2.50

THE ART NOUVEAU STYLE, edited by Roberta Waddell. 579 rare photographs, not available elsewhere, of works in jewelry, metalwork, glass, ceramics, textiles, architecture and furniture by 175 artists—Mucha, Seguy, Lalique, Tiffany, Gaudin, Hohlwein, Saarinen, and many others. 288pp. 8⅜ x 11¼.
23515-7 Pa. $6.95

CATALOGUE OF DOVER BOOKS

THE PHILOSOPHY OF HISTORY, Georg W. Hegel. Great classic of Western thought develops concept that history is not chance but a rational process, the evolution of freedom. 457pp. 5⅜ x 8½. 20112-0 Pa. $4.50

LANGUAGE, TRUTH AND LOGIC, Alfred J. Ayer. Famous, clear introduction to Vienna, Cambridge schools of Logical Positivism. Role of philosophy, elimination of metaphysics, nature of analysis, etc. 160pp. 5⅜ x 8½. (Available in U.S. only) 20010-8 Pa. $1.75

A PREFACE TO LOGIC, Morris R. Cohen. Great City College teacher in renowned, easily followed exposition of formal logic, probability, values, logic and world order and similar topics; no previous background needed. 209pp. 5⅜ x 8½. 23517-3 Pa. $3.50

REASON AND NATURE, Morris R. Cohen. Brilliant analysis of reason and its multitudinous ramifications by charismatic teacher. Interdisciplinary, synthesizing work widely praised when it first appeared in 1931. Second (1953) edition. Indexes. 496pp. 5⅜ x 8½. 23633-1 Pa. $6.00

AN ESSAY CONCERNING HUMAN UNDERSTANDING, John Locke. The only complete edition of enormously important classic, with authoritative editorial material by A. C. Fraser. Total of 1176pp. 5⅜ x 8½. 20530-4, 20531-2 Pa., Two-vol. set $14.00

HANDBOOK OF MATHEMATICAL FUNCTIONS WITH FORMULAS, GRAPHS, AND MATHEMATICAL TABLES, edited by Milton Abramowitz and Irene A. Stegun. Vast compendium: 29 sets of tables, some to as high as 20 places. 1,046pp. 8 x 10½. 61272-4 Pa. $12.50

MATHEMATICS FOR THE PHYSICAL SCIENCES, Herbert S. Wilf. Highly acclaimed work offers clear presentations of vector spaces and matrices, orthogonal functions, roots of polynomial equations, conformal mapping, calculus of variations, etc. Knowledge of theory of functions of real and complex variables is assumed. Exercises and solutions. Index. 284pp. 5⅝ x 8¼. 63635-6 Pa. $4.50

THE PRINCIPLE OF RELATIVITY, Albert Einstein et al. Eleven most important original papers on special and general theories. Seven by Einstein, two by Lorentz, one each by Minkowski and Weyl. All translated, unabridged. 216pp. 5⅜ x 8½. 60081-5 Pa. $3.00

THERMODYNAMICS, Enrico Fermi. A classic of modern science. Clear, organized treatment of systems, first and second laws, entropy, thermodynamic potentials, gaseous reactions, dilute solutions, entropy constant. No math beyond calculus required. Problems. 160pp. 5⅜ x 8½. 60361-X Pa. $2.75

ELEMENTARY MECHANICS OF FLUIDS, Hunter Rouse. Classic undergraduate text widely considered to be far better than many later books. Ranges from fluid velocity and acceleration to role of compressibility in fluid motion. Numerous examples, questions, problems. 224 illustrations. 376pp. 5⅝ x 8¼. 63699-2 Pa. $5.00

THE SENSE OF BEAUTY, George Santayana. Masterfully written discussion of nature of beauty, materials of beauty, form, expression; art, literature, social sciences all involved. 168pp. 5⅜ x 8½.   20238-0 Pa. $2.50

ON THE IMPROVEMENT OF THE UNDERSTANDING, Benedict Spinoza. Also contains *Ethics, Correspondence,* all in excellent R. Elwes translation. Basic works on entry to philosophy, pantheism, exchange of ideas with great contemporaries. 402pp. 5⅜ x 8½.   20250-X Pa. $3.75

THE TRAGIC SENSE OF LIFE, Miguel de Unamuno. Acknowledged masterpiece of existential literature, one of most important books of 20th century. Introduction by Madariaga. 367pp. 5⅜ x 8½.
20257-7 Pa. $3.50

THE GUIDE FOR THE PERPLEXED, Moses Maimonides. Great classic of medieval Judaism attempts to reconcile revealed religion (Pentateuch, commentaries) with Aristotelian philosophy. Important historically, still relevant in problems. Unabridged Friedlander translation. Total of 473pp. 5⅜ x 8½.   20351-4 Pa. $5.00

THE I CHING (THE BOOK OF CHANGES), translated by James Legge. Complete translation of basic text plus appendices by Confucius, and Chinese commentary of most penetrating divination manual ever prepared. Indispensable to study of early Oriental civilizations, to modern inquiring reader. 448pp. 5⅜ x 8½.   21062-6 Pa. $4.00

THE EGYPTIAN BOOK OF THE DEAD, E. A. Wallis Budge. Complete reproduction of Ani's papyrus, finest ever found. Full hieroglyphic text, interlinear transliteration, word for word translation, smooth translation. Basic work, for Egyptology, for modern study of psychic matters. Total of 533pp. 6½ x 9¼. (Available in U.S. only)   21866-X Pa. $4.95

THE GODS OF THE EGYPTIANS, E. A. Wallis Budge. Never excelled for richness, fullness; all gods, goddesses, demons, mythical figures of Ancient Egypt; their legends, rites, incarnations, variations, powers, etc. Many hieroglyphic texts cited. Over 225 illustrations, plus 6 color plates. Total of 988pp. 6⅛ x 9¼. (Available in U.S. only)
22055-9, 22056-7 Pa., Two-vol. set $12.00

THE ENGLISH AND SCOTTISH POPULAR BALLADS, Francis J. Child. Monumental, still unsuperseded; all known variants of Child ballads, commentary on origins, literary references, Continental parallels, other features. Added: papers by G. L. Kittredge, W. M. Hart. Total of 2761pp. 6½ x 9¼.
21409-5, 21410-9, 21411-7, 21412-5, 21413-3 Pa., Five-vol. set $37.50

CORAL GARDENS AND THEIR MAGIC, Bronsilaw Malinowski. Classic study of the methods of tilling the soil and of agricultural rites in the Trobriand Islands of Melanesia. Author is one of the most important figures in the field of modern social anthropology. 143 illustrations. Indexes. Total of 911pp. of text. 5⅝ x 8¼. (Available in U.S. only)
23597-1 Pa. $12.95

SECOND PIATIGORSKY CUP, edited by Isaac Kashdan. One of the greatest tournament books ever produced in the English language. All 90 games of the 1966 tournament, annotated by players, most annotated by both players. Features Petrosian, Spassky, Fischer, Larsen, six others. 228pp. 5⅜ x 8½. 23572-6 Pa. $3.50

ENCYCLOPEDIA OF CARD TRICKS, revised and edited by Jean Hugard. How to perform over 600 card tricks, devised by the world's greatest magicians: impromptus, spelling tricks, key cards, using special packs, much, much more. Additional chapter on card technique. 66 illustrations. 402pp. 5⅜ x 8½. (Available in U.S. only) 21252-1 Pa. $3.95

MAGIC: STAGE ILLUSIONS, SPECIAL EFFECTS AND TRICK PHO- TOGRAPHY, Albert A. Hopkins, Henry R. Evans. One of the great classics; fullest, most authorative explanation of vanishing lady, levitations, scores of other great stage effects. Also small magic, automata, stunts. 446 illus- trations. 556pp. 5⅜ x 8½. 23344-8 Pa. $5.00

THE SECRETS OF HOUDINI, J. C. Cannell. Classic study of Houdini's incredible magic, exposing closely-kept professional secrets and revealing, in general terms, the whole art of stage magic. 67 illustrations. 279pp. 5⅜ x 8½. 22913-0 Pa. $3.00

HOFFMANN'S MODERN MAGIC, Professor Hoffmann. One of the best, and best-known, magicians' manuals of the past century. Hundreds of tricks from card tricks and simple sleight of hand to elaborate illusions involving construction of complicated machinery. 332 illustrations. 563pp. 5⅜ x 8½. 23623-4 Pa. $6.00

MADAME PRUNIER'S FISH COOKERY BOOK, Mme. S. B. Prunier. More than 1000 recipes from world famous Prunier's of Paris and London, specially adapted here for American kitchen. Grilled tournedos with anchovy butter, Lobster a la Bordelaise, Prunier's prized desserts, more. Glossary. 340pp. 5⅜ x 8½. (Available in U.S. only) 22679-4 Pa. $3.00

FRENCH COUNTRY COOKING FOR AMERICANS, Louis Diat. 500 easy-to-make, authentic provincial recipes compiled by former head chef at New York's Fitz-Carlton Hotel: onion soup, lamb stew, potato pie, more. 309pp. 5⅜ x 8½. 23665-X Pa. $3.95

SAUCES, FRENCH AND FAMOUS, Louis Diat. Complete book gives over 200 specific recipes: bechamel, Bordelaise, hollandaise, Cumberland, apri- cot, etc. Author was one of this century's finest chefs, originator of vichyssoise and many other dishes. Index. 156pp. 5⅜ x 8. 23663-3 Pa. $2.50

TOLL HOUSE TRIED AND TRUE RECIPES, Ruth Graves Wakefield. Authentic recipes from the famous Mass. restaurant: popovers, veal and ham loaf, Toll House baked beans, chocolate cake crumb pudding, much more. Many helpful hints. Nearly 700 recipes. Index. 376pp. 5⅜ x 8½. 23560-2 Pa. $4.00

"OSCAR" OF THE WALDORF'S COOKBOOK, Oscar Tschirky. Famous American chef reveals 3455 recipes that made Waldorf great; cream of French, German, American cooking, in all categories. Full instructions, easy home use. 1896 edition. 907pp. 6⅝ x 9⅜. 20790-0 Clothbd. $15.00

COOKING WITH BEER, Carole Fahy. Beer has as superb an effect on food as wine, and at fraction of cost. Over 250 recipes for appetizers, soups, main dishes, desserts, breads, etc. Index. 144pp. 5⅜ x 8½. (Available in U.S. only) 23661-7 Pa. $2.50

STEWS AND RAGOUTS, Kay Shaw Nelson. This international cookbook offers wide range of 108 recipes perfect for everyday, special occasions, meals-in-themselves, main dishes. Economical, nutritious, easy-to-prepare: goulash, Irish stew, boeuf bourguignon, etc. Index. 134pp. 5⅜ x 8½. 23662-5 Pa. $2.50

DELICIOUS MAIN COURSE DISHES, Marian Tracy. Main courses are the most important part of any meal. These 200 nutritious, economical recipes from around the world make every meal a delight. "I . . . have found it so useful in my own household,"—*N.Y. Times.* Index. 219pp. 5⅜ x 8½. 23664-1 Pa. $3.00

FIVE ACRES AND INDEPENDENCE, Maurice G. Kains. Great back-to-the-land classic explains basics of self-sufficient farming: economics, plants, crops, animals, orchards, soils, land selection, host of other necessary things. Do not confuse with skimpy faddist literature; Kains was one of America's greatest agriculturalists. 95 illustrations. 397pp. 5⅜ x 8½. 20974-1 Pa. $3.50

A PRACTICAL GUIDE FOR THE BEGINNING FARMER, Herbert Jacobs. Basic, extremely useful first book for anyone thinking about moving to the country and starting a farm. Simpler than Kains, with greater emphasis on country living in general. 246pp. 5⅜ x 8½. 23675-7 Pa. $3.50

HARDY BULBS, Louise Beebe Wilder. Fullest, most thorough book on plants grown from bulbs, corms, rhizomes and tubers. 40 genera and 335 species covered: selecting, cultivating, naturalizing; name, origins, blooming season, when to plant, special requirements. 127 illustrations. 432pp. 5⅜ x 8½. 23102-X Pa. $4.50

A GARDEN OF PLEASANT FLOWERS (PARADISI IN SOLE: PARADISUS TERRESTRIS), John Parkinson. Complete, unabridged reprint of first (1629) edition of earliest great English book on gardens and gardening. More than 1000 plants & flowers of Elizabethan, Jacobean garden fully described, most with woodcut illustrations. Botanically very reliable, a "speaking garden" of exceeding charm. 812 illustrations. 628pp. 8½ x 12¼. 23392-8 Clothbd. $25.00

HISTORY OF BACTERIOLOGY, William Bulloch. The only comprehensive history of bacteriology from the beginnings through the 19th century. Special emphasis is given to biography-Leeuwenhoek, etc. Brief accounts of 350 bacteriologists form a separate section. No clearer, fuller study, suitable to scientists and general readers, has yet been written. 52 illustrations. 448pp. 5⅝ x 8¼.   23761-3 Pa. $6.50

THE COMPLETE NONSENSE OF EDWARD LEAR, Edward Lear. All nonsense limericks, zany alphabets, Owl and Pussycat, songs, nonsense botany, etc., illustrated by Lear. Total of 321pp. 5⅜ x 8½. (Available in U.S. only)   20167-8 Pa. $3.00

INGENIOUS MATHEMATICAL PROBLEMS AND METHODS, Louis A. Graham. Sophisticated material from Graham *Dial*, applied and pure; stresses solution methods. Logic, number theory, networks, inversions, etc. 237pp. 5⅜ x 8½.   20545-2 Pa. $3.50

BEST MATHEMATICAL PUZZLES OF SAM LOYD, edited by Martin Gardner. Bizarre, original, whimsical puzzles by America's greatest puzzler. From fabulously rare *Cyclopedia*, including famous 14-15 puzzles, the Horse of a Different Color, 115 more. Elementary math. 150 illustrations. 167pp. 5⅜ x 8½.   20498-7 Pa. $2.50

THE BASIS OF COMBINATION IN CHESS, J. du Mont. Easy-to-follow, instructive book on elements of combination play, with chapters on each piece and every powerful combination team—two knights, bishop and knight, rook and bishop, etc. 250 diagrams. 218pp. 5⅜ x 8½. (Available in U.S. only)   23644-7 Pa. $3.50

MODERN CHESS STRATEGY, Ludek Pachman. The use of the queen, the active king, exchanges, pawn play, the center, weak squares, etc. Section on rook alone worth price of the book. Stress on the moderns. Often considered the most important book on strategy. 314pp. 5⅜ x 8½.
20290-9 Pa. $3.50

LASKER'S MANUAL OF CHESS, Dr. Emanuel Lasker. Great world champion offers very thorough coverage of all aspects of chess. Combinations, position play, openings, end game, aesthetics of chess, philosophy of struggle, much more. Filled with analyzed games. 390pp. 5⅜ x 8½.
20640-8 Pa. $4.00

500 MASTER GAMES OF CHESS, S. Tartakower, J. du Mont. Vast collection of great chess games from 1798-1938, with much material nowhere else readily available. Fully annoted, arranged by opening for easier study. 664pp. 5⅜ x 8½.   23208-5 Pa. $6.00

A GUIDE TO CHESS ENDINGS, Dr. Max Euwe, David Hooper. One of the finest modern works on chess endings. Thorough analysis of the most frequently encountered endings by former world champion. 331 examples, each with diagram. 248pp. 5⅜ x 8½.   23332-4 Pa. $3.50

CATALOGUE OF DOVER BOOKS

THE STANDARD BOOK OF QUILT MAKING AND COLLECTING, Marguerite Ickis. Full information, full-sized patterns for making 46 traditional quilts, also 150 other patterns. Quilted cloths, lame, satin quilts, etc. 483 illustrations. 273pp. 6⅞ x 9⅝. 20582-7 Pa. $3.95

ENCYCLOPEDIA OF VICTORIAN NEEDLEWORK, S. Caulfield, Blanche Saward. Simply inexhaustible gigantic alphabetical coverage of every traditional needlecraft—stitches, materials, methods, tools, types of work; definitions, many projects to be made. 1200 illustrations; double-columned text. 697pp. 8⅛ x 11. 22800-2, 22801-0 Pa., Two-vol. set $12.00

MECHANICK EXERCISES ON THE WHOLE ART OF PRINTING, Joseph Moxon. First complete book (1683-4) ever written about typography, a compendium of everything known about printing at the latter part of 17th century. Reprint of 2nd (1962) Oxford Univ. Press edition. 74 illustrations. Total of 550pp. 6⅛ x 9¼. 23617-X Pa. $7.95

PAPERMAKING, Dard Hunter. Definitive book on the subject by the foremost authority in the field. Chapters dealing with every aspect of history of craft in every part of the world. Over 320 illustrations. 2nd, revised and enlarged (1947) edition. 672pp. 5⅜ x 8½. 23619-6 Pa. $7.95

THE ART DECO STYLE, edited by Theodore Menten. Furniture, jewelry, metalwork, ceramics, fabrics, lighting fixtures, interior decors, exteriors, graphics from pure French sources. Best sampling around. Over 400 photographs. 183pp. 8⅜ x 11¼. 22824-X Pa. $5.00

Prices subject to change without notice.

Available at your book dealer or write for free catalogue to Dept. GI, Dover Publications, Inc., 180 Varick St., N.Y., N.Y. 10014. Dover publishes more than 175 books each year on science, elementary and advanced mathematics, biology, music, art, literary history, social sciences and other areas.